AF531401

ENERGY SECURITY

ENERGY SECURITY

Editors

Dr. Digumarti Bhaskara Rao

Ms. Digumarthi Harshitha

"Academy of Communication Culture Education Science and Service"
1-22-10 Srinivasa Nagar
Guntur 522006
Andhra Pradesh
India

2001

DISCOVERY PUBLISHING HOUSE

NEW DELHI—110 002

Published by :
Discovery Publishing House
4831/24, Ansari Road, Prahlad Street
Darya Ganj, New Delhi—110 002 (INDIA)
Phone : 327 92 45
Fax : 91–11–3253475

First Published—2001

ISBN 81–7141–598–9

Laser Typeset by :
Allied Computers,
Karnal (Haryana)

Printed at:
Arora Offset Press
Laxmi Nagar, Delhi 110 092.

Preface

Everyone needs energy in different forms to lead life comfortably. The existing energy sources, except some forms of renewable energy, may not last for a long time. Energy security, hence, has become a part of all human activities.

This book on energy security deals with several aspects covering Energy and Geopolitics, Global Energy Perspectives and Trends, Regional Energy Demands and Supply Options, Environmentally Secure Energy Options and Technologies, Power Generation Technologies for Future, and Bio-climate Architecture. The authors of the articles included in this book are eminent personalities in the field of energy.

I am thankful to Prof. Pierre Lasserre, Director, UNESCO-ROSTE, Venice, Italy for granting permission to reproduce articles presented in the Forum of the UNESCO International School of Science for Peace on "Energy Security in the Third Millennium: Scientific and Technological Issues" held at Villa Olmo-Como, Italy during 14-16 May, 1998. I am also thankful to R. Santesso, Assistant Information Officer, UNESCO, Venice; UNESCO International School of Science for Peace; and Landau Network, Centro Volta, Como, Italy.

I hope that there contributions may generate vision and vigour in the people and personnel involved in energy security.

D. Harshitha

Contents

Contributors

Al Jaber Abdul Malek, President, Palestinian Energy Centre, Nablus, PALESTINE

Andreini Pierangelo, Faculty of Engineering, Polytech of Milan, Piazza L. da Vinci, 32 20133 Milan, ITALY

Arungu Olende S., Dept. for Economic and Social Affairs (DESA), Div. for Sustainable Development, United Nations, DC2-2278, 2 UN Plaza, NEW YORK, N.Y. 10017, USA

Aslanian Garegin, Centre for Energy Policy, Off. 125, 14 Petrovka, 103031 Moscow, RUSSIAN FEDERATION

Baguenier Henri, CEEETA, Rua Gust. De Matos, Sequeira 28 1ºdtº, 1200 Lisbon, PORTUGAL

Berkovski Boris, Director, Division of Engineering Sciences, UNESCO, 1 rue Miollis, Paris 75015, FRANCE

Bosseboeuf Didier, ADEME, Paris, FRANCE

Botta Alberto, Mayor, City of Como Via Vittorio Emanuele, 97, 22100 Como, ITALY

Canepa Margherita, Assistant Secretary General, Landau Network–Centro Volta, Via Cantoni 1, 22100 Como, ITALY

Canobbio Ernesto, Fusion Division, European Commission, DGXII, 200 rue de la Loi, B-1049 Brussels, BELGIUM.

Carettoni Tullia, President, Italian National Commission for UNESCO, Piazza Firenze 27, 00186 Rome, ITALY

Casana François, Unit Energy Cooperation with Third Countries, European Commission, DGXVII-Energy, 200 rue de la Loi, B-1049 Brussels, BELGIUM

Casati Giulio, Scientific Coordinator, Centro "A. Volta", Via Cantoni 1, 22100 Como, ITALY

Citterio Marco, President, Chamber of Commerce, Industry Artisans and Agriculture, Via Parini 16, 22100 Como, ITALY

Cumo Maurizio, Department of Engineering, University of Rome, "La Sapienza", Piazzale Aldo Moro 5, 00186 Rome, ITALY

De Santis Paolo, City Councilor for Culture, Mobility and Traffic, Urbanisation Water & Rods, Via Vittorio Emanuele 97 22100 Como, ITALY

Di Lorenzo Aldo, Director, Istituto Motori–CNR, Via G. Marconi 8, 80125 Naples, ITALY

Dufour Angelo, Vice Director, Ansaldo Ricerche Srl, Corso F.M. Perrone 25, 16161 Genoa, ITALY

Farinelli Ugo, ENEA, Studies and Strategies of Development Unit, Lungotevere Thaon de Revel 76, 00196 Rome, ITALY

Favorsky Oleg, Head, Dept. of Physics for Technical Problems of Energetics, Russian Academy of Sciences, Moscow, RUSSIAN FEDERATION

Gherardi Giuseppe, Deputy Director for Advanced Systems Engineering, ENEA, Via Martiri Di Monte Sole 4, 40129 Bologna, ITALY

Ghirlando Robert, University of Malta, Dept. of Manufacturing Engineering, Msida MSD 06, MALTA

Gutermuth Paul-Georg, Head, Division of Renewable Energies, BMWi, Federal Ministry for Economics, Villemombler St. 76, 53107 Bonn, GERMANY

Ibrahim Hassan, Director of the ASEAN-EC, Energy Management Training and Research Centre, Jakarta, INDONESIA

Kagramanian Vladimir, Planning and Economies Studies Sector, Division of Nuclear Power, International Atomic Energy Agency (IAEA), Wagramer Strasse 5, P.O. Box 100, A-1400 Vienna, AUSTRIA

Karekezi Stephen, Director, AFREPREN/FWD, Nairobi, KENYA

Khalatnikov Isaak, Hon. Director, Landau Institute for Theoretical Physics, Russian Academy of Sciences, Kosygina Str. 2, 117940 Moscow V-334 GSP-1, RUSSIAN FEDERATION

Körbitz Werner, Austrian Biofuels Institute, Graben 14/2, P.O. Box 97, A-1014 Vienna, AUSTRIA

Kouzminov Vladimir, Chief, UNESCO Venice Office, 1262 A Dorsoduro, 30123 Venice, ITALY

Lantieri Antonino, ENEA, ERG-FISS-FIRE, Via Martiri di Monte Sole 4, 40129 Bologna, ITALY

Lavrencic Cannata Devana, Consultant, ENEA, Viale Lieig 58, 00198 Rome, ITALY

Lion Robert, President, Energy 21, Paris, FRANCE

Martellini Maurizio, Secretary General, Landau Network-Centro Volta and Dep. Physics, University of Insubria at Como, Via Cantoni 1, 22100 Como, ITALY

Mastepanov A., Ministry of Fuel and Energy, Moscow, RUSSIAN FEDERATION

Mourogov Victor, Deputy Director General, Head Dept. Nuclear Energy, International Atomic Energy Agency (IAEA), Wagramer Strasse 5, P.O. Box 100, A-1400 Vienna, AUSTRIA

Nadejdin Eugeni, Regional Advisor of Sustainable Energy Division, UN European Economic Commission, Palais des Nations, 8–14 Ave.de la Paix, 1211 Geneva 10, SWITZERLAND

Nicoletti Manfredi, University of Rome "La Sapienza", Via di San Simone 75, 00186 Rome, ITALY

Nuno Vitorino, CEEETA, Rua Gust. De Matos Sequeira, 28 1ºdtº, 1200 Lisbon, PORTUGAL

Pagliano Lorenzo, Dept. of Energy, Polytech of Milan, Piazza L. da Vinci 32, 20133 Milan, ITALY

Pico Ponte Efrain, Director of International Relations, Ministry of Energy and Mines of Venezuela, Caracas, VENEZUELA

Priebe Klaus-P., Managing Director, PINA Gmbh, Technologie Zentrum, Emil-Figge Str. 76, D-44227 Dortmund, GERMANY

Roturier Jacques, University of Bordeaux 1, C.E.N.B.G., B.P. 120 Le Haut-Vigneau, F-33175 Gradignan, Cedex, FRANCE

Sala Tiziana, Provincial Administration/Culture Sector, Via Borgovico, 148, 22100 Como, ITALY

Sanguini Armando, Director General for Cultural Relations, Ministry of Foreign Affairs, Piazzale della Farnesina 1, 00194 Rome, ITALY

Silberg Peter, EUS Gmbh Munscheldstrasse 14, D-448861 Gelsenkirchen, GERMANY

Tirone Livia, Tirone-Nunes Urbanismo Lda, Herança do Pinheiro Ral 2710, Sintra, PORTUGAL

Ullrich Christian, Director INFU, University of Dortmund, D-44221 Dortmund, GERMANY

Valette Pierre, European Commission, DGXII-Investigations Energy Strategy, 200 rue de la Loi, Brussels, BELGIUM

Velikhov Evgenij, President, I. Kurchatov Research Centre, Kurchatov Sq. 1, 123182 Moscow, RUSSIAN FEDERATION

Zaleski C. Pierre, Deleque General CEGMP University of Paris-Dauphine 75116 Paris, FRANCE

PART — 1
ENERGY AND GEOPOLITICS

1
Energy and Geopolitical Issues

P. ZALESKI
France

I. Geopolitics

The majority of International commerce in strategic commodities is controlled by national governments, which means that the classical theories of economics cannot be applied. Indeed, the inequality in the world distribution of "production factors"—that is, the ensemble of natural resources, manpower, know-how, technology level, availability of investment capital and other elements which determine wealth-renders incorrect one of the generally admitted economic hypotheses: the maximization of collective interest results from the maximisation of individual satisfaction.

In addition, the behaviour of economic players, even if it may be considered as rational, does not follow conventional economic "rationality"; rather, these players try to establish a position of strength which will then dominate international exchanges.

Geopolitics is the study of the "relations of strength" which largely shape international exchanges and the behaviour of economic players. Geopolitics aims to evaluate a political situation and its possible development. In view to do this, the geopolitician must analyse different domains that contribute to the overall position of strength: geography, demography, economy, finance, technology, and politics including military capabilities.

II. Geopolitics of Energy

Geopolitics of energy must analyse the positions of strength for the complete energy chain, from exploration and production to consumption, through long-distance transportation and distribution. However, the application of geopolitics to energy or to other commodities is justifiable only for what we will call "strategic commodities". This term is difficult to define with precision, and the strategic importance of a given commodity varies over time. We may, however, indicate two main attributes of a strategic commodity; it is essential for the economy of a country; and the supply is uncertain or may present a risk.

To illustrate this notion, let me show you a figure taken from a book of the geopolitics of oil and gas by the late Andre Giraud, former French Minister of Industry and then Defence, and also the founder of our center for geopolitics of energy at University of Paris-Dauphine, and my former boss and friend. (Figure 1). This book, published in 1986, stressed the strategic importance at that date of oil, to which we may add today natural gas.

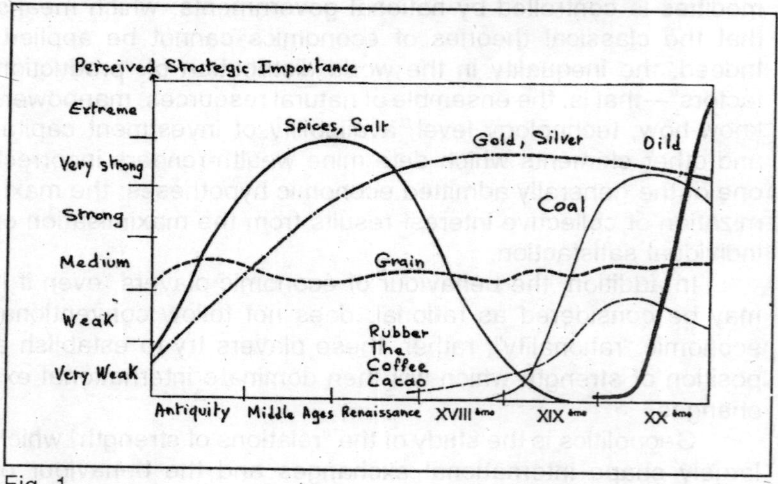

Fig. 1

III. The Strategic Importance of Persian Gulf Oil

Here I will present a figure which I borrowed from Mr. William F. Martin, former U.S. Deputy Secretary of Energy and currently Chairman of the think tank, Washington Policy and Analysis. Mr. Martin presented the figure in testimony to the U.S. Senate in October of last year. (Figure 2).

We can see in this figure that Mr. Martin suggests that the next disruption in the supply of oil from the Persian Gulf may occur in the near future. This may come true or not. We should note that in this region, the United States, with some backing of its OECD allies, has a real position of strength. This position has

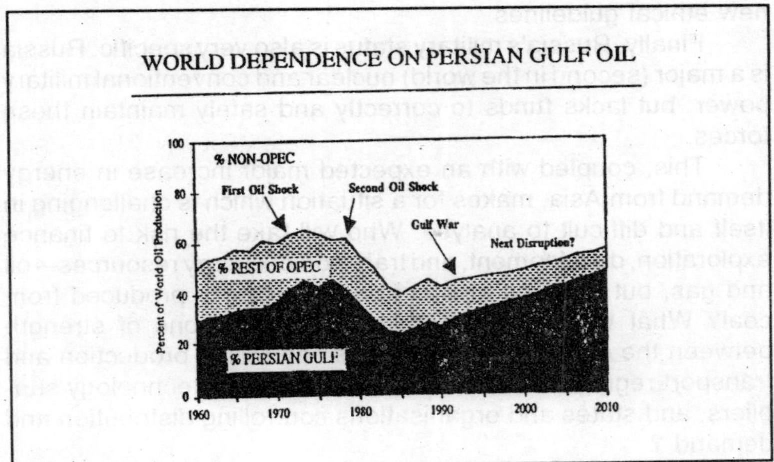

Fig. 2

a cost—some have estimated its cost at $ 50 billion a year, including the cost of the Sixth Fleet—but certainly has a real benefit as well: the low cost of energy for OECD countries, allowing continued favourable economic development. Certainly, the situation of Persian Gulf oil is an interesting geopolitical case.

IV. Recent Developments of Strategic Importance

The increased importance of oil resources, and even more of gas resources, from Siberia and the Caspian Sea region, coupled with the large increase in Asian energy demand, pose intriguing geopolitical problems, Russia, the state which owns the major part of these resources and controls a large part of the territory necessary for the transport of resources from other countries in transition, is potentially an extremely rich country, perhaps one of the richest in the world. Indeed, it has vast natural resources—both mineral and land for agriculture; large and skilled human resources; a high quality of technology and scientific knowledge. But on the other hand it presently has major problems with stability of political and legal structures, with development of a dynamic economy, and especially with the financing of this development. There are also major social problems to be resolved, and perhaps even more fundamentally, the transition from a Marxist philosophy to a more market-oriented society has not yet been accompanied by formation of new ethical guidelines.

Finally, Russia's military status is also very specific. Russia is a major (second in the world) nuclear and conventional military power, but lacks funds to correctly and safely maintain these forces.

This, coupled with an expected major increase in energy demand from Asia, makes for a situation which is challenging in itself and difficult to analyse. Who will take the risk to finance exploration, development, and transport of energy resources—oil and gas, but perhaps also coal and electricity produced from coal? What will be the interplay of the positions of strength between the different actors: states controlling production and transport regions, resource owners, investors, technology suppliers, and states and organisations controlling distribution and demand ?

V. Transition from Conventional Oil and Gas to New Liquid and Gaseous Fuels

At the end of last year, the International Energy Agency published a report on World Energy Prospects to 2020. It indicates that the progressive exhaustion of conventional oil resources will lead to a production peak between 2010 and 2020, followed by a progressive decline, or perhaps a plateau and then a progressive decline, in production. For conventional natural gas, the IEA report predicts the decline beginning some years later. Potential replacements exist: non-conventional oil, or other gaseous or liquid fuels (for example coal gasification, biomass). The geopolitics of these new fuels will be quite different.

Some questions may also be asked: what will be the cost of producing these new fuels ? What investment will be required to obtain them? And what will be their environmental impact ?

Another interesting question is, will the transition to these new fuels be smooth, or—because of delays and difficulties in implementing major new fuel development projects, combined with a more demanding attitude of conventional oil and gas suppliers—will there be turbulence in the world's supply of liquid and gaseous fuels?

VI. New Players in Energy Geopolitics

The final issue I would like to bring to your attention is the emergence of a new factor in the energy geopolitics game—the "position of strength" coming from activists in the public. This emergence is linked to increased affluence, shorter working times, a higher level of general knowledge, and the explosion of means of communication— not only radio, television and print media, but also the Internet. This new trend concerns mostly the more affluent fraction of the public, in the most developed parts of society, because these people have time, and usually money, to devote to these activities. Using available legal avenues and political pressure, they may bring into the decision-making process some long-term issues which earlier did not receive significant attention. The public and environmentally oriented associations must, however, avoid falling into the trap of opposing a given option on emotional grounds, without a systematic and objective analysis of the relative risks of different options.

The central new long-term issue for energy introduced by public opinion is a requirement for "sustainable development". To satisfy this requirement, one of the keys is efficient use of energy. But this does not come easily, in particular due to societal inertia and competition from powerful energy producing companies. To illustrate this difficulty, let me mention the recent (March, 1998) joint report from the International Energy Agency and the Energy Charter Secretariat, entitled, "Energy Investment". This report indicates that according to a study by the European Commission and the EBRD, there is a market of $40 billion for energy efficiency projects in countries in transition with a playback of less than three years, while the present level of investment in energy efficiency projects is less than 10 per cent of that sum.

Another key is the substitution of fossil fuels by other energy sources, primarily because of the impact of fossil fuels on the environment—in particular, the "greenhouse effect".

Because of technological progress, which extends the estimate of fuel reserves, the formerly important issue of exhaustion of resources no longer mobilises public opinion, even though for the longer term it is certainly an issue for sustainable development.

Let us look in particular at the issue of the greenhouse effect. By this we mean, of course, the discovery that, for the first time in the history of mankind, human activity may perturb the global environmental balance, leading to an increase of atmospheric greenhouse gas concentration and eventually an increase in average world temperature. The perception, right or wrong, but quite general among the activist fraction of the public in developed countries, that this increase may lead to catastrophic consequences has induced some politicians to propose a limitation of greenhouse gas emissions. Late last year, this concern was reflected in the Kyoto Agreement to control greenhouse gas emissions in industrialised countries.

It is not clear today if and how the Kyoto Agreement will be implemented. (Figure 3). This figure, taken from William Martin's testimony, shows projections of world carbon dioxide emissions to 2010 which are clearly not consistent with the Kyoto commitment. We cannot today say how strong is the position of the "new player" public opinion. But if the greenhouse effect issue becomes a real constraint, it will modify the balance between different energy resources, limiting the use of fossil fuels even before the problem of resource exhaustion appears, and the

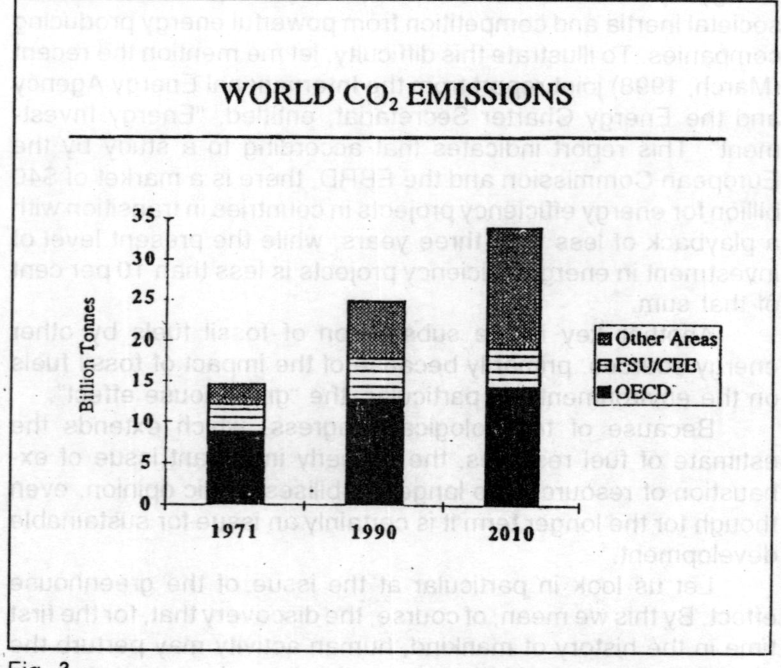

Fig. 3

geopolitics of energy will be affected.

The real test of the strength of public pressure in limiting greenhouse gas emissions will be its confrontation with the will of developing countries like China, India, Brazil and others to develop as fast as possible with available means. The importance of these other countries in CO_2 emissions can be seen in Figure 3. When this confrontation occurs, will industrialised countries, under the pressure of public opinion, take into account not only some arbitrary level of greenhouse gas emissions to which nations should restrict themselves (1990 level), but at the same time a contribution to the admissible quota based on the right of each human being to emit a certain quantity of greenhouse gases, and therefore on national populations?

If this combined system prevails, will it lead, by a mechanism like emissions permits or joint implementation, to transfer of money and technology from industrialised to developing countries in view to allow the latter to choose, without compromising their development, not to use less expensive fossil fuels, but rather some likely more expensive, but hopefully not too much more expensive, option that does not emit greenhouse gases, like renewables or nuclear energy?

VII. Conclusion

During this meeting, we will hear many interesting presentations, in particular about new technology developments that may contribute to solving some of today's problems. We should not, however, forget the very great inertia of the world energy system, the huge financial requirements of the energy sector, and the time needed to implement new solutions and impose new trends. Above all, to achieve energy security in the Third Millennium, we need stable and cooperation-oriented political systems in all regions of the world.

PART — 2

Global Energy Perspectives and Trends

2

Climate Technology Strategy within Competitive Energy Markets

P. VALETTE
Belgium

Technology Story

- World base line 2030
- Impacts of new technologies

General Policies

- Internalisation of external costs
- Flexibility instruments

The Five Technological Scenarios

- **The nuclear scenario**
  Assumes a breakthrough in nuclear technology in terms of cost and safety.
  Has an influence on standard large LWRs but especially on a **new evolutionary nuclear design.**
- **The clean coal scenario**
  Involves major improvements in solid fuel burning technologies and especially effects the technical-economic characteristics of Super Critical Coal Plants, Integrated Gasification Combined Cycles and Advanced Thermodynamic Cycles.
- **The Gas Technology Scenario**
  Assumes enhanced availability of natural gas and introduces major technical-economic improvements for gas turbine combined cycles and combined heat and power plants.
  Some improvements assumed for all gas turbine related technologies.

"Technology Story" Scenarios

Objective

To examine the possible impact on CO_2 emissions and world energy markets of eventual breakthroughs, in specific energy technology fields (2030).

Methodology

- Identify possible generic technological breakthroughs.
- Using expertise in technological perspectives, translate these breakthroughs into a more optimistic trajectory for the technical-economic characteristics of specific clusters of technologies.
- Use the POLES world energy model to simulate in a consistent framework the impact of the scenarios by comparing to a reference case study taking care to include:
  - — Competition between Technologies
  - — Secondary effects regarding fuel prices
- Internalize the "Environment costs" and re-evaluate the impacts of the scenarios.

The Fuel Cell Scenario

Retains the assumptions of the gas technology scenario but additionally involves a breakthrough in fuel cell technologies in the latter part of the forecast horizon affecting proton exchange membrane fuel cells for fixed applications, solid oxide fuel cell with cogeneration and a hydrogen fuel cell car.

The Renewable Scenario

Employing a major effort and breakthroughs in renewable technologies, notably wind power, biomass gasification, solar thermal power plants, small hydro and photovoltaic cells.

The different blocks in the POLES 2 model

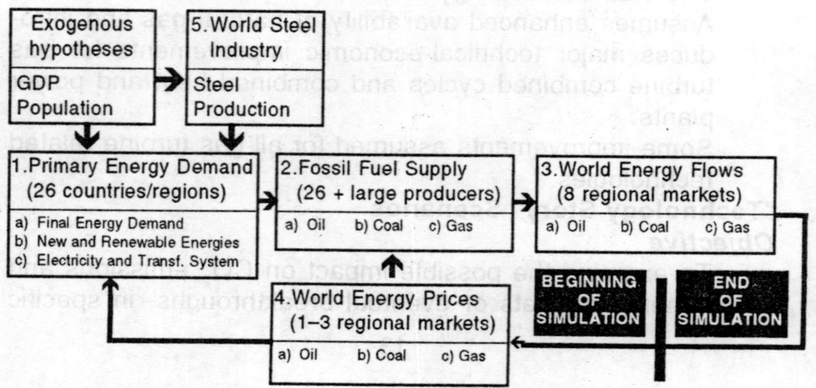

Reference Case: World Energy Balance

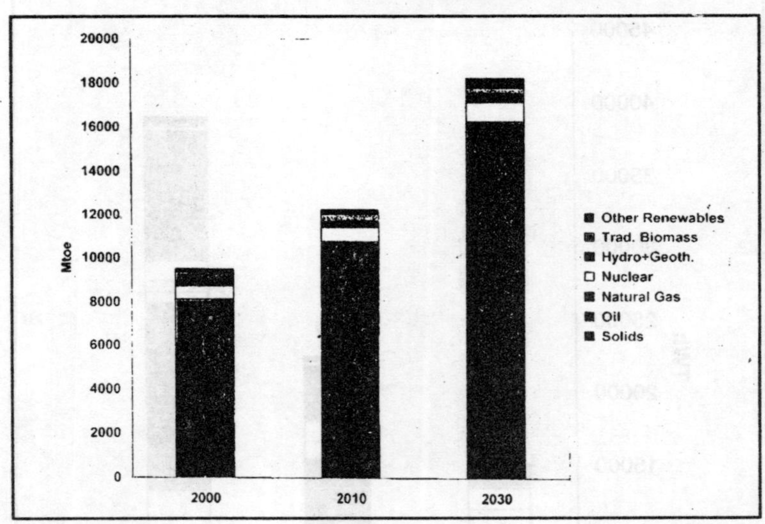

Reference Case: World CO_2 Emissions

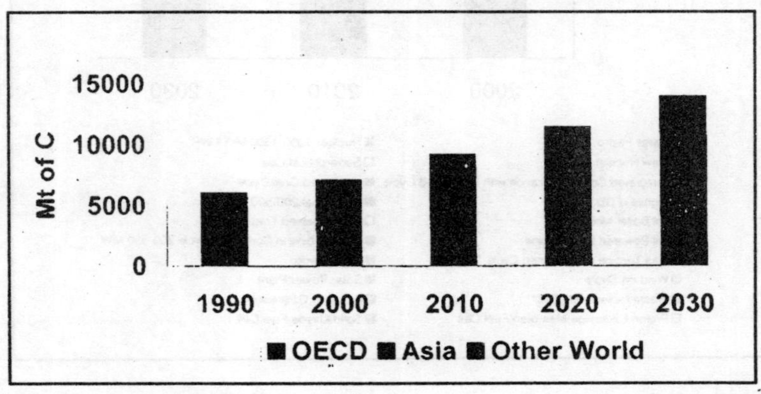

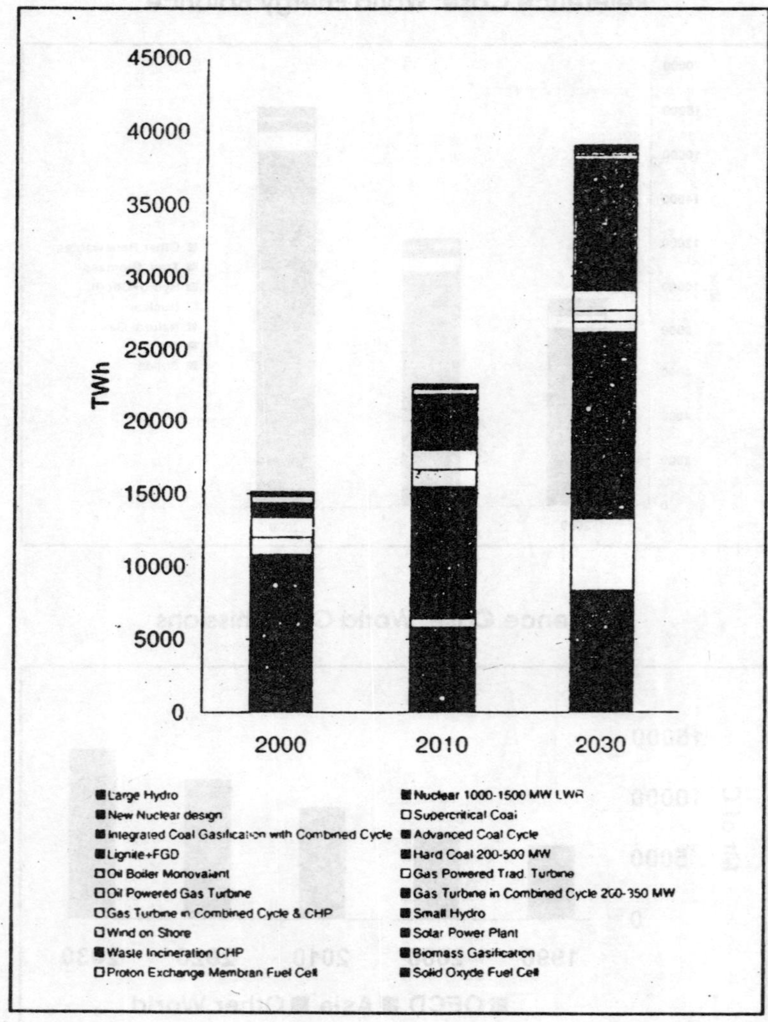
Reference Case: World Electricity Production
TWh
45000
40000
35000
30000
25000
20000
15000
10000
5000
0
2000
2010
2030
Large Hydro
New Nuclear design
Integrated Coal Gasification with Combined Cycle
Lignite+FGD
Oil Boiler Monovalent
Oil Powered Gas Turbine
Gas Turbine in Combined Cycle & CHP
Wind on Shore
Waste Incineration CHP
Proton Exchange Membran Fuel Cell
Nuclear 1000-1500 MW LWR
Supercritical Coal
Advanced Coal Cycle
Hard Coal 200-500 MW
Gas Powered Trad. Turbine
Gas Turbine in Combined Cycle 200-350 MW
Small Hydro
Solar Power Plant
Biomass Gasification
Solid Oxyde Fuel Cell

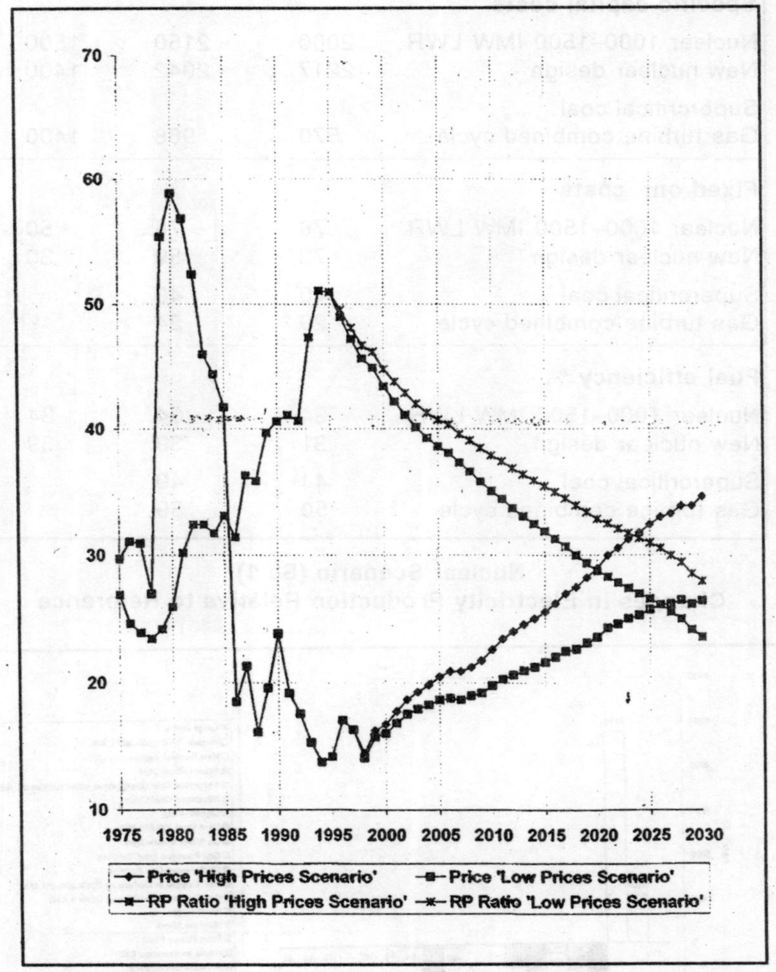
World Crude Oil price & RP rations
IEA-EU
70
60
50
40
30
20
10
1975 1980 1985 1990 1995 2000 2005 2010 2015 2020 2025 2030
Price 'High Prices Scenario'
Price 'Low Prices Scenario'
RP Ratio 'High Prices Scenario'
RP Ratio 'Low Prices Scenario'

Nuclear Technologies Alternative Scenarios			
ECU 90/KW	***1995***	***2030 reference case***	***2030 scenario***
Specific capital costs			
Nuclear 1000–1500 IMW LWR	2000	2150	1500
New nuclear design	2217	2042	1400
Supercritical coal			
Gas turbine combined cycle	570	968	1400
Fixed om costs			
Nuclear 1000–1500 IMW LWR	76	76	50
New nuclear design	73	59	30
Supercritical coal	40	40	
Gas turbine combined cycle	29	24	
Fuel efficiency %			
Nuclear 1000–1500 IMW LWR	34	34	34
New nuclear design	31	39	39
Supercritical coal	44	49	
Gas turbine combined cycle	50	59	

Nuclear Scenario (Sc 1)
Changes in Electricity Production Relative to Reference

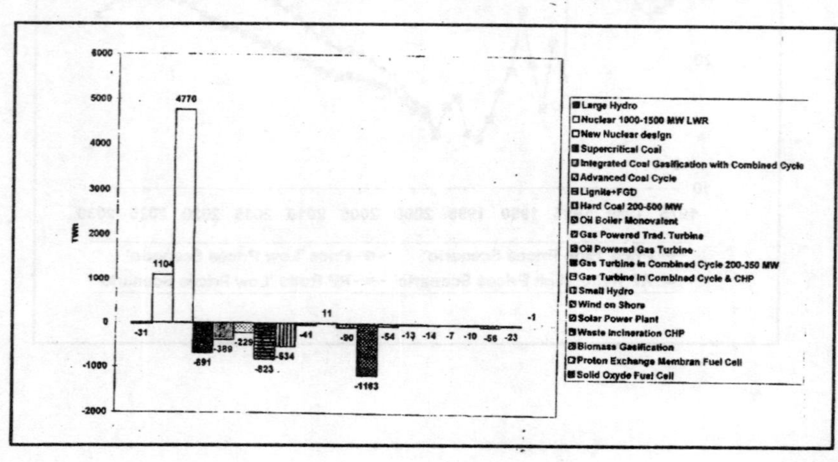

Nuclear Scenario (Sc 1)

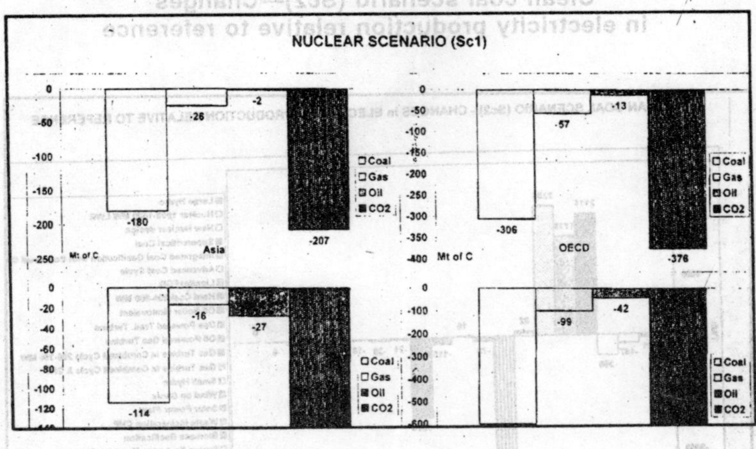

Changes in CO_2 Relative to Reference in 2030

ECU 90/KW	*1995*	*2030 no. tech. progress*	*2030 refer. case*	*2030 Scenario*
Specific capital costs				
Hard coal (conven.)	980	928	835	760
Supercritical coal		1268	968	750
Integr. gasif. combinated cycle		1370	1090	820
Advanced coal cycle (direct. comb.)		1200	1000	750
Gas turbine combined cycle	570		470	
Fixed om costs				
Hard coal (conven.)	46	46	46	46
Supercritical coal		40	40	40
Integ. gasif. combinated cycle		70	60	60
Advanced coal cycle (direct. comb.)		45	35	30
Gas turbine combined cycle	29		24	
Fuel efficiency %				
Hard coal (conven.)	38	38	39.6	39.6
Supercritical coal		44	49	49
Integ. gasif. combinated cycle		46	49.8	54
Advanced coal cycle (direct. comb.)		46	50	52
Gas turbine combined cycle	50		59	

Clean coal scenario (Sc2)—Changes in electricity production relative to reference

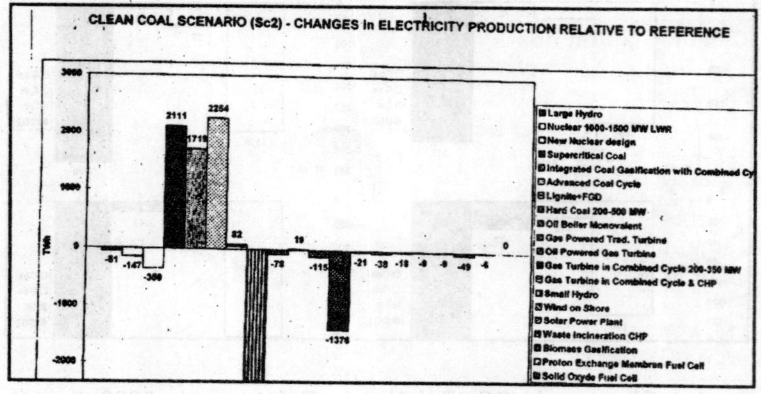

Clean coal scenario (Sc2)—Changes in CO_2 relative to reference in 2030

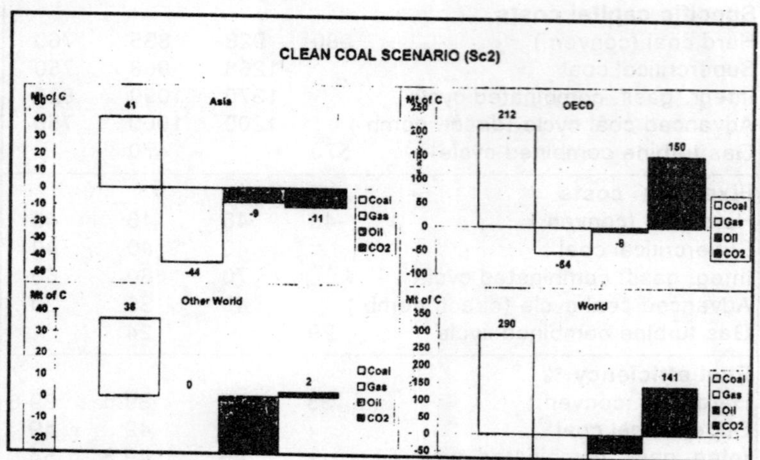

Enhanced Gas Technologies Alternative Scenarios			
ECU 90/KW	*1995*	*2030 refer. case*	*2030 scenario*
Specific capital costs			
Gas trubines in combined cycle			
200-350 MW	570	470	410
CHP	900	814	640
Adv. coal cycle		1000	780
Supercritical coal		968	
Fixed om costs			
Gas trubines in combined cycle			
200-350 MW	29	24	10
CHP	45	40	33
Adv. coal cycle		35	30
Supercritical coal	40	40	
Fuel efficiency %			
Gas trubines in combined cycle			
200-350 MW	50	59	63
CHP (electricity)	28	40	42
CHP (steam)	45	45	47
Adv. coal cycle		50	52
Supercritical coal	44	49	

Enhanced gas scenario (Sc3)—Changes in electricity production relative to reference

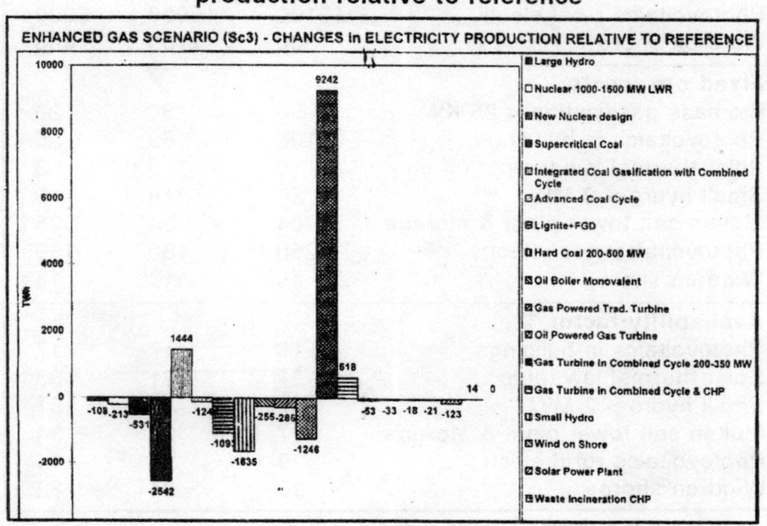

Enhanced gas scenario (Sc3)—Changes in CO_2 relative to reference in 2030

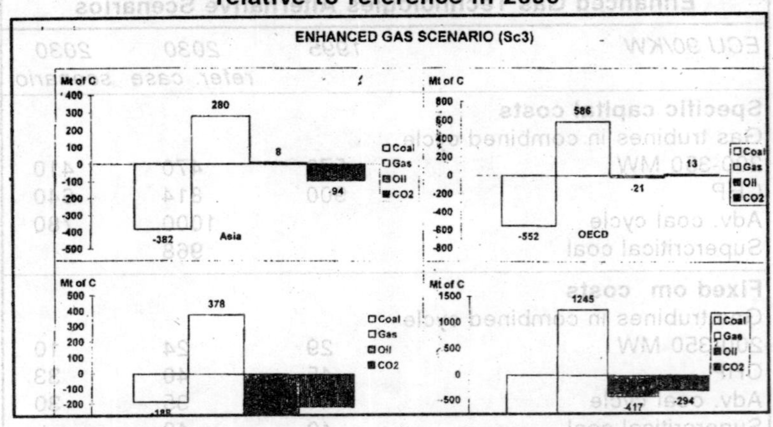

Renewable Technologies Alternative Scenarios			
ECU 90/KW	*1995*	*2030* *refer. case*	*2030* *scenario*
Specific capital costs			
Biomass gasification < 25 KW	1767	1767	960
Photovoltaics in buildings	12000	4000	2000
Solar thermal law temp.	1200	750	500
Small hydro > 2 MW	2000	1850	1000
Molten salt tower plant & storage	2500	1603	1131
Photovoltaics rural electr.	13100	10000	5000
Wind on shore	1300	1100	460
Fixed om costs			
Biomass gasification < 25 KW	50	30	20
Photovoltaics in buildings	100	50	25
Solar thermal law temp.	10	5	3
Small hydro > 2 MW	20	16	16
Molten salt tower plant & storage	64	34	25
Photovoltaics rural electr.	250	180	160
Wind on shore	16	14	14
Availability factor %			
Photovoltaics in buildings	15	17	17
Solar thermal law temp.	27	31	33
Small hydro > 2 MW	57	57	57
Molten salt tower plant & storage	27	32	34
Photovoltaics rural electr.	9	9	14
Wind on shore	20	24	32

A Reference Case—B Renewable Breakthrough Scenario (Sc5)

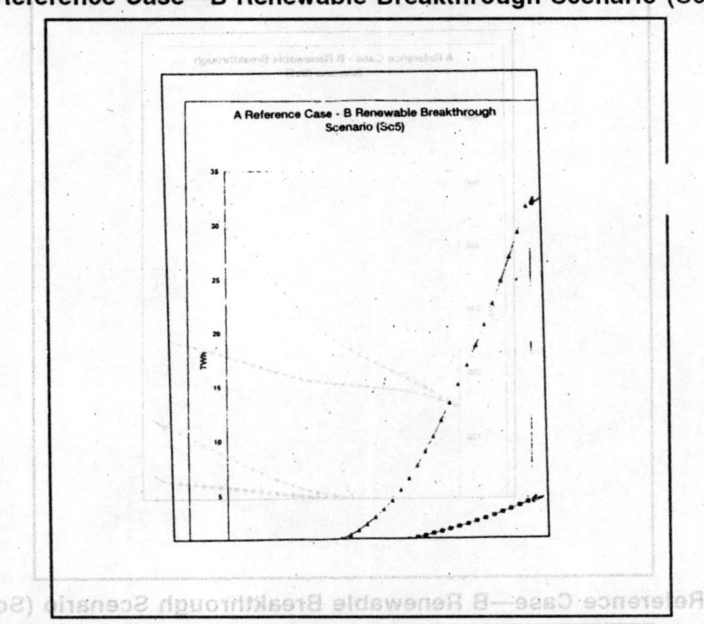

A Reference Case—B Renewable Breakthrough Scenario (Sc5)

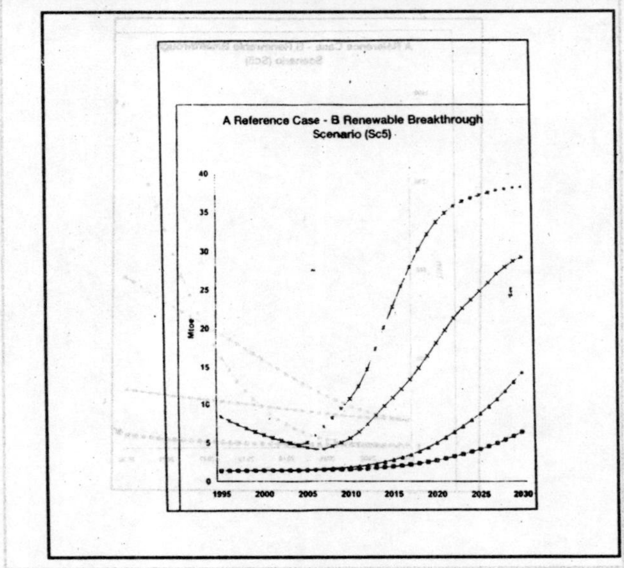

A Reference Case—B Renewable Breakthrough Scenario (Sc5)

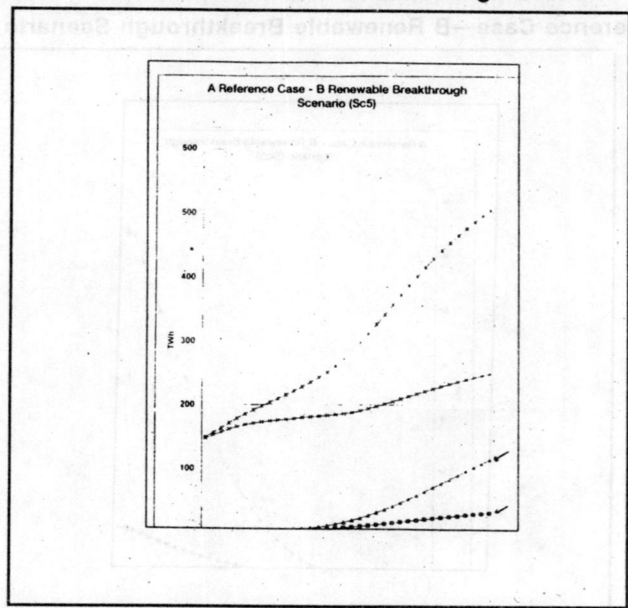

A Reference Case—B Renewable Breakthrough Scenario (Sc5)

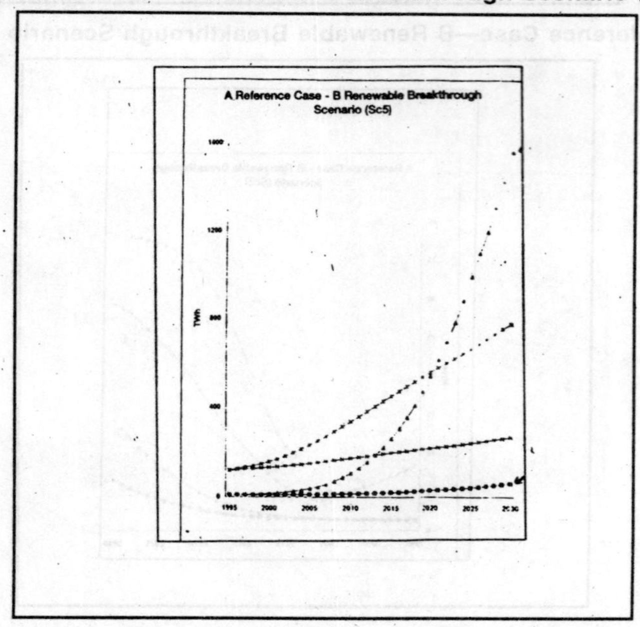

Renewable Breakthrough Scenario (Sc5)
Changes in Electricity Production Relative to Reference

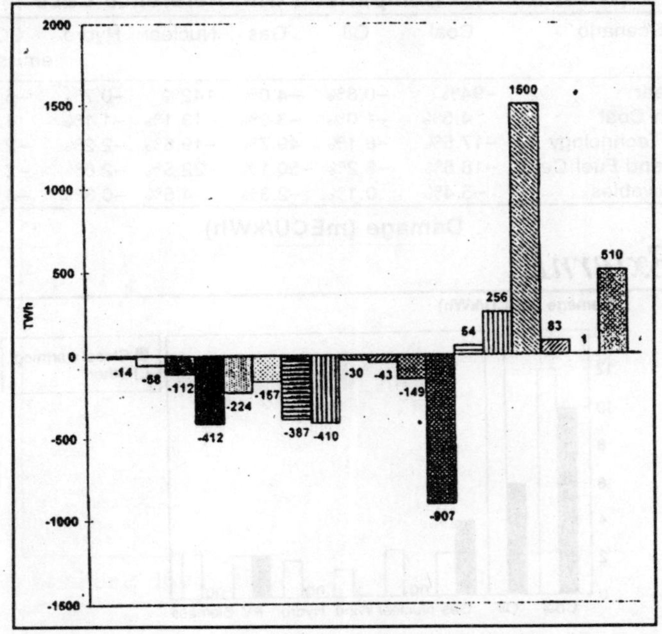

Renewable Breakthrough Scenario (Sc5)
Changes in CO_2 Relative to Reference in 2030

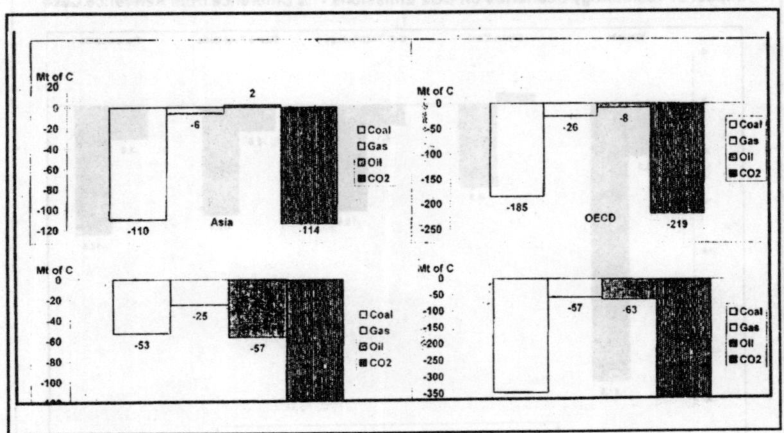

Summary Table of the impact of technology scenarios (World 2030) – Comparison with Reference Case

Scenario	Coal	Oil	Gas	Nuclear	Hydro	CO_2 emissions
Nuclear	–94%	–0.8%	–4.0%	142%	–0.7%	–5.3%
Clean Coal	4.5%	–1.0%	–3.9%	–13.1%	–1.1%	1.0%
Gas Technology	–17.5%	–8.1%	49.7%	–19.6%	–2.2%	–2.1%
Gas and Fuel Cells	–18.6%	–8.2%	–50.1%	–22.5%	–2.6%	–2.5%
Renewables	–5.4%	0.1%	–2.3%	–4.5%	–0.3%	–3.3%

Damage (mECU/kWh)

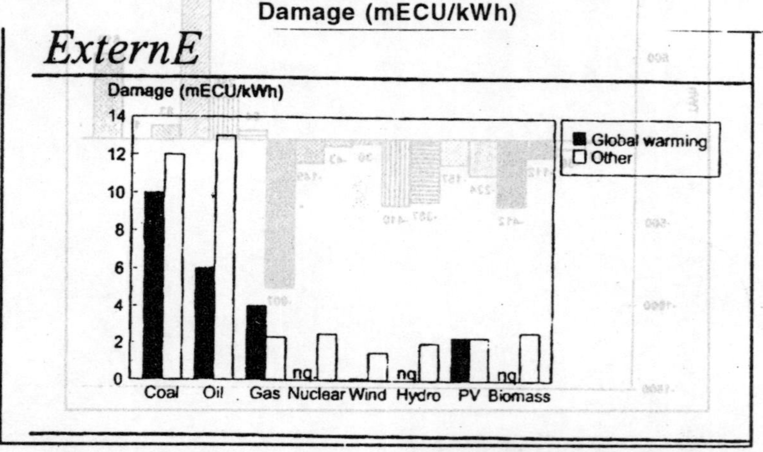

Impact of Technology Scenarios on CO_2 Emissions % Difference from Reference Case

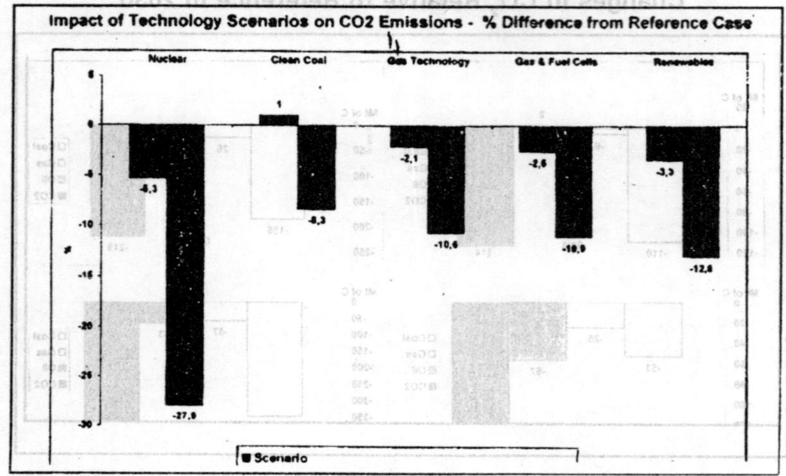

Impact of Technology Scenarios on Total Electricity Consumption–Difference from Reference Case

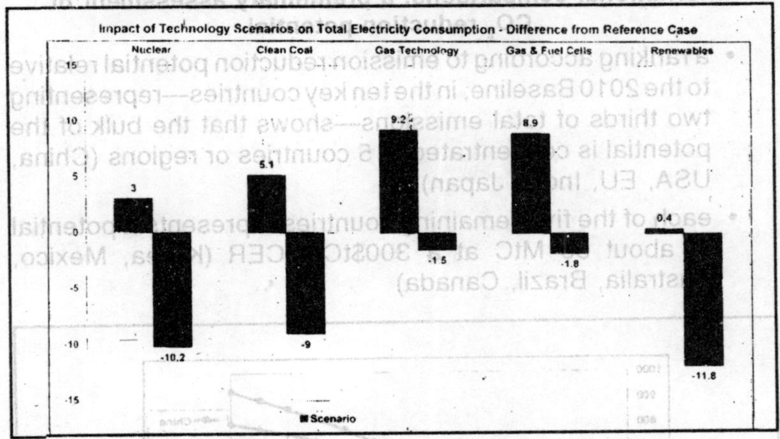

Marginal Cost of Emission Reduction Using Shadow Carbon-Taxes in the POLES Model 11

International comparisons: the debate among Annex I countries, common targets or equal MCER ?—1

- because of differences in : Baseline dynamics, energy price structure and primary fuel-mix, the main partners in "Annex I" show quite different MCER curves

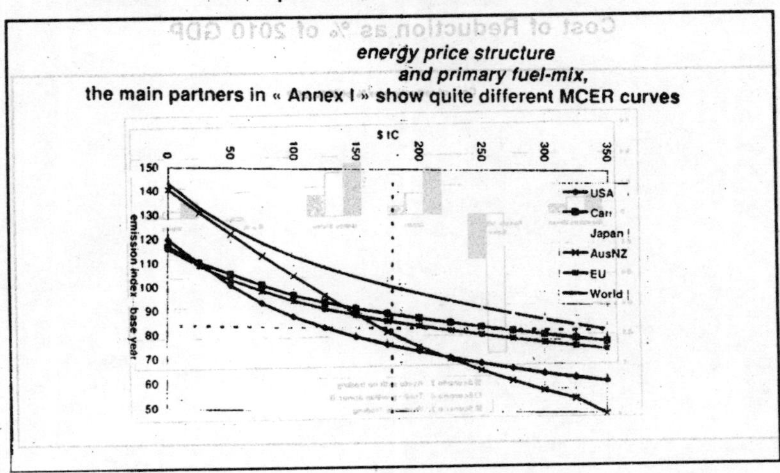

Marginal Cost of Emission Reduction Using Shadow Carbon-Taxes in the POLES Model 14

International comparisons: a preliminary assessment of CO_2 reduction potential

- a ranking according to emission reduction potential relative to the 2010 Baseline, in the ten key countries—representing two thirds of total emissions—shows that the bulk of the potential is concentrated in 5 countries or regions (China, USA, EU, India, Japan)
- each of the five remaining countries represents a potential of about 50 MtC at a 300$tC MCER (Korea, Mexico, Australia, Brazil, Canada)

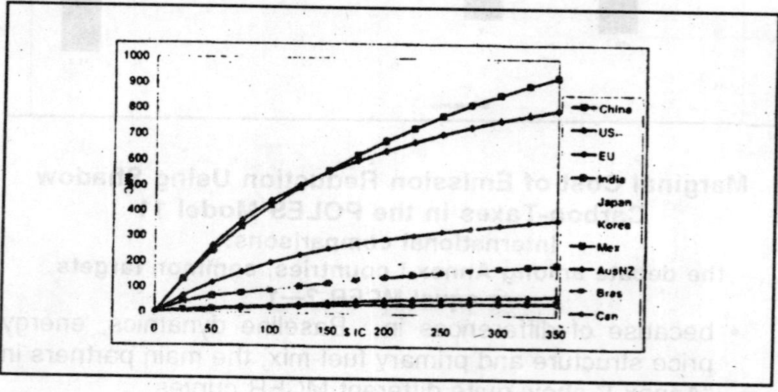

Cost of Reduction as % of 2010 GDP

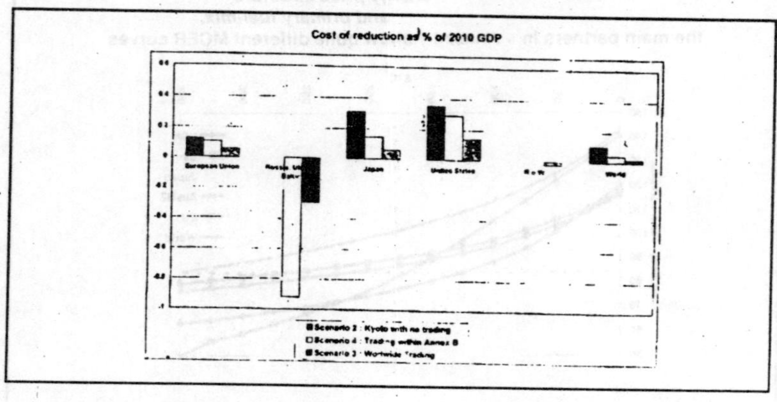

IMPACTS OF EMISSIONS TRADING

3 Scenarios:

Scenario

- KYOTO with NO TRADING

Scenario

- WORLDWIDE TRADING

Scenario

- TRADE within ANNEX B

Cost of Reduction as % of 2010 GDP

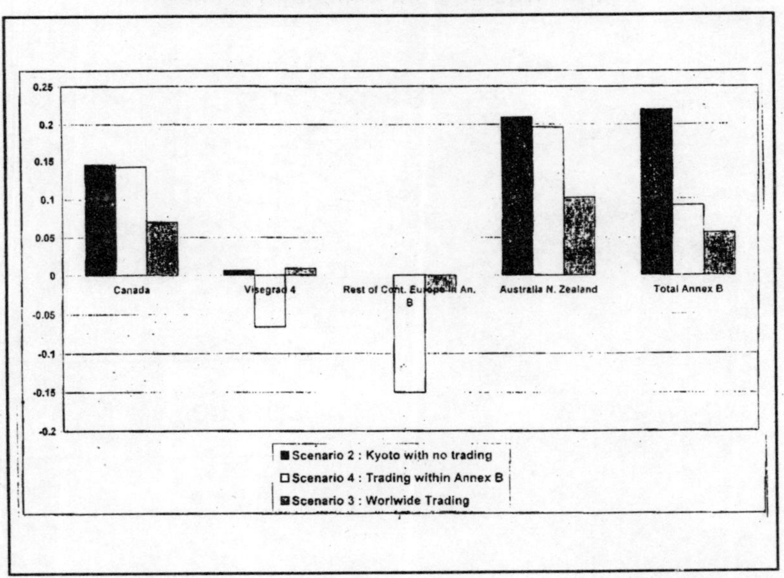

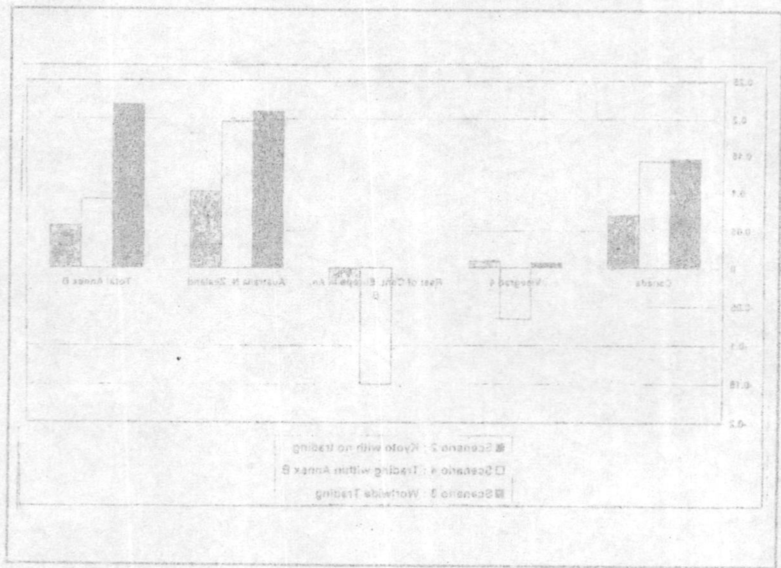

PART—3

THE REGIONAL ENERGY DEMANDS AND SUPPLY OPTIONS

3

On the Road Towards Fusion Energy: an International Perspective

E. CANOBBIO
Belgium

REFLECTIONS ON THE QUEST FOR FUSION

A. The Challenge

B. Fusion Grand Scales
[an historical travel shot]

- Geopolitics
- A big mistake: to cry wolf
- A bad discovery: anomalous losses

C. ITER at Present

D. Lessons for the Future

Annual Fuel Requirements for a Town of 5,00,000 People

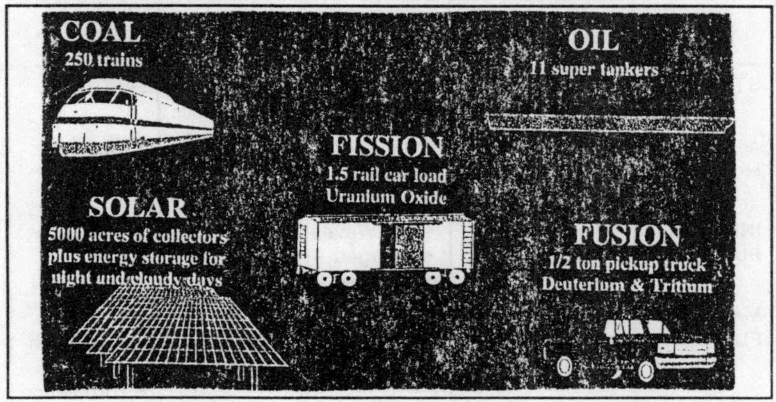

1932	Discovery of the neutron
1933	Fermi's weak nuclear forces
1935	Yukawa's strong nuclear forces
1938	Protons fusion in stars
	Discovery of fission
1942	Fermi's Chicago Reactor
1945	Little Boy and Fat Man
1949	QED
1950	Alfven's book
1952	US H-Bomb
1953	CERN Convention
	USSR (true) H-Bomb
	Eisenhower's "Atoms for Peace"
1954	Oppenheimer "unfit to serve his country"
	Death of Fermi
1956	US (true) H-Bomb
	Suez crisis
	Spitzer's book
	Khrushchev (+ Kurchatov) in the UK
1957	The Treaties of Rome
	Lawson criterion
	IAEA creation
1958	2nd UN "Peaceful Uses" conference
1959	Birth of EU Fusion Programme (Palumbo)

THERMONUCLEAR FUSION

	Nuclear Process	*Ignition Driver*	*Equilibrium*	*Thermal Insulation*
STARS	Weak Forces → γ's	Gravity	Gravo-Hydrostatic → Sphere	By Opacity
H-BOMB	Strong Forces → α & n	A-Bomb	None	None
INERTIAL FUSION	Strong Forces → α & n	• EM Waves • Charged Particles	None	None
MAGNETIC FUSION	Strong Forces → α & n	• EM Waves • Neutral Particles	Magneto-Hydrostatic → Torus	Limited by • Bremstrahlung • Turbulence

1961 1st IAEA Conference
Berlin Wall
1963 De Gaulle "No" to UK
1965 QED Nobel Prices
1968 Tokomak results
Invasion of Czechoslovakia
1969 Thomson scattering on T-3
1973 UK accedes to EU
Death of Artsimovitch
Kippur war
1974 Hawkin's Black Holes radiation
IEA creation
1975 Jet objectives definition
1976 "Anomalous" transport discovery
1977 JET site decision
1978 Iran revolution
1980 CCFP creation ("single voice")
1982 Net Team creation
JET operational
1984 Transport scaling (Goldstone)
Fusion → Big Science
Nobel Prize to Rubbia
1985 Gorbachev's (+ Velikhov)
Summits initiative:
Fusion cooperation on global scale

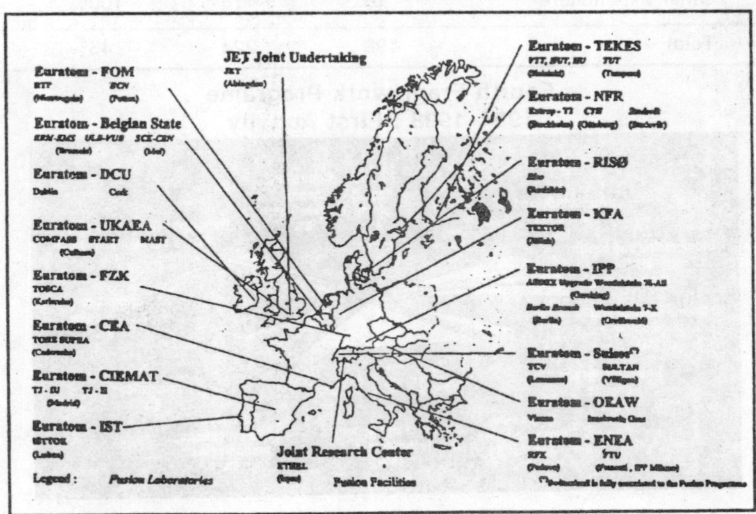

1986 Sakharov's release
Chernobyl
1987 UNO's "our common future"
1988 ITER CDA Agreement
TORE SUPRA operational
Diplomatic relationships EU-USSR
1989 Fall of the Berlin Wall
"Cold Fusion"
1990 Reunification of Germany
USSR collapse
1991 Fusion energy production in JET
1992 ITER EDA Agreement
Fusion record on JET
Japan's financial austerity measures (1998–2001)
Superphoenix shut-down decision
Kyoto Global Warming Protocol
1998 ITER EDA Extension (till 2001)
ITER mission/cost reconsidered

	Expenditure	*Cummunity Share*	*Community Share*
Associations	350	100	29%
JET	80	62	77%
NET/ITER	40	34	85%
JRC	12	12	100%
Other expenditure	16	16	100%
Total	498	224	45%

Fourth Framework Programe
1994–1998 : First Activity

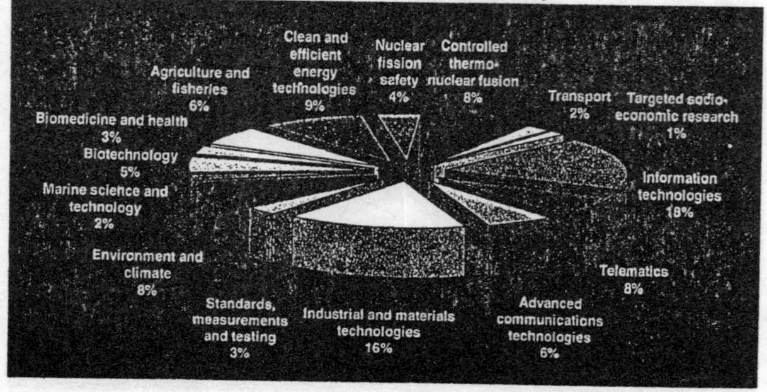

Strategy towards the first Electricity Producing Fusion Device (DEMO)

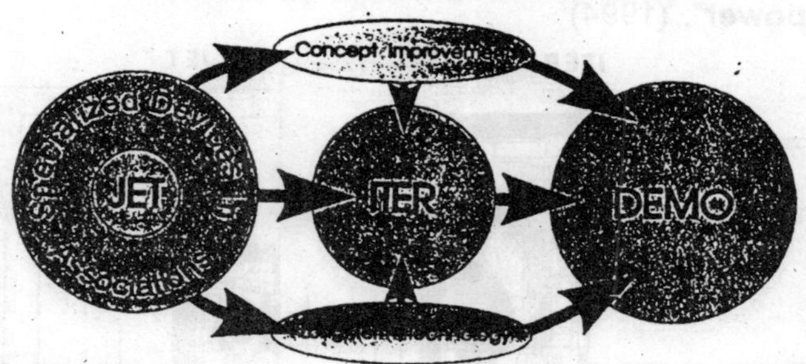

ITER MISSION

The overall programmatic objective of ITER is to demonstrate the scientific and technological feasibility of fusion energy for peaceful purposes.

ITER would accomplish this objective by demonstrating controlled ignition and extended burn of deuterium-tritium plasmas, with steady-state as an ultimate goal, by demonstrating technologies essential to a reactor in an integrated system and by performing integrated testing of the high-heat-flux and nuclear components required to utilise fusion energy for practical purposes. (1992)

Crude Oil Prices and US DoE Fusion Budget, 1970–1998

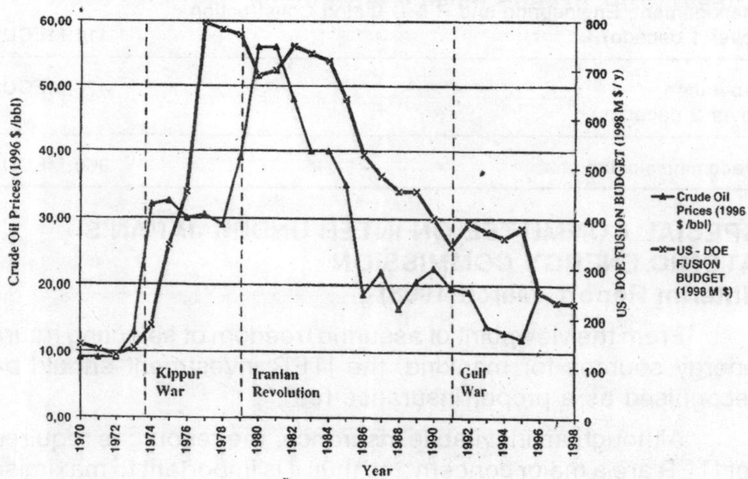

"ITER should be designed **to operate safely and to demonstrate the safety and environmental potential of fusion power"**. (1994)

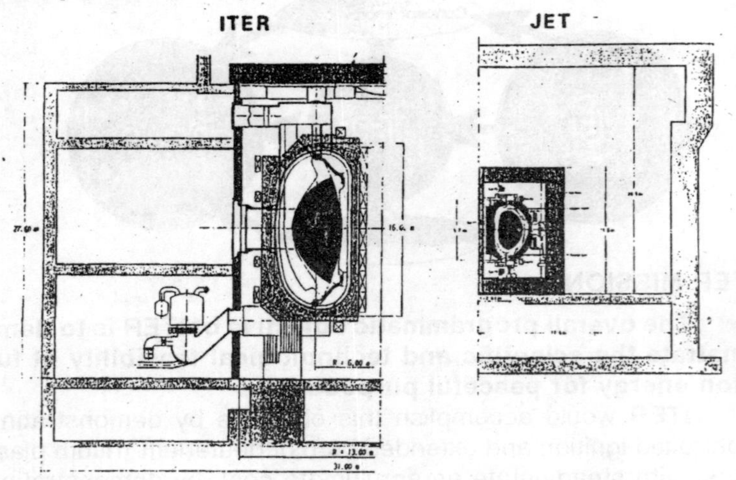

ITER REFERENCE DESIGN COST

1998 Cost Estimates including Contingencies

Construction (over 1 decade)	6,550 MECUs
Management, Engineering and R & D during Construction (over 1 decade)	1,116 MECUs
Operation (over 2 decades)	420 MECUs
Decommissioning	360 MECUs

SPECIAL COMMITTEE ON INTER UNDER JAPAN'S ATOMIC ENERGY COMMISSION
(Interim Report, March 1998)

"From the viewpoint of assuring freedom of selecting future energy sources for mankind, the ITER investment should be recognised as a proper insurance fee".

"Although an inevitabile insurance, the resources required for ITER are a major concern and thus it is important to maximise the positive outcome as well as to minimise the expense by balancing the technical objectives, the technical margins and the cost of ITER".

"Each generation should seek to satisfy the needs of its present without jeopardising the capacity of the future generations to cover their own needs"

G. Bruntland (1998)

"We are never fully the contemporary of our present time. History advances masked: it comes on the stage with the mask of the previous scene and we are bewildered by the play".

R. Debray (1969)

4

World Energy in the Next Century: International Analytical Study

E. VELIKHOV
Russian Federation

The report is prepared by the Energy Centre of the World Federation of Scientists with the objective to present the status of the world energy sector and its long-term perspectives.

The analysis is conducted in three interrelated facets: regional, technological and environmental.

When viewing the results of the study, one should be aware that the solution of the energy problem is closely linked with the general problem of resource supply for man (water, air, food etc.) and with the problem of the attitude of man to nature. The report considers only a part of this resource problem, i.e. the part related to the supply of energy. It is assumed that in the next century there will not be revolutionary changes in energy supply such as fast, voluntary and drastic self-limitation of energy consumption on the Earth or a sudden discovery of a new, environmentally clean and easily available energy resource. At the same time, it is assumed that mankind is capable of evolutionary improvement of the consumption of energy resources. The authors consider this the most probable development of the present situation.

The report is intended for the scientific community, governments and the media.

EXECUTIVE SUMMARY

If the current trends continue, the world annual energy consumption, which is at present about 8000 Mtoe, will increase 4–6 times by the end of the next century and will reach 35000–50000 Mtoe/year. The majority of the growth will occur in developing countries of Asia, Africa and Latin America.

Natural gas that will cover the major part of the increase in energy consumption at least in the first half of the next century will be the main fuel of the world energy system. The share of natural gas in the world energy balance that is at present about 23 per cent will reach 30 per cent in the first quarter of the XXI century and it may continue to grow.

Notwithstanding the dominance of natural gas in covering the increase in energy consumption, world energy systems will retain their multi-component character: coal, oil, nuclear energy, renewable resources will continue to supply energy. The share of renewable energy sources will increase as compared with their current share. However, these resources will not become quantitatively important for energy consumption in the world scale. Specific outlook of energy systems will greatly vary from region and from country to country.

The main constraints of the growth in energy supply will be the costs of the extraction of energy resources, the costs of the production and transport of energy and the emission of greenhouse gases. The importance of controlling greenhouse gas emissions will be a new key factor of energy planning in the XXI century.

The problems of energy supply to the growing and developing world under environmental-constraints cannot be timely and adequately solved by any country taken separately nor by any business structure. Only joint international efforts would give a chance of solving the global energy problems. For moving forward in this direction it is necessary to develop a joint plan of the solution of energy problems including the creation of the necessary climate for private investments. It is necessary to complement the Final Protocol of the recent international conference in Kyoto by concrete control mechanisms including international political and economic measures. In view of the threat of global warming and with fossil fuels gradually coming to an end, growth of nuclear energy is prognosticated to recommence in the first quarter of the XXI century as well as the expansion of the use of renewable resources. For the second part of the century the start of commercial use of nuclear fusion is expected.

At the beginning of the XXI century the present phase of the development of nuclear power will terminate. During this phase it was proved that nuclear energy can be used and the key problems that would determine the role of nuclear energy in the future were identified. The principal solutions for these problems are found and with their implementation the start of a new phase in the development of nuclear power—a large-scale development of nuclear energy—can become possible in the next century. This phase will be characterised by a closed character of the nuclear energy system with reprocessing and breeding of nuclear fuel, practically complete limitation of the emissions beyond the limits of the system, expansion of the applications of nuclear energy and radical decrease of the risk of proliferation of nuclear materials. For a gradual transition to the second phase it is necessary to develop a special strategy for the period of transition including technological advancements necessary for long-term solutions of the problems of nuclear power.

In line with the development of large-scale nuclear power plants (NPPs) it is desirable to consider the possibility to cover local demands for autonomous energy sources (up to 50 MWe) by small nuclear reactors on the basis of new technologies of fuel reprocessing and waste disposal with due regard to non-proliferation requirements for nuclear materials and technologies. Availability of an autonomous and safe source of nuclear energy will facilitate the creation of conditions for economic development of many regions that are experiencing development difficulties to a large extent due to the absence or insufficiency of energy supply.

For gas technologies, it is necessary to accelerate the development of advanced technologies of gas conversion into the liquid form, to advance the technologies for gas transportation, to increase the efficiency of use, and to sharply decrease methane losses and CO_2 emissions to the atmosphere.

In addition to the existing oil and gas regions of the world, the Arctic is becoming a large oil and gas province and is entering the international economic exchange.

Of the non-renewable fossil fuels, coal is the most abundant resource. The world coal deposits amount to 1000 billion metric tonnes (1 billion=10^9) as compared with 140 billion tonnes of oil and 130 trillion (1 trillion 10^{12}) m^3 of natural gas. Due to the abundance, coal will retain its role in the world energy system. The extraction and use of coal is an important task of public and private structures. It is necessary to concentrate efforts on the development and implementation of technologies for effective and environmentally acceptable use of coal.

Nuclear fusion remains a promising solution of the energy problem. Due to the exceptional complexity of mastering this energy source, the transition to commercial use of nuclear fusion would be possible only under the condition of close and effective international cooperation. As a first practical step, it is necessary to make the decision to build the International Thermonuclear Experimental Reactor (ITER).

Energy conservation policy is today's must for both developed and developing countries. The development and implementation of energy conservation technologies should be one of the priorities in the energy policy. Timely implementation of energy efficient technologies is especially important for developing countries in view of the expected fast growth of energy consumption.

Most of strategic energy problems are of long-term character. Participation of private investors in their solution is hindered by the global character of the problems, uncertainty of technical solutions and elevated economic risk, in particular in view of the on-going global privatisation of the power sector. Therefore, the formulation and implementation of a responsible energy policy is a task for national governments. An important part of such policy is the creation of a climate facilitating the participation of private capital in the solution of energy problems.

Russia possesses a considerable part of world resources of fossil fuels, natural gas first of all, also resources of hydraulic energy and all existing technologies for energy extraction, transportation and conversion. Up to 40 per cent of world resources of natural gas, 13 per cent of oil and about 30 per cent of coal are located in Russia with the existing energy conservation potential up to 0.5 billion tonnes of coal equivalent. It opens wide opportunities for both increase in useful consumption and the expansion of energy export. Thus, Russia is a key partner in the global solution of the energy and the climate problem.

A most important task is a timely creation of a reliable and effective global infrastructure of energy transportation, because energy sources are located at considerable distances from energy consumers. Russia has created in this century unique and largest in the world transportation systems for electricity and gas, and the termination of the cold war and the conversion of military industries open opportunities for the use of powerful Russian industries for the development of both energy facilities and transportation systems.

The creation of a technical and economic system that would link Russian fuel resources and Russian energy technologies with energy consumers in Asia is an important aspect of the solution of energy problems in the XXI century.

5

The Moscow G 8 Energy Ministerial Meeting: Conclusions and Recommendations

A. MASTEPANOV
Russian Federation

The representatives of the world's leading energy companies who held their meeting in Moscow to discuss the most important issues facing the global energy sector over the next two decades, have identified the following important messages for the G8 Governments in their deliberations tomorrow. The business community welcomes this initiative on the part of G8 Governments to dialogue directly with the energy industry in order to address together the challenges facing the global energy sector.

GLOBAL ENERGY PERSPECTIVES

1) Studies on long-term energy development indicate a significant rise of global energy use, for example WEC studies point towards a 55 per cent increase in global energy consumption between 1998 and 2020. This development is driven by:
    - world population growth;
    - economic development;
    - continued urbanisation, industrialisation and increasing mobility.
2) In terms of energy resource requirements, this could mean that in 2020:
    - 90m barrels of oil could be consumed globally each day, an increase of 25m, or more than the entire current OPEC production today;
    - Coal output could double to nearly 7 billion tonnes annually, or more than twice the UK's known coal reserves;

- Gas production could reach 4 trillion cubic meters, or almost as much as the USA's present gas reserves.
- more electricity generating capacity will probably have been built over the period 1990 to 2020 than was built in the previous century.

3) Identified global energy resources are more than adequate to meet this growth in demand. Fossil fuels will continue to dominate energy supply over the period, and all sources of energy—fossil, nuclear and renewables—will need to be exploited in order to satisfy demand. But accessing remoter resources, lengthening supply lines, increasing environmental impacts such as potential climate change, growing dependence on trade of energy, and the changing geographical patterns of use, pose political, institutional and financial as well as technological challenges. Failure to meet these challenges risks geo-political instability.
4) Developing countries which today account for approximately 35 per cent of global energy demand will account for 50 per cent by 2020, and for at least 70 per cent by 2100. Concentrated within developing countries, 2 bn people (i.e. 35% of all people in the world today, 800 million of them in South Asia, and 400 million in sub-Saharan Africa) have no access to commerce energy supplies.
5) To enable the world to cope smoothly with the consequences of the growing and changing demands for energy resources will require:
   - *international dialogue and agreements to avoid arbitrary political interventions and consequent market instability;*
   - *removal of barriers to international trade;*
   - *diversifying energy supplies so far as practical in terms of energy types and regions where national resources are exploited and the means by which they are transmitted and distributed to consumers;*
   - *timely action:* many energy projects still tend to have long lead-times and new technologies can take decades to penetrate energy markets on any major scale. The current low price of energy supplies should not be allowed to defer early action on the grounds that there appear to be no early problems and government and

industry together should seek to take a long view and avoid short-term volatilities;

- *mobilisation of the necessary finance;*
- *sustained attention to environmental issues* so that energy can be supplied in an increasingly sustainable way.

ENERGY SECTOR LIBERALISATION AND THE EVOLVING ROLE OF GOVERNMENT

6) The main trends in the energy sector in the last decade have been liberalisation, deregulation, the development of competitive markets, the withdrawal of Governments from the role of producers of energy or reliance on State-owned monopolies and the emergence of a wide range of global energy supply companies. While it is still early, days to reach final conclusions on the benefits and deficiencies of energy sector liberalisation and privatisation, some clear messages are emerging:

- diversity of energy supply in a market-oriented environment seems to deal adequately with worries about security of supplies in many instances;
- *Whilst each Government will need to find its own balance between the sometimes conflicting objectives of competitiveness, security of supply, environmental protection and public service, Governments can increasingly focus on their core responsibilities: managing the economy to maintain stable, healthy economic conditions; ensuring an adequate framework of law and regulation suited to their particular needs; and building or maintaining the institutions required to enable the energy sector to provide consumers with cost-efficient, acceptably clean and user-friendly energy services. Government activity in particular should concentrate on providing stable, transparent, and non-discriminatory legal and fiscal regimes; fair, independent and effective regulation that finds a balance between necessary regulatory intervention and freedom of energy companies to act; minimisation of bureaucracy and opportunities for arbitrary government intervention; and abolition of national preferences. These concepts*

are embedded in the Energy Charter Treaty whose ratification remains important, and adoption of these concepts seems fundamental to enabling developing countries and economies in transition to overhaul their deficiencies in meeting energy demand.

7) For markets to work efficiently and effectively, it is important that proper economic prices are signalled both to give customers a clear basis for their decisions and to ensure optimal resource allocation. *In many energy markets the wrong signals are being sent out by government controlled agencies through subsidising energy prices for a wide range of consumers instead of targeting direct financial support to those who need subsidies.* The result is huge inefficiency in the supply and use of energy, the waste of energy resources, adverse environmental effects, and gross distortion of the operation of markets. This is particularly true of capital markets. *Governments* accordingly *need to pursue policies that phase out energy subsidies and encourage market prices*, particularly to ensure the proper working of domestic capital markets. Where temporary hardship results, financial support can be channeled to those in need and should not be broadcast to users indiscriminately.

ENERGY FINANCING

8) Meeting growing energy demand in general, and the needs of the 2 bn now without access to commercial energy in particular, will require the mobilisation of the necessary capital. The WEC estimates that the cumulative energy supply investment requirement for the period 1990–2020 could be of the order of US $ 20–30 trillion, about 3-4 per cent of world GDP, if the consequential infrastructural and environmental mitigation investments are included. Internal cash generation by energy enterprise is a first source of finance, hence the importance of enterprise to be efficient in all aspects of its operations, including collecting payments from customers. Aid and multilateral institutional finance can only provide a fraction of the necessary capital, through that finance can be useful in reinforcing appropriate market reforms. Global capital markets are, in principle,

more than adequate to provide all the necessary finance. However, mobilising private capital resources may prove difficult, especially in those countries which do not satisfy investors' criteria for a stable economy, market transparency, a well-founded legal framework that includes safeguard of property rights, properly regulated financial markets, and an appropriate balance of risk and reward. Capital will naturally flow to the better-organised markets and more attractive projects: countries who are slow to recognise and respond to these points will continue to lose out in the competition for funds.

9) *Governments have a crucial role in creating and maintaining the institutions and the framework necessary to encourage the mobilisation of private domestic capital and attract international investors. Again, the keywords are stability, transparency, non-discriminatory, well-founded law and regulation, independence from political intervention.* Decisions on projects, in particular, approvals, permits and certification need to be taken in a timely and consistent way, using procedures and standards consistent with international practice. This remains a problem in many of the developing countries and economies in transition, and Governments may like to consider how to encourage, e.g. the Asian countries to establish an Asia Energy Charter on the lines of the Energy Charter.

ENERGY AND ENVIRONMENT

10) The exploitation of any energy resource gives rise to some form of environmental problem:
   - local, represented by mining spoils, "end of pipe" emissions, accidents, loss of amenity;
   - regional, represented by deforestation and soil impoverishment where commercial energy is unavailable and fuel wood and dung are the only available fuels, and by acid deposition where abatement measures are not taken;
   - global, in particular possible climate change as a result of enhanced global warming.

Whilst particular environmental problems call for particular solutions and technology continues to make significant advances in reducing environmental impacts and exploiting the opportunities that environmental challenges also offer, there are some common prudent measures of universal benefit that Governments should seek to encourage:

- *greater efforts (whether by incentives, education, restructuring of enterprise) to increase energy efficiency* in production, conversion and use. Huge potential can be found in the under-performing assets in many countries: This will involve *phasing out subsidies and including external environmental costs in consumer prices* in-so-far as they are measurable and not reflected in market prices;
- *educational efforts* and *public information campaigns* about costs and benefits of all energy forms;
- wider commercialisation of renewable energies, particularly with a view to providing energy for rural areas remote from existing energy infrastructure;
- *increasing public confidence in the development of nuclear power, which can offer major benefits in terms of potential climate change as well as securing a balance of energy sources.* This may require further attention to nuclear waste management and proliferation issues.
- faster diffusion of the most-efficient technologies to developing countries.

Governments should base legal/regulatory decisions on full life-cycle costs and benefits (on which more research is required), establish stable, clear and non-discriminatory environmental regulations, including at an international level where appropriate; and consult with industry and consumers as well as scientists and technologists to establish realistically achievable standards applicable to all.

11) Government interventions in support of sustainable (including environmentally acceptable) supplies of energy should observe the following principles:
   - *they should be realistic*. If they are not, they will become discredited.
   - *any regulation should be aimed at the objectives sought—not the means of achieving them*, which are best left to the workings of competitive markets and diversity of approach;

- where possible, economic instruments should be preferred to legislation and regulation (e.g. emissions trading where this is directed at reducing future emissions), and Joint Implementation programmes and projects. Again, Governments should dialogue with industry to work out suitable and practical systems, and should not be shy of inviting industry itself to come up with practical solutions to meet the objectives set by Governments.
- *fiscal interventions* (e.g. to promote R&D in relation to renewable energies) should be *tightly targeted and time-limited*.

RUSSIA

12) The meeting took note of the importance of Russian energy resources for the global energy economy and of the encouraging and necessary steps being taken by the Russian Government and Russian energy industries to establish a legislative and independent regulatory framework within which to increase the efficiency of those industries, encourage inward investment and facilitate the trading of Russian energy.

6

Private Finance for Sustainable Energy

R. LION
France

"Sustainable Energy" is the addition of two large energy paths, which are contributors to Sustainable Development:

1. energy efficiency, and
2. "new renewables", including biomass and co-gen.

Sustainable Energy is a major element for security:

a) it relies on local sources (and this involves energy efficiency since the cheaper energy resource is the energy which is not used);

b) it reduces the global environmental risks, such as climate change, pollutions, deforestation.

Attracting private finance to Sustainable Energy is:

1. Indispensable;
2. Tricky;
3. Possible.

1. PRIVATE INVESTMENT IS INDISPENSABLE

A) Whatever the global need for investment in energy through the next century, public money will be short.

Energy scenarios present very different pictures about the global energy demand (in GToe):

	1990	2050	2100
WEC IIASA A	9	25	45
WEC IIASA CI	9	14	21
NOE-CNRS	9	11.5	12

...and many others, including the Shell "Dematerialisation scenario" and IPPC scenarios.

Six months after Kyoto, it becomes more and more likely the world will follow on the longer term a low-consumption scenario. Still, for decades, from now on, the needs will grow, because of growth and demography in emerging countries, including very heavily populated ones such as India and China. Also because the over 2 Bn people having no access to commercial electricity have, sooner or later, to be served.

The volume of capital needed is estimated by the WEC "business as usual" or "A scenario" around 1,000 Bn US $/year in the decade 2020, versus around 450 Bn US $/year presently. It is generally considered that such a volume is not sustainable unless one accepts terrible constraints on other major issues—which usually rely on public investment—such as education, health, fighting poverty.

The greatest shortness will be that of subsidies of all kinds, coming from governments, development aid, multilateral or regional institutions. Fiscal constraints have become, and will remain, universal.

B) **The shift to private investment** is already spectacular for conventional energy. Ten years ago, in most countries investments in power generation and energy transportation were made by public entities. Now it is the opposite: gas pipes, power plants (IPPs), central heating systems are awarded to private investors. This is the case even in China, and of course in countries in transition. Privatisation of power distribution is also more and more frequent.

This has to happen to energy efficiency investment and, to a large extent, to renewable energy projects. For many technologies here, the learning period is over, the competitiveness is acceptable—especially when you consider bringing electricity, through renewables such as solar, wind, micro-hydro, in very remote areas, far away from any grid. For many investments hertz, the pay back is rapid and high—especially in energy savings.

There is still some need for government capital, for instance when:

a) the end-user's solvency is low (P.V. in poor countries)... but the end-user's money, which is private savings, must be in any case called upon. Granting power is not a sustainable solution;
b) one wants to develop renewables where power generated by conventional fuel or nuclear is sold at low prices, which

is the case in France... but here is it subsidies you need, or higher prices from the utility purchasing the "green" power ?

Still the broad picture for Sustainable Energy is: time has come to stop being bottle-fed with public funding. So private investment in this area is indispensable.

Note: *Investment of private capital can be done in many ways: the end-user's payment; the local savings through local private institutions; international funding through the corporate sector or capital markets.*

C) **And is it happening**. Private funding is already flowing in investments such as:

- private sector energy savings: ;making more efficient and less costly industrial plants, office facilities, transportation systems etc.
- in heating, cogeneration;
- in building materials, home appliances (low consumption lamps, refrigerators,...) motor vehicles.

I want to mention here the fast-growing number of funds in North America and Europe to invest in (or purchase) micro-hydro facilities, wind farms, bio-mass power plants with no government subsidies.

It is also very likely that the Kyoto Protocol Act 17 mechanisms, such as joint implementation and/or treadable permits will foster private investments in this area.

Certainly there are gaps. Private capital is not attracted much by energy savings in public buildings (even if through third party financing, paid back on energy savings, the profitability can be high), sustainable town planning or architecture, etc. Private investment is also quite low in energy efficiency investments in developing countries (ex: unfrequent search for efficiency in office or hotel buildings, in transportation, in many sectors of manufacturing industry: a paper-mill in South Asia will use 3 times more energy than the same plant in the E.U.).

Finally, there is little private investment in renewables, with significative exceptions:

- micro-hydro in Western/Central Europe,
- wind farms, even in emerging countries (Morocco),—biomass power generation.

2. PRIVATE INVESTMENT HERE IS TRICKY

A) **For specific reasons**

The problem with Sustainable Energy is not a technical problem: technologies are available, and often competitive; market growth is year after year reducing their costs. Specific problems, especially with renewables, are:

- dispersion, small-size, no institutional backing;
- frequent low expertise of the promoters, in particular financial/management knowledge;
- high upfront costs, compared to low maintenance costs, and to the end-users' resources;
- end-user low solvency for instance in remote and poor rural areas; lack of micro-credit systems;
- unfamiliarity, for banks and investors; short track record, skepticism from a large portion of the financial community.

B) **Private Capital's Requirements**

Considering developing countries and countries in transition (but could I say: everywhere), capital compares investing here with competing investment possibilities:

- return on investment ? Capital looks for high yields, specially when the sector is not familiar or the country's stability is problematic (risk premium);
- liquidity for the equity invested, or likeliness of good servicing if it is debt?
- in the case of energy: certainty of supply (there could be problems with bio-mass, co-gen)? power purchase agreements (PPA)? long term price guarantee?

Let us add that private investment has not always been successful in the energy sector as a whole. There are sometimes negative impacts: too high prices for the consumer, burden for the local utility which has been obliged to purchase high-priced power. There are problems when the country endures recession, or social disruption (South East Asia). But this is not proper to energy. So it is not yet a very easy sort of investment.

3. PRIVATE INVESTMENT IS POSSIBLE

A) **A good regulatory and political Environment**

There is a need for a level playing field, breaking with situations where conventional and nuclear fuel has been, and still is, directly or indirectly subsidised. The imbalance is nearly everywhere against renewables, co-gen, energy savings, demand-side management, etc. This must and can be changed through:

- reconsidering perverse incentives;
- proper regulations, such as purchase obligations (the UK NF-FO system), price guarantees; the role of utilities is here essential;
- capacity building,
- full cost pricing, taking into account externalities or, more simply, enforcing carbon taxes as the Netherlands do.

And then leadership in the country is required. Investors need confidence, and a credible and well established energy policy. It is finally a question of politics.

B) **Pro-active Partners**

Financiers. Good news: more and more financiers are convinced this is an emerging, and possibly profitable, market. They need to find and build good partnerships, including utilities. They have to apply here, as they are beginning to do, good and eventually innovative financial techniques[1].

A new instrument is becoming serious: Green Funds, following or not tax incentives such as in the Netherlands. More large investors, such as pension funds, want now to have at least part of their assets in "SER investments"[2]. Even ratings (of Funds and Companies) are appearing, both in North America and Europe, to certify the "SER" quality of investment. This of course is not going to channel money only to sustainable energy, but it will help this activity on a large scale. It could mean billions of US $ a year.

As already mentioned, Tradeable Permits will also flow corporate sector investment to sustainable energy in Third World Countries.

Developers. Indeed, they are pro-active. They have to understand better the private capital's requirements (see above) and speak its language.

They have to take benefit of the large market the "Green power" is catching (as "bio-products do) in industrialised coun-

---

1. See Proceedings of the Seminar "Attracting Private Finance to Sustainable Energy", organised in Brussels by Energy 21, October, 1996, under the sponsorship of E.U. DG12.
2. Socially and Environmentally Responsible".

tries: in several countries, the end-user is ready for "Green pricing", which is paying the Kwtt up to 10 per cent more (15 per cent in the UK) if its generation by "clean energy" is certified.

Funding Institutions, such as Government Agencies, aid mechanisms, or multilateral banks have to play a leverage role, or a driving role. They must better understand, and venture on a regular basis with, private finance when sustainable energy projects are at stake.

Possibly, this should be the role of "mainstream" loans from institutions such as the Word Bank. This would be better than leaving such initiatives to "Focused" Funds (several have been launched in the Word Bank Group, with no great success), or the Global Environment Facility (GEF). This could concern as well the E.U.'s aid policies. "Mainstreaming" green concerns in their core business will make large agents powerful leverages, and attract flows of private funds.

C) Lower Needs

Facing the capital shortage, a good response for whom wants an easier funding of the investments really needed, is : less energy production. This as I showed requires investment savings, efficiency, low consumption goods; it means producing "Negawatts", which is generally the cheapest Energy source, and is usually profitable enough to make private investors happy. It is also a job-intensive activity, when conventional generation of power is only capital intensive.

So we are on a **win-win-win-win issue**. Correct private investment is good here:

- for the economy,
- for the environment,
- for jobs creation, and
- for the private capital itself.

There are indeed many reasons to turn to private capital in the area of Sustainable Energy, if a good balance is struck between deregulation, and proper regulation. The latter is very much needed to ensure that power is affordable for all categories of population, wherever they live.

There are indeed reasons to be optimistic.

7
The Energy Challenge for Developing Countries

F. CASANA
Belgium

The developing countries will face in the next decades a strong increase of their population and a strong economic development.

The Asian crisis could lower prospects further, however even at the most conservative estimates it is expected that Asian GDP will rise over 160 per cent from 2000 to 2020.

The GDP of Latin America will increase by 105 per cent in the same period.

From an energy point of view this evolution will result in two challenges for developing countries in the future:

— Increase in energy demand;
— Sustainability of the development.

1. DEVELOPING COUNTRIES AND INCREASE IN ENERGY DEMAND

In its study "*European energy to 2020—A scenario approach*" the European Commission forecasts an important increase in the share of developing countries in world energy demand.

From 1990 to 2020 this share will increase from 28,2 per cent to 41 per cent, while the share of OECD countries will decrease from 51.2 per cent to 45 per cent.

The annual average growth of energy demand will be 6 per cent in developing countries against 1 per cent only in OECD countries. Globally, the energy demand of developing countries will more than double and will represent, in 2020, 90 per cent of the demand of the OECD countries. Half the increase in the world energy demand will be represented by developing countries.

The same trend remains whatever scenario is taken. The share of developing countries in the world energy demand in 2020 is expected to be between:

— 40.7 per cent in a "battlefield" scenario where contractions and instabilities in the global system make economic integration very difficult. Globalisation is seen as too ambitious. The geopolitical system fragments into blocks, with tensions and friction between and within blocks;
— 41.8 per cent in a "forum" scenario where the process of global economic integration produces new imperatives for public action. National, European and International institutions are gradually restructured so as to be able to deal more effectively with broader, more complex shared problems and interests;
— 42.2 per cent in a "hypermarket" scenario where global economic integration is self-reinforcing and continued. The force driving this scenario is continued application of the market mechanism, which is seen as the best way to produce wealth and to handle complexity and uncertainty. Liberalisation and privatisation deliver results and produce new market entrants.

When it comes to regional trends some important elements appear:

— China alone represents more than 20 per cent of the developing countries increase and 10 per cent of the world increase;
— Asia together with China represents nearly 65 per cent of the developing countries increase;
— Africa represents 20 per cent of this increase, but with a particular concentration in North Africa;
— The rest is represented by Latin America.

The increase will be dramatic for China, India, Brazil and North Africa.

This evolution will result in three major consequences:

1) Important Financing Needs

The investment required in the energy supply sector to meet the anticipated increase in demand will be extremely high.

In electricity supply alone the 2020 Study estimates total EC power plant investment requirements to be 543 billion ecu per year.

In developing countries this investment is estimated at 80 billion ecu per year between 2000 and 2020.

2) Pressure on the World Energy Market

Global energy supply is likely to grow from an estimated 8.5 btoe in 1995 to nearly 12.6 btoe in 2020, which is 1.6 per cent per year. This is a little less than the growth seen over the past twenty years.

Natural gas is forecast to be the fastest growing fuel with a growth rate of around 2.7 per cent per year.

Oil would retain the largest share of global energy, albeit falling from 37 per cent today to around 30 per cent by 2020. This is slightly more than natural gas.

The supply of solid fuels will grow mainly in the developing world, albeit with its share of world energy falling from 27 per cent in 1990 to 24 per cent in 2020. The main region for solid fuel consumption will be Asia, which will absorb about 50 per cent of world production in 2020 against 32 per cent in 1990. This is associated essentially with the expansion of the power sector.

Nuclear energy will grow the least due to the decline in the developed world.

Renewables are not seen as having a growing share unless there are specific incentives or there is a rise in energy prices.

This fuel mix shows a good prospect for security of supply with more diversification both in terms of fuels and geographically. However, the environmental concerns could rise.

On the another hand, an eventual rise in prices due to this increase in the demand for energy would particularly affect developing countries, which have fewer resources than developed ones.

3) Increase in Emissions of Greenhouse Gases

Without going into details, as this is a well-known debate, it is worth noting only two points:

— from 2000 to 2020 the level of emissions from developing countries will represent 60 per cent of the world emissions increase in Pre-Kyoto assumptions, and 70 per cent in Post-Kyoto assumptions. China, India and Brazil will together represent 25 per cent of world emission increases.
— In this context the objectives are clearly stabilisation or a decrease in emissions from the developed world, and the

monitoring of the increase in emissions from the developing world.

In order to avoid problems for developing countries, such as power shortages, impeding economic growth, large scale pollutions problems or having sections of the population without access to energy, several international cooperation actions should be implemented.

A) The first action should be an *increase in international private investments* in production, transport and the distribution of energy. Public resources are scarce, the debt of many state companies in developing countries makes new financial loans impossible. It is therefore important to implement new financing schemes. On the another hand, the private sector appears to be more efficient than public sector companies.

B) In order to facilitate these private investments a process of opening liberalisation and privatisation of the energy sector is needed. This process is well advanced in Asia (not including China), progressing quickly in Latin America (Chile, Argentina and now Brazil and beginning in Africa.

The second action would be to increase the use of local energy resources. *Integration and open regionalism* could speed up this evolution. Resource, which are not needed at national level in one country due to insufficient demand or which cannot be exported overseas could be used in neighbouring countries if proper regional integration existed. The use of the export of Argentinian and Bolivian gas productions to supply to growing markets in Chile and Brazil is one of the main examples. *The increase in the use of renewables* could also be part of this process. Many renewable technologies could be used locally: wind, solar, photovoltaic, geothermal etc. The use of this local energy in some remote areas could be advantageous and even bring more benefits than a mere extension of the grid.

The third action would be an *increase in energy efficiency*. If in European countries the promotion of energy efficiency was the consequence of the oil shocks, and now is one tool of mitigation of greenhouse gases emission in developing countries, it could contribute to a reduction in the number of new plants and thereby a reduction in the financial needs for new investments.

The last action would be a *transfer of technology* that could respond to the needs of developing countries once these technologies have been proven in developed countries.

2. SUSTAINABILITY AND ENERGY

Energy plays a very important role in sustainable development, particularly for poverty, gender issues, population, health, the environment, investment, foreign exchange and security.

The United Nations Development Programme (UNDP) took a strong interest in this topic with a recent publication "*Energy after Rio: prospects and challenges*". More recently again, a common work co-funded by the European Community and UNDP addresses this topic in Africa, Caribbean and Pacific countries.

Eradication of Poverty

Eradication of poverty is both a social and an economic issue. Development is proceeding in diversified ways and at very different speeds in the various countries and within each country. While some developing countries, have achieved considerable rates of economic growth and impressive reductions in poverty, the process of development and globalisation has left other countries increasingly marginalised and suffering continuing deterioration in their living standards. In Sub-Saharan Africa and Small Islands Developing States the per-capita GNP has decreased by 18 per cent from 1980 to 1995, generally due to an increase in the population.

Energy is a key element for apportioning resources to the population in the agricultural sector and remote areas. Even in a fast developing and widely liberalised country such as Argentina it is estimated that 2.5 million inhabitants out of an overall population of 14 million have no access to the grid.

Due to the liberalisation process, itself linked with the increase in financing needs for new investments and the increase in prices for common users, the poor population in urban regions could have access to energy denied.

Environmental Dimension

Environmental effects of the energy systems represent a heavy burden for developing countries, even when very little useful energy is actually consumed. These problems will worsen as energy consumption increases. Indoor air pollution resulting from cooking with firewood or agricultural residues on primitive stoves in poorly ventilated dwellings is a major health concern, especially for women and children.

Air quality in many of the large urban conglomerations in developing countries is generally worse than in the metropolitan areas of industrialised countries. This problem is particularly relevant in Asia and Latin America where energy demand is growing quickly.

Energy systems are important contributors to the pollution of soil and especially of water, an essential and often scarce resource. Indirect, but important, environmental effects of the depletion of wood resources, contribute eventually to deforestation and desertification.

Energy systems account for the largest share of the emission of greenhouse gases, which threaten the stability of global climate. In the above-mentioned study, the European Commission estimates that from 2000 to 2020, 60 per cent of world emission increases will be represented by developing countries.

Energy is also an important element in the development of health, social and education services that will grow with the increase in population. The energisation of public services is one of the key elements of rural energisation as illustrated, for example, by the Argentinean and South-African examples.

The measures to be taken in order to face these problems consist of several elements:

First the use of Sustainable Energy Technologies

There are three main components of a more sustainable energy system:

— improvement in the efficiency of final energy users;
— development and diffusion of renewable energy sources;
— more efficient utilisation and transformation of final fuels.

A continuing effort in research and development on sustainable energy technologies is necessary to find new solutions, better ways to apply them, improved economics and performance. An increase in cooperation between research and development in developed and developing countries is needed, with a special emphasis on the needs of developing countries.

Transfer of technology is also a vital element, but with regard to developing countries, it has to obey the following rules:

— it should consider transfer of proven technologies. Developing countries could not "test" new technology;
— it must be adapted to the needs of developing countries as many technologies are more adapted to the conditions prevailing in developed countries;
— the international cooperation should focus not only on the investment phase, but also on the maintenance phase in order to keep the material equipment functioning in good conditions. Training linked to the investment is a key element of success;
— there is a gap between market interest and countries' interest. The main example is the Small Island States where renewable technologies are extremely important but there is not the market to attract foreign private investors.

The main technologies available are:
— on the supply side: gas fuelled dual cycle power stations, combined heat and power generation, micro-generation units, fuel cells, gasification of coal, new methods for the exploration and exploitation of oil and gas fields: advanced prospection methods, horizontal drilling, transport of two-phase mixtures and field stimulation technologies, gasification of tars in refineries, underwater pipelines. In renewable energy continuous improvements are taking place, but at a slower pace than expected. Photovoltaics and wind energy are the two most reliable technologies today. The use of biomass through direct combustion and through gasification is also progressing, while less work has gone into the production of liquid fuels from biomass.
— On end use technologies: improvement of energy efficiency through the use of new materials and components in buildings, increasing efficiency of vehicles, fuel cell vehicles, new services: monitoring and targeting, third party financing....

Another point is *to identify and remove legislation and regulatory barriers* (due to norms, regulations and the institutional environment) to innovations and investments in the energy field. Regional integration is also a main element as it allows countries facing similar problems to tackle them together, eventually helped by international assistance, and find and implement joint solutions.

International cooperation and assistance has to be sustained and in particular take into consideration energy as a tool for sustainable development. More often, energy is considered as only a problem of infrastructure and important investments. Its role in the aspects of sustainable development should be enhanced. Moreover, the decrease in oil price, must not lead to a decrease in international aid in the field of energy.

The work of the UNDP and EU concerning "*Energy as a tool for sustainable development*" in the above mentioned ACP countries, as well as the work implemented by the Interamerican Development Bank and also co-funded by the EC "*sustainable markets for sustainable energy*" in Latin America, are good examples of thee new orientations.

In conclusion, the energy situation in developing countries will need important attention from the world community in the next decade. Increased cooperation, transfer of technology and flows of capital are necessary in order to allow the satisfaction of the fast growing demand. The evolution of the energy sector concerning this point has more to do with the "forum" rather than the "battlefield" or "hypermarket" scenarios.

8

Problems of Energy Supply in a Small Island State

R. GHIRLANDO
Malta

1. INTRODUCTION

In the last few years, the United Nations has recognized that Small Island Developing States (SIDS) have special problems of their own. The first such recognition was made as far back as 1972 during UNCTAD III (ref. 1). In 1994, the "UN Global Conference on the Sustainable Development of Small Island States" was held in Barbados. The small island states have themselves set up an organisation called the Alliance of Small Island States (AOSIS); a list of the members of AOSIS is given in appendix 1. It has been recognised, for example, that the economic development of such states cannot be judged by their GDP, since these micro-states are very vulnerable to external forces. One of Malta's economists has developed the concept of a "vulnerability index" that tries to measure this vulnerability by taking into account the imports and exports of a country as a percentage of GDP (ref 1). Another factor of vulnerability is the proneness of small island states to natural disasters.

The problems of satisfying the energy demands of small island states arise partly from their insularity and geography, and are similar to those suffered by other small islands that are far enough from the mainland for this to cause problems to their energy supply. For small island states, the problems are made worse by their being fully fledged sovereign states and therefore totally responsible for their own well-being. This paper discusses the problems of satisfying the energy needs of a small island state, using Malta as a case study.

With a land area of 312 sq. km. and 375,000 inhabitants, to which must be added another 30,000 tourists that would be visiting Malta on any one day, the Maltese archipelago is some 100 km. from the nearest "mainland", i.e. the island of Sicily. Its energy requirements are met by a power station of 466 MW installed capacity and fuel imports of approximately one million metric tonnes of petroleum products, including fuel for the generation of electricity (ref. 2).

2. EXPECTATIONS AND PERCEPTIONS

In an island state such as Malta with its European aspirations, the standards expected by its population are nothing less than European. Thus the cost of electricity, the reliability of its supply, the quality of the fuels sold on the local market, must be comparable to those on mainland Europe, and yet few people realise that the conditions are totally different. Being an island state poses problems and costs of a different nature from those found on the mainland.

Also, whenever recourse is made to foreign consultants, these bring with them their experiences based on solutions that are correct for large country systems. Whether these solutions are also right in the context of a small island state is hardly ever questioned. One of the worst problems caused by the importation of such foreign advice has been the loss of age-old traditions in the way buildings were designed in Malta, to be substituted by solutions meant for colder climates. Thus thick walls, tiled floors, high ceilings, wooden louvered shutters have been abandoned in favour of thinner walls, low ceilings with soffits, carpeting, large steel and glass facades, etc. which then require energy-hungry, air-conditioning to make the space bearable in summer.

At the local personnel level, the small size of the country also poses a problem. In a large country, there is bound to be mobility of personnel, especially at the higher levels, between different operators in the sector. This encourages the transfer and development of technology and know-how. When the country is small, on the other hand, most of the personnel will have gained their knowledge in-house, which is not very healthy. One way round this problem is for the energy company to forge strategic alliances with large overseas organisations that can provide training and opportunities for work experience for its personnel.

3. ELECTRICITY GENERATION

1) Peak Demands

One of the problems of providing electricity is the difficulty of storing it; the supply (and hence generation) of electricity must follow instantaneously the continuous variation in demand. If the number of consumers is small, then the demand curve will have considerable fluctuations. On the contrary, a large number of consumers will produce a smoother and flatter curve. The larger the number and the greater the area served by the network, the smoother and flatter will the demand curve be. Hence the effects of variation in electricity demand are more pronounced for small isolated islands than for the mainland. Malta in terms of population is not more than a small town in Europe, but whereas the latter obtains its electricity from the European grid, Malta stands alone. Thus it must have enough generating capacity to cope with its peak demands even if for most of the time much of this capacity is not needed. Clearly this additional equipment adds to the cost, especially since some of the machines have to be switched off at night and started again early in the morning. Generating machines are not particularly happy to operate in such a cyclical manner. Malta uses as much as 25 per cent of its electricity to produce fresh water from the sea. Some preliminary studies have been conducted to see whether it is possible to operate the Reverse Osmosis (RO) Plants at night in an attempt to balance the load, especially now that there is enough spare RO capacity to operate them this way. Unfortunately, these (RO) plants suffer even more from cyclical operation than generating plant.

2) System Reliability

Another advantage of being connected to a large grid is the increased reliability of supply. In a relatively small system, the output of a single generator is a large proportion of the total output of the system. In such a system, when a single unit fails, the sudden loss of power is considerable in proportion to the power being generated (and therefore) demanded at that point in time. Because it is a relatively high proportion, the other machines find it difficult to make up for the lost machine and to take up themselves the load of the unit that has failed, even if at that moment in time they do have the spare capacity. Their speed of response does not match the speed at which the power is lost from the failed unit. Hence there is only one thing that the controller of the system can do and that is to shut down part of the load. And since the unit lost is a relatively high proportion of the supply at that instance, a substantial part of the country would be plunged into darkness. In a much larger system, however, the loss of a single generator of the same size would hardly be felt. The other machines would be able to absorb the extra load quite easily, since it would represent only a very small increase. It must be appreciated that even a stoppage of one per year means a machine availability of 99.989 per cent. On the other hand, the aviation industry is well known for the high level of reliability that it manages to achieve. Hence the attainment of high reliability may be more a problem of cost rather than a purely technical matter.

A study carried out a few years ago showed that it was technically feasible to connect electrically Malta to Sicily via a 100–200 MW DC link (ref. 3). The study envisaged three options for this link: (i) as spare capacity to meet reserve requirements during the winter and occasional exchanges between the two systems, (ii) for regular base load imports from Sicily and (iii) full supply. The study did not consider, however, whether this link could serve to improve the reliability of Malta's electricity supply, although it is clear from the study that such a utilisation, even if technically feasible, would be very costly.

3) Size of Generating Units

One question regarding the generation of electricity is the size of the units that should be installed. It has long been the theory that the larger the generating unit the lower the costs of generation. Malta's needs could be supplied with just one machine but clearly that is not an option since it would be a very high risk indeed. The question then arises: how big should the units be? On the one hand, large size means lower costs, but on the other hand, smaller size means greater reliability of the system (as was discussed in a previous paragraph) and smaller extra capacity required for breakdowns and maintenance. Spare capacity is needed not only to meet peak demand but also to make up for breakdowns and maintenance. Thus it is necessary to have at least as much spare capacity as may be required for the regular maintenance of the machines. This will depend on the periodicity of maintenance and how much fit can be achieved between the maintenance plan and the seasonal demand curve. Further, spare capacity equivalent to at least the largest machine is also required so that if this breaks down, there is enough to spare. Clearly, the bigger the largest machine and the higher is the spare capacity required, with all that entails in terms of additional capital and running costs. So a compromise must be reached between having a few large machines and many smaller units. At one stage Malta had opted for 60 MW units, but even this size, although small by normal standards has proved too large. The latest machines being installed now are gas turbines of 37.5MW each. A theoretical calculation of the ideal size would have to take into account the initial capital cost, the running costs based on the cycle of operation (which in turn depends on the demand curve), the machine's efficiency curve, the assessment of the risk of the unit failing and the impact of such a failure, maintenance, etc.

4. ENVIRONMENTAL PROBLEMS

A problem of microstates is their limited land area. In the case of Malta, this is made worse by the high population density of 1200 inhabitants per sq. km. Malta is not only an island state, it is a city state, has very little land space, and no hinter-land. This lack of space had caused problems when a suitable site for a second power station was being sought. Malta has had to abandon the use of coal and opt for more expensive fuels due to the lack of space for disposing of the coal ash, besides other environmental problems associated with coal. Space limitations also limit the utilisation of renewable energy sources on a grand scale. A few years ago, a German consultancy firm had made a study that proposed the setting up of a wind farm on Malta's southern coast where the cliffs are at their highest point. Such a proposal would have had a very negative visual impact that would have spoilt a relatively large proportion of Malta's panoramic coast.

5. MARKET CONSIDERATIONS

1) The Present Scenario

In the last few years, the energy sector in various parts of the world has been revolutionised and turned into a proper market, with energy being traded as a marketable product. Malta has still to restructure its energy sector, but how far it is wise to do so is questionable. It is true that notions that were almost sacred only a few years ago, such as that electricity is a natural monopoly, have since fallen like tenpins, but the question still remains whether a market as small as Malta's, if one can cali it a market at all, can be restructured in the same way as in most Western economies. Ever since the introduction of electricity in Malta in 1896, the provision of electrical power has always been a public utility provided by the State, in the same way as water; indeed the two have often gone together. Today the generation and supply of electricity is still a state monopoly. The general perception in Malta is that this is how it should be. Gas followed more or less the same story. Petroleum fuels on the other hand have only been a state monopoly since the oil crisis of 1973. Up to that date the local market had been supplied by Esso, BP and Shell who had been operating on the island since early this century. All three companies had their own installation for receiving and storing fuel as well as their own distribution system. Even today, notwithstanding the pressure on land use, one may still find areas with three petrol stations right next to each other, reminiscent of the days when the three companies competed with each other. During the 1973 crisis, the companies could not or would not guarantee continuity of supply and the Government of the day nationalised their operations. Eventually electricity, gas and petroleum were incorporated into one state corporation, Enemalta. When the last of the British forces left Malta in 1979, the Maltese Government also inherited superb underground storage facilities that had been built by the British for military purposes.

2) Global Trends

The world events of the last few years have shown that privatisation and the liberalisation of markets are not just fashionable political ideology but that they make economic sense, since the competition so generated brings about economic efficiency to the benefit of consumers and tax-payers. These considerations cannot be ignored for much longer in Malta. Indeed the telecommunications sector has led the way and is being liberalised. It should be clear however that what is needed for economic efficiency is competition. Simple privatisation that turns a public monopoly into a private monopoly will not achieve that.

The European Commission's White Paper "An Energy Policy for the European Union" proposes that the energy policy for the EU should be based on three objectives (ref. 4):

— overall competitiveness
— security of energy supply
— environmental protection

In the 1996 annual report of Malta's State Energy Corporation, Enemalta, the same three objectives were considered as essential conditions to be satisfied in any restructuring of the energy market in Malta (ref. 5). These were restated as follows:

"The over-riding considerations must be three: to ensure security of supply; to ensure that the service to the consumer is of the right price and quality; and that the environment is respected."

The question then is: Can the market for electricity, petroleum products and gas in such a small economy as Malta be opened up and still achieve the above conditions? Each will be considered in turn.

3) The Electricity Market

It is not difficult to envisage that if Enemalta had to lose its monopoly for the generation and distribution of electricity, there would be no rush of private investors wanting to build another power station. And even if there had to be one, it would have difficulties finding a suitable site. Hence one can possibly foretell that the wholesale generation and distribution in Malta would still remain a *de facto* monopoly even if the *de jure* condition had to change. It is also not difficult to see that if the monopoly had to abolished, there could be some establishments that might be tempted to generate their own electricity, particularly if this could be combined with providing heat either for hot water, process steam or space heating. One can envisage the larger hotels wanting to do this, as well as the hospital, the national swimming pool and the reverse osmosis plants, especially if combined with desalination using low-grade heat. In this respect, it is interesting to note the provision in the Italian Energy Plan (ref. 6) which enjoins the designer to consider co-generation in certain buildings and to include his evaluation with the documentation that is required for planning permission. Clearly, such small individual generating plants would still need to be carefully regulated not only to ensure safety and environmental standards, but also to ensure that their coupling to the national grid for extra or emergency power would not overload the public utility. The way would also be open for photovoltaic operation by the private sector although at the moment this still appears economically unfeasible, unless the photovoltaic panels are used as an alternative way of cladding buildings. Cladding by photovoltaic panels can work out cheaper than marble, with free electricity as an added bonus.

4) The Market for LPG

Since 1971, when the network of piped town gas was phased out as being too expensive to maintain, LPG has been supplied to Maltese consumers in cylinders. Gas is imported and bottled by the state corporation which also takes customer orders. The actual distribution is handled by a number of private contractors, each of whom covers a particular area and is paid by commission on sales. There would seem to be no reason why the importation, bottling and distribution of gas could not be carried out by a small number of private companies. Clearly, the size of the market does not permit to have too many companies, and yet one must avoid the monopolisation of the market. Some pilot projects could also be initiated to introduce piped networks in apartment blocks and small self-contained residential areas. There would be the need for proper safety rules which would have to be strictly enforced. Even then, the biggest obstacle to such projects could be the fear that the Maltese have of gas.

5) The Market for Petroleum Products

Petroleum products are imported into Malta for supplying fuel to vehicles, aircraft, ships and industry, and, except for the bunkering of ships, is carried out exclusively by the state monopoly.

Because of its strategic location in the middle of the Mediterranean, Malta has been an important bunkering station for many years. At the height of the British Empire, it was an important coaling station for steamships and this made an important difference to the island's economy. Bunkering was completely liberalised in 1995, but even before then the state corporation was not the only organisation providing this service.

Petrol, diesel, kerosene, jet fuel and other fuel oils are imported and stored solely by the state corporation (monopoly) which then also distributes and sells them to the various consumers. When these products were being imported by the three companies already mentioned, the actual sale of petrol and diesel to vehicles at the filling station was already being carried out by private individuals, although the actual pumps and sometimes the stations themselves were provided by the three companies. This system remained in operation until last year when the responsibility for the pumps was passed over to the individuals. We can now say that at least part of the operation has been privatised, although not entirely liberalised since the owners of filling stations belong to a strong syndicate that would not look too favourably on the opening of new stations. Distribution of kerosene and other fuel oils for heating purposes to domestic consumers and industry has also been in private hands for some time now, subject to a system of licences and zoning that need to be reformed for proper liberalisation. However, the distribution of fuel to filling stations and to aircraft remains entirely in the hands of the state corporation. It is not difficult to envisage the privatisation of the distribution of petrol and diesel to filling stations. That would then leave only the importation and storage of fuel as a state monopoly.

At the start of this discussion on the liberalisation of the energy market in Malta, mention was made of three conditions that must be satisfied in any case. One of these, relating to the right of the consumer to have energy products at the right price and quality, would surely be better assured if there were to be truly market competition with some rules to ensure that the competition is fair and just. Another condition referred to the environment. In the immediate future, one can only see the environment being safeguarded through strict regulation. One can only hope that sometime in the future, the culture will have changed to the extent that safeguarding the environment becomes second nature to our civilisation. The need to ensure the security of supply, the third condition, gives rise to a certain hesitancy to a complete liberalisation of the importation and storage of fuels in a market as small as that of Malta. The issue of security cannot be taken lightly. Malta's GDP is smaller than the turnover of many of the larger oil companies. Should Malta's economy then be placed at the mercy of a few oil companies? Obviously the situation would be totally different if Malta were a member of the European Union, since one assumes that the EU would be able to put pressure on the oil companies if they did not play fair with the Maltese market. But that may now not happen for some time yet. The situation could also change if the European Energy Charter Treaty, of which Malta is a signatory, proves to be as strong an instrument for guaranteeing free and fair trade as the promoters of the treaty would have hoped for.

Even assuming that the security of supply is assured, there are also serious doubts whether there is the physical space on the islands to have more than one operator in this sector. Even the current single operator, the state corporation, is having problems with one of its location and will probably have to dismantle it due to urban pressure. That would not leave it with many more alternative storage sites apart from the previously mentioned "British-built" reservoirs.

Given that the importation and storage of fuels remains an activity of the state corporation, there is still a question. How can the ordinary consumer be assured that the state corporation is indeed procuring fuels at the best possible prices, given that state corporations are notoriously known to be rather poor at business?

6. CONCLUSION

The above discussion has used Malta as a case study to highlight some of the problems of energy supply in a small island state.

The paper has presented the problems of providing a reliable and economical source of electrical energy as well as the problems associated with the liberalisation of the energy market in a small island state. Arguments have been put forward to show that it might be politically and practically difficult to open out completely. And yet, is there no way that there can be some competitive element introduced into the system to ensure at least economic efficiency? When the author of this paper was

Chairman of Enemalta, he had proposed in the Corporation's 1996 annual report that in this age of virtuality, it might be possible to have *virtual* competition as the next best alternative to *real* competition (ref. 5). He had later suggested that a theoretical exercise was necessary to estimate the "theoretical" cost of producing electricity in Malta. Actual costs are known and often quoted but what is not known is what that cost *should* be. If that "ideal" cost were known it could be the measure against which to compare the performance of the corporation, i.e. a sort of virtual competition. There may well be other possibilities of introducing some degree of economic efficiency in the energy sector through such as arrangements as, for example, BOT (build, operate and transfer), leasing agreements for facilities, etc.

This paper has discussed the problems of supplying energy to a small island state that is Malta. A number of these problems, particularly those relative to the generation and supply of electrical energy are problems that one would find on any isolated area, such as small islands some distance away from the mainland. Other problems, such as finding the appropriate structures for the energy market are not only due to the geography of small island states, but are more a consequence of their state as a "State", i.e. their sovereignty.

Although this paper has presented problems without outlining solutions, a clear statement of the problems may well be the first step towards finding adequate solutions. Some creative thinking should lead to some imaginative solutions. It will certainly be interesting to see how the energy sector will continue to evolve, to influence and be influenced by developments in the new socio-economic order that we see evolving before us on a daily basis (ref. 7).

7. REFERENCES

1. Briguglio Lino, "Small Island Developing States and their Economic Vulnerabilities", World Development, vol. 23, no. 9, pp. 1615–1632, Elsevier Science Ltd., Great Britain, 1995.
2. Enemalta Annual Report 1997.
3. "Pre-feasibility Study of a Malta-Sicily Interconnection", prepared by Electricite de France and presented at the

Third Biennial Conference of the Malta Council for Science and Technology, June 1996, Malta.

4. White paper "An Energy Policy for the European Union", 1995.
5. Enemalta Annual Report, 1996.
6. "Norme per l'attuazione del piano energetico nazionale in materia di uso razionale dell'energia, di risparmio energetico e di sviluppo delle fonti rinnovabili di energia. Legge 9 gennaio 1991, n. 10" and "Regolamento recante norme per la progettazione, l'installazione, l'esercizio e la manutenzione degli impianti termici degli edifici ai fini del contenimento dei consumi di energia, in attuazione dell'art.4, comma 4, della legge 9 gennaio 1991, n. 10. Decreto del Presidente della Repubblica 26 agosto 1993, n.412".
7. "Mediterranean Energy Policy", a draft energy policy paper, Document no. 4, Euro-Mediterranean Energy Forum Secretariat, 23/7/97.
8. Appendix 1. The Alliance of Small Island States Atlantic:

Cape Verde
Guinea Bissau
Sao Tome and Principe

Carribean:

Antigua and Barbuda
Bahamas
Barbados
Belize
Cuba
Dominica
Grenada
Guyana
Jamaica
St Kitts and Nevis
St Lucia
St Vincent and the Grenadines
Surinam
Trinidada and Tobago

Indian Ocean:

Comorros
Maldives
Mauritius

Seychelles

Mediterranean:

Cyprus

Malta

Pacific:

Cook Islands

Federal States of Micrones.

Fiji

Kiribati

Marshall Islands

Nauru

Papua New Guinea

Samoa

Solomon Islands

Tokelau

Tonga

Vanuatu

South China Sea:

Singapore

Observers:

American Samoa

Guam

Netherlands Antilles

Niue

US Virgin Islands

9
The African Energy Sector: Status and Prospects

S. KAREKEZI
Kenya

This paper examines key characteristics and trends of the African energy sector based on the findings of research work of over 100 African energy researchers and policy makers—members of the African Energy Policy Research Network (AFREPREN). The paper identifies fundamental changes that have defined and shaped the African energy sector. Using an innovative differentiated approach, the paper provides a comprehensive framework for assessing the highly diverse national energy sector found on the African continent—from the advanced high-tech power sector of South Africa to the biomass-dominated traditional energy sector of the low-income countries of the Sahel. The paper highlights key trends that will drive future energy demand and supply in Africa. It concludes by proposing an innovative and self-sustaining initiative for developing and maintaining an African energy database and information service that can provide the basis for accurate assessment of the current status and future prospects of the African energy sector.

KEY PAST TRENDS IN THE AFRICAN ENERGY SECTOR

With an estimated total population of just under 700 million, Africa's total modern energy consumption in 1992 was estimated to be 217 Mtoe (IEA, 1995). North and South Africa account for a large share of the modern energy consumption. North Africa with a population of about 120 million accounted for 37 per cent of total modern energy consumption while South Africa with about 40 million inhabitants accounted for 43 per cent of the continent's consumption of modern energy. The balance of 20 per cent was accounted for by sub-Saharan African countries (IEA, 1995).

In energy terms, Africa can be divided into the aforementioned three regions, namely: South Africa, North Africa and sub-Saharan Africa. North Africa is predominantly an oil and gas producer region while South Africa is largely dependent on coal. Sub-Saharan Africa is mainly reliant on traditional bio-mass energy which often accounts for over 75 per cent of many sub-Saharan African countries' total energy supply (Karekezi and Ranja, 1997). In electricity terms, the contrast between the three regions is even more striking. South Africa consumed an estimated 50 per cent of the continent electricity with North Africa and sub-Saharan Africa accounting for 30 per cent and 20 per cent, respectively.

Region	*Population in 1992 (Million)*	*% of Africa's Total Modern Energy Consumption (1992)*	*% of Africa's Total Electricity Consumption (1992)*
South Africa	40	43	50
North Africa	120	37	30
Sub-Saharan Africa	503	20	20

Source : IEA, 1995

The energy intensity of African economies is a subject of some debate because of the wide divergence among countries. While available data on energy intensity of African economies vary widely, both the World Bank and the IEA figures indicate that South Africa is likely to be the most energy intensive country in Africa with an estimated energy intensity of 1.13 toe for every 1,000 of 1987 US Dollars at market exchange rates (IEA, 1995; World Bank, 1996). The next most intensive region is North Africa followed by sub-Saharan Africa. The reliability of available data on the energy intensity of sub-Saharan Africa economies largely depend on the accuracy of traditional and biomass energy estimates. Available data indicates that the inclusion of traditional and biomass energy intensity calculations could make sub-Saharan Africa the second most energy intensive region after South Africa.

Reliance on traditional biomass energy is particularly high in sub-Saharan Africa (Karekezi and Ranja, 1997). Even oil rich sub-Saharan African countries continue to rely on biomass energy to meet the bulk of their household energy requirements. In Nigeria, the dominant sub-Saharan African oil producer, it is estimated that about 91 per cent of household energy needs were met by biomass (IEA, 1995). Existing estimates of biomass energy use vary widely. The IEA estimates that traditional energy, which is virtually all in the form of biomass, could amount to 150 Mtoe per year most of it consumed in sub-Saharan Africa. This would bring the total continental energy consumption to just under 400 Mtoe, equivalent to about 4–5 per cent of world total energy consumption in 1992 (IEA, 1995; World Bank, 1996).

Reliance on inefficient traditional bio-fuels has two main environmental drawbacks. The first drawback and perhaps the most serious is indoor air pollution from unvented bio-fuel cookstoves. There is growing evidence that indoor air pollution from smoke could be a major contributor to the incidences of respiratory diseases, a major cause of death in the region (Karekezi and Ranja, 1997).

Although the felling of trees for construction and the clearing of land for agricultural activities, rather than the increasing demand for energy, have proved to be the major causes of deforestation and land degradation in sub-Saharan Africa, the role played by bio-fuel demand, particularly in the form of charcoal, cannot be underestimated. For example, a comprehensive bio-mass energy assessment of Zambia urban areas indicated that charcoal demand in Lusaka, the capital city of Zambia, was a major cause of deforestation (Karekezi and Ranja, 1997).

The electricity sector is largely dominated by South Africa which in 1992 generated 44 GW. The rest of Africa accounted for 51 GW (IEA, 1996). Most of South-Africa's electricity was from coal (93.5%) while the rest of Africa produced electricity from a balance of hydro (35.9%), oil (33.5%) and gas (27.1%). Most of the electricity generated in North Africa and oil producing sub-Saharan African countries is from either oil-fired or gas-fired power stations. The rest of sub-Saharan Africa is largely reliant on hydro-based electricity although the situation is beginning to change.

Total oil production in Africa for the year 1991 was 329.7 million tonnes, found mostly in North Africa and in the Western

Installed Electricity Generation Capacity in Africa, 1992

	S. Africa	Rest of Africa	Africa Total
Installed Capacity	44GW	51GW	95GW
Of which sourced from:			
Coal	93.5%	-------	43.3%
Nuclear	4.4%	-------	2.0%
Hydro	-------	35.9%	19.3%
Oil	-------	33.5%	18.0%
Gas	-------	27.1%	14.5%
Others	2.1%	3.5%	2.9%
	100.0%	100.0%	100.0%

Source : IEA, 1996

coastline. Although oil consumption in 1991 was about 2.0 million barrels per day, projections indicate that by the year 2010, oil consumption in Africa will be 75 per cent higher than the 1991 figure (Welham, 1994:10). Estimates by the IEA project a doubling of oil products consumption in Africa by year 2010 (IEA, 1995). Although the African continent has about 15 million km^2 of sedimentary basin (17.5 per cent of the total sedimentary basin in the world), only about 4 per cent of the total world petroleum exploration/production expenditure to date has been in Africa. The extent and pace of petroleum exploration/production in Africa is at present decreasing rather than increasing (ECA, 1995:6). Consequently, in the near term, oil production is unlikely to rise very rapidly but this situation could radically change in the medium to long-term.

South Africa, the dominant coal producer and consumer in Africa, accounts for 92 per cent of Africa's coal production in 1992 (IEA, 1996). The SADC (Southern African Development Community) region has by far the largest share of Africa's known reserves of coal. It is estimated that 90 per cent of the continent's proven and economically attractive coal reserves and a substantial portion of its uranium deposits are located in South Africa, Zimbabwe and Namibia (World Bank, 1994).

The average solar radiation in Africa is between 5-6 KWh/m^2. Exploitable hydropower capacity in Africa is massive (Johansson et al, 1992:9) of which less than 4 per cent has been harnessed.

Africa has an estimated proven geothermal potential of 9,000 MW and only 45 MW in Kenya has been exploited. Currently, plans are underway to increase production from geothermal energy, from other areas of the Kenyan Rift Valley.

It is difficult to separate the energy constraints with the overall macro-economic, environmental and social-political problems that face Africa particularly sub-Saharan Africa (AFREPREN, 1990). Insecurity of access to adequate levels of energy services is one of the key problems that faces Africa. Energy imports constitute a major drain on the diminishing convertible currency reserves of many countries in the region. In spite of low world market oil prices, fossil fuel imports routinely account for up to a third of export earnings of many African countries.

The paradoxical heavy and growing burden of fossil fuels imports in an era of low oil prices is a result of a combination of factors that include:

i) Falling export earnings reflecting the persistent unfavourable terms of trade that faces regions such as Africa that still rely on the export of unprocessed raw materials;
ii) Growing demand arising from high population growth;
iii) Changing demand patterns largely driven by rapid urbanisation;
iv) Failure to exploit the advantages that increased intra-regional trade in energy would offer;
v) High fossil fuel transportation costs that faces many African countries which are land-locked (compounded by preference for costly road transport instead of low-cost rail transportation); and,
vi) Inadequate capability and skills in international energy trade negotiations.

The majority of access to sufficient supplies of energy continues to be inadequate. Energy for rural areas is increasingly scarce and costly (Karekezi, 1992). As shown in the Table overleaf, electrification levels are still woefully inadequate.

The rapid decline of the vegetative cover continues to constrain energy supplies to the overwhelming majority of the population of the region, most of whom reside in the rural areas (Pasztor and Kristoferson, 1990; Karekezi and Ranja, 1997).

Many of the problems faced by the energy sector in Africa can be traced back to developments within the oil and gas and

Country	% of Electrified Households	
	Urban	Rural
Malawi	11.00	0.32
Tanzania	13.00	1.00
Lesotho	14.00	4.00
Mozambique	17.05	0.66
Zambia	17.85	1.39
Namibia	26.00	5.00
Botswana	26.48	2.09
Swaziland	42.00	2.00
Zimbabwe	64.72	0.60
Average	25.78	1.70

Source : Karekezi et al, 1996 and Karekezi et at, 1997a

electricity sub-sectors which have accounted for a large proportion of energy investments in the region. A distinctive feature of the region's energy sector in the wake of independence was, with the exception of the nationalised petroleum and power utilities, the absence of a coherent and all-embracing institutional infrastructure for energy policy formulation, analysis, monitoring and implementation.

Coordination of activities within the energy sector was almost non-existent. Biomass energy which was, and still is, used by the overwhelming majority received virtually no policy attention (Bachou, 1990; Karekezi and Ranja, 1997). Due to the prevailing low oil prices, little attention was given to energy policy and related environmental considerations.

In the 1960s, the development of the energy sector was confined to increasing the supply of conventional fuels such as petroleum and electricity. Major refineries and electric generating stations were built in several African countries based on the optimistic assumption that Africa was set to experience substantial growth in its industrial and agricultural sectors. In brief, a supply-oriented energy strategy demonstrated by the rapid growth of investment in the oil and gas and electricity sub-sectors dominated this era. It is, however, noteworthy that many of the large power sector investments made in this period have proven to be very productive assets that continue to provide essential supplies of modern energy to many African countries.

Since then, the key events of the energy sector can be grouped into three distinct phases mainly linked to developments in the oil and gas sub-sector. The first phase coincides with the response by African countries to the first oil price rise in 1973. The second oil price rise in 1979 marks the beginning of the second phase. The era of low oil prices combined with a growing economic crisis starting in the mid-1980s to date, marks the third phase.

1. The First Oil Crisis

In the first phase, oil prices rose in real terms by a factor of three. Excluding oil exporting countries, the region's oil import bill jumped from an average of 10 per cent of the region's total export earning to over 20 per cent (World Bank, 1989). Most African countries resorted to external borrowing to pay for the rapidly rising energy import bills. This was compounded by the low prices that the region's exported commodities fetched in the international market which further reduced the exports earnings of Africa. Prices of imports rose dramatically. Paradoxically, this period coincided with high Government spending on inappropriate and non-productive projects largely financed by large deficits and external loans.

The first oil shock undermined the validity of African countries' development plans for the 1970s which were based on the rapid growth experienced in the 1960s. The region's unpreparedness left it no choice but to move from planning for future development to struggling for survival. The initial response of the region to the first oil shock was to restrict oil imports with little recourse to long-term demand management. Ministries of Energy were created to address energy issues and coordinate relevant activities on behalf of the Government. Unclear objectives and inadequately defined institutional and legal backing arrangements, made the work of many of these newly-created Ministries of Energy difficult.

This period also saw the rapid advent of the state in the energy sector. Utilities that were not already state-owned were nationalized. The oil and gas sector was brought under the control of the state with Government taking control of national refineries; importation of oil products; and, fixing of the prices of petroleum products. While these measures significantly increased the power of the state of the energy sector, there was no discernible improvement in the performance of key constituents parts of the energy sector. On the contrary, it appears that this period marked the beginning of the steady deterioration of the energy sector in the continent.

External assistance was largely confined to support for oil imports as a balance of payment management tool and for oil exploration. A much smaller proportion of this assistance was allocated to development of renewable energy technologies and energy efficiency (Energy Research Group, 1986).

South Africa, the largest coal producer and consumer in Africa, was then still under *apartheid* era sanctions. In response to the oil sanctions and increases in oil prices, South Africa continued to invest in synthetic coal-based production of fuel at its Sasol complex. Coal is first gasified and the result is transformed into a range of liquid fuels and feedstocks for the petrochemical industry (IEA, 1996). In the wake of the 1973 oil price rise, South Africa embarked on an extensive expansion of Sasol and constructed a new plant, Sasol II (IEA, 1996). With the continent's most extensive coal-based electricity generation and distribution system, the development of a complementary coal liquifaction industry was aimed at making South Africa self-sufficient in energy and immune to growing oil sanctions.

Year	*Coal Production in S. Africa (million tons)*	*Coal Exports from S. Africa (millions tons)*
1956	33.6	0.5
1960	38.2	0.9
1964	45.0	1.2
1968	51.7	1.2
1972	58.4	1.2

Source : Eberhard and Trollip, 1994

For oil exporting countries such as Nigeria, Algeria and Libya, higher crude oil prices translated into a burgeoning increase in oil revenues used to underwrite massive investments. In the case of Algeria and Libya, a substantial portion of this increased revenue was used to rapidly expand their respective national oil and gas infrastructure which continues to serves these countries well. In the case of Nigeria, a large proportion of the oil revenues were diverted to current consumption and poorly though-throught investment.

On the whole, the region's response to the first oil shock was characterised by ad hoc and short-term measures such as fuel rationing. This contrasts sharply with the long-term coordinated approach of the Organisation for Economic Cooperation and Development (OECD) countries which diversified their energy resource base and emphasized energy efficiency. These measures reduced the reliance of OECD countries on oil imports and triggered innovations such as demand management planning and energy efficiency technologies that effectively delinked energy consumption from economic growth. The total per capita energy use of OECD countries decreased by six percent between 1973 and 1985 while their per capita GDP growth increased by 21 per cent (Kubursi and Naylor, 1985). Because of Africa's weak science and technology base and absence of an early warning system that could marshal local resources to respond to crisis, the first oil shock aptly demonstrated the paralysis in the region's innovative capacity.

2) The Second Oil Crisis

The second oil crisis in the late 1970s found Africa no better prepared than before. Oil import bills jumped from an average of 20 per cent of the region's export earnings to over 50 per cent for a number of low-income countries (World Bank, 1989). The resultant heavy oil import bill and the continued price collapse of raw material exports increased Africa's debt load. Many countries failed to meet their external loan repayment obligations and were forced into debt re-scheduling. As before, the region's response was largely confined to physical restrictions on oil consumption.

Country	*Energy imports as a % of merchandise exports*	
	1960	**1981**
Keny	18	63
Senegal	8	77
Morocco	9	50
Burkina Faso	38	71
Ghana	8	52
Tunisia	15	31

Source : World Bank, 1984

Most of the energy investment flowing to the region targeted the conventional electricity and petroleum sectors. In aggregate terms, however, the level of new investment in the energy sector slowed down significantly with very few major large-scale energy investments. A significant proportion was credit and was channelled towards rehabilitation and modest expansion investments.

A more positive development in this period was the initiation of national energy surveys which began the difficult process of building a critical mass of base line energy data and concomitant analytical expertise for the development of energy master plans in certain countries. In addition, substantial local expertise in energy research particularly in the renewable energy and biomass energy sub-sectors was developed in this period. With external support, more substantive programmes on afforestation, improved bio-fuel cookstoves and renewable energy technologies were initiated (Bhagavan and Karekezi, 1992).

The main intervention of the United Nations was the organization of the United Nations Conference on New and Renewable Sources of Energy in 1981 in Nairobi, Kenya. Partly due to lack of substantive financial support, the performance of UN follow-up activities to this Conference were almost all below expectations. With grant support from bilateral donor agencies, the UNDP/World Bank Energy Sector Management Assistance Programme (ESMAP) carried out national energy assessment studies which provided some useful base line energy data on several countries in the region. Follow-up activities to the ESMAP studies were, in many cases, not as successful as expected largely due to limited local involvement. The then European Economic Community (EEC) provided substantial support to the region particularly in energy planning and energy policy research.

Some bilateral donor agencies such as CIDA (Canada), the Commonwealth Science Council, DANIDA (Denmark), FINNIDA (Finland), GTZ (Germany), IDRC (Canada), the Netherlands Government, NORAD (Norway), SAREC (Sweden), SIDA (Sweden) and USAID (USA) initiated major regional renewable energy and energy efficiency programmes in selected African countries. In spite of substantial investment, the envisaged benefits of these programmes have not been widely replicated as it was originally envisaged.

Although the above activities indicated increased interest in biomass, renewable energy and energy efficiency, the bulk of energy investment in the region still went to conventional energy activities such as those of the electricity and oil and gas sub-sectors. Environmental considerations within this period were limited to general platitudes concerning the need to develop a sustainable biomass energy resource base and the environmentally-benign character of renewable energy technologies and energy efficiency. With the exception of a few major biomass energy studies, no major efforts were made to initiate viable environmentally-sound energy activities.

In a few African countries, however, there was increased interest in large-scale investment in renewables. Notable achievements include the sugar-based ethanol plants established in Kenya, Malawi and Zimbabwe plus substantial investment in co-generation of electricity in the large sugar industry of Mauritius.

South Africa, the continent's dominant energy giant, continued to strengthen both its electricity industry and synthetic fuels sub-sector. In 1980, the commercial energy consumption of South Africa was estimated to be close to 60 per cent of sub-Saharan Africa's total energy consumption (World Bank, 1986).

South Africa also embarked on an extensive destabilisation campaign in the southern African region with the objective of crippling the economies of the independent nations on its borders. Energy installations were a primary target. Angola and Mozambique suffered heavy losses. The oil-rich Angolan enclave of Cabinda was targeted for destabilisation with South Africa chanelling support to local secessionist rebels. In Mozambique, the giant Cabora Bassa dam was also targeted with its main transmission lines crippled by South-Africa backed rebels.

Oil producing African countries continue to enjoy growing oil revenues. In 1980, the World Bank estimates that Nigeria's export earnings totalled close to US$ 26 billion, most of it consisting of revenues from oil exports. North African countries continued to expand their local and regional electricity and oil and gas networks. Sub-Saharan oil-producing African countries led by Nigeria used a large proportion of the oil revenues to expand current consumption at the expense of investment in their traditional cash-crop sectors which lay the ground work for the deep-seated economic problems that many of the oil producing African states currently face.

3) The Era of Low Prices and Growing Economic Problems

Due to the region's chronically weak and disjointed economies, Africa has been largely unable to benefit from the low oil prices that emerged in the mid-1980s to date. The region's debt burden (in part, driven by continued relatively heavy oil imports bill) rose to a new peak and prices of Africa's raw materials exports continued to decline in real terms while the cost of manufactured imports kept on increasing. Many countries began to experience serious socio-political instability which had a negative impact on their respective energy sectors. Over-manning, poor maintenance and excessive line losses characterised conventional energy utilities in the region.

The World Bank, the leading lender to Africa's electricity industry, started to push for more radical reform of the continents power sector. No longer satisfied with modest reforms of the national electricity utilities (such commercialisation, corporatisation and establishment of arm-length relationship between utilities and Government), the World Bank started to call for complete privatisation of national power utilities.

Refineries regularly recorded poor performances. Many African refineries were limited to first stage distillation which severely constrained the range of oil products that could be obtained from imported crude oil (Nyoike, 1993). In addition, their capacity was, in most cases, under-utilised. As in the case of the electricity sub-sector, the World Bank called for radical rationalisation of the region's refineries. The World Bank was particularly keen to see the closure of small national refineries and commercialisation and privatisation of the larger refineries that served large regions of Africa.

As a result of increased population and declining performance of the conventional energy utilities such as national electricity and petroleum companies, pressure on the biomass energy resource base continued to increase. International interest in renewable energy, biomass energy and energy efficiency faded with declining oil prices. Consequently, many of the renewable, energy efficiency and biomass energy initiative that started in the late 1970s and early 1980s were gradually wound-down.

Status of Refineries in Selected African Countries in 1990s

Country	*No. of Refineries*	*% Capacity Utilization*
Ethiopia	1	82.4
Tanzania	1	60.8
Angola	1	58.4
Cameroon	1	55.7
Ghana	1	47.5
Nigeria	4	46.5
Congo	1	46.2
Cote D'Ivoire	1	44.5
Kenya	1	43.4
Gabon	2	42.1
Zaire	1	36.5
Madagascar	1	23.3

Source : Karekezi et al, 1997b

On a more positive note, small and medium-scale renewable energy enterprises started to record encouraging growth. Initial success was recorded in informal-sector based improved cook-stoves initiatives followed by household solar home PV systems. At a later stage, solar water heaters, wind pumps and small-scale hydro technologies (particularly hydro ram pumps) began to show signs of success.

Improved Stoves Disseminated in Selected Countries

Country	**Rural**	**Urban**	**Total**	**Year**
Kenya	180,000	600,000	780,000	1993
Tanzania	—	54,000	54,000	1991
Uganda	—	52,000	52,000	1995
Ethiopia	22,000	23,000	45,000	1993
Sudan	1,380	26,580	27,960	1992
Zimbabwe	10,000	10,880	20,880	1994
Malawi	—	3,700	3,700	1991
Botswana	—	1,500	1,500	1994
Lesotho	800	670	1,470	1993

Source : Karekezi and Ranja, 1997

With the exception of the co-generation industry of Mauritius which continued to operate successful and is now set for substantial growth, many of the large-scale investments in renewables began to face problems. Most notable were the fuel-blending ethanol initiatives in Kenya and Malawi which were gradually wound-down due to lower world oil prices and resistance from local subsidiaries of transnational oil conglomerates.

With the advent of majority rule, oil sanctions facing South Africa were dismantled which made the massive investments in costly coal-to-fuel technology and even more expensive gas-to-fuel technologies uneconomic. The South African energy sector is still grappling with the implications of these massive *apartheid* era energy investments.

The dismantling of *apartheid* in South Africa effectively put a halt on the destabilisation campaign aimed at its neighbours. Mozambique was the first beneficiary of this welcome development. Its transmission system linking its gain Cabora Bassa electricity generation dam with neighbouring countries was rehabilitated and is now operational. Massive investments aimed at exploiting Mozambique's ample gas reserves are now being actively pursued.

On a more positive note, the dismantling of *apartheid* in a rapidly-reforming South Africa resulted in the launching of a large electrification program aimed at bringing the benefits of modern energy services to the previously disadvantaged communities. The country's largest utility Eskom committed itself to the continent's most ambitious electrification target. Eskom committed itself to increasing the percentage of electrified homes in South Africa from 40 per cent to 70 per cent by the year 2000. Preliminary results of ESKOM's electrification programme indicates that it is on target.

Falling oil prices had a negative impact on the economies of African oil exporting countries compounded by the misguided investments of oil revenues in the 1970s and early 1980s. Largely due to lower oil prices and the limited development of other export sub-sectors, Nigeria's annual total exports earnings declined precipitately from US$ 28 billion in 1980 to US$ 9.9 billion in 1995. Algeria also experienced a rapid fall in its export earning which declined from US$ 14.5 billion in 1980 to US$ 10.9 billion in 1995. Many analysts link this rapid decline in export earnings combined with misguided investment and economic policies to the socio-economic problems faced by many oil-producing countries in Africa. Algeria, Angola, Congo, Libya and

Nigeria have slid into debilitating political instabilities ushered by internal civil strife and international economic sanctions in the case of Libya.

Nigeria's energy sector has since slid into chaos. Signs of the impending chaos were visible as early as 1982 when the country's power system experienced total collapse on 23 occasions between 5 and 27 April, 1982. Actual measures of average station effectiveness in a sample period of 1982 indicated an effectiveness of only 37.9 per cent (World Bank, 1983).

Average Weekly Generating Station Efficiencies

Period in 1982	*Total Installed Capacity (MW)*	*% Available*	*Effectiveness*
12–18 July	2440	60.0	40.0
19–25 July	2440	57.6	39.3
25–31 July	2440	58.2	39.0
01–07 August	2440	54.5	38.2
08–14 August	2440	54.5	38.6
15–21 August	2440	52.0	38.0
22–18 August	2585	40.7	31.1
29 Aug. – 04 Sept.	2585	46.7	38.2
05–11 September	2585	53.9	39.0
Average		53.0	37.9

Since then, the situation has sharply deteriorated with Nigeria's national utility, NEPA carrying an excessive reserve margin of 33 per cent and recording an average transmission and distribution losses of 25 per cent (Iwayemi, 1994). Throughout the 1980s, NEPA continues to record mounting losses as shown in the following table;

Operating Surplus or Loss of NEPA: 1981–1985 (Naira Millions)

Year	*Surplus or Loss*
1981	135.36
1982	81.09
1983	−37.27
1984	−11.92
1985	−58.63
1986	−26.73
1987	−81.04

The above developments have shaped the current principal characteristics of the energy sector of sub-Saharan Africa discussed in the next chapter that evaluates that current status of the African energy sector.

CURRENT STATUS OF THE AFRICAN ENERGY SECTOR

Key features that have important implications for the future of Africa's energy sector and discussed in the next chapter include the following:

— Low consumption of modern fuels.
— Limited links within the energy sector.
— Energy insecurity.

1) Low Consumption of Modern Energy

Although, Africa has substantial energy resources—the World Bank estimates that the continent's petroleum reserves could last for 120 years at current consumption rates (World Bank, 1989)—sub-Saharan Africa has one of the lowest per capita consumption of modern energy in sub-Sahara Africa was estimated to be 249 kgs. of oil equivalent (kgoe). This was fractionally less than half the corresponding figure for all low and middle income economies (686 kgoe). By 1994, the situation had remained largely stagnant in per capita terms with sub-Saharan Africa consuming an average of 237 kgoe per capita compared to an average of low and middle income economies of 739 kgoe. Not only was the sub-Saharan Africa figure now lower than the average of low-income economies, it was less than a fifth of the world average of 1,433 kgoe (World Bank, 1997).

Regions	*Per Capita Modern Energy Use (kgoe)*	
	1980	*1994*
Sub-Saharan Africa	249	237
East Asia and Pacific	378	593
South Asia	123	222
Middle East and N. Africa	825	1,220
Low and Middle Income	686	739
World	1,419	1,433

Source : World Bank, 1997

The abundant hydro capacity of the region is still largely unexploited (US Department of Interior, 1976; World Energy Conference; and, World Resource Institute, 1986). Estimates of current exploitation of available hydro energy resources range from 4 per cent to 6 per cent, one of the world's lowest figures. While this is a symptom of stunted industrial growth and the stagnation of the modern-sector of the economy, it also represents a widow of opportunity. The continent could eschew the traditional energy-intensive and environmentally-harmful modernisation path of the North and develop a low-energy and ecological-sound path to sustainable development.

As mentioned earlier, the energy sector in Africa continues to be dominated by traditional and inefficient biomass fuels in both the supply and demand sub-sectors. For certain countries, bio-mass accounts for 70–90 per cent primary energy supply and up to 95 per cent of the total consumption. With modern energy supplies not keeping pace with increasing demand, the biomass energy source base is facing growing pressure particularly from rising demand in urban consumption of charcoal and increased use of fuelwood in industries and institutions.

Importance of Biomass Energy in Selected African Countries (late 1980s and early 1990s)

Country	*% Contribution of Biomass Energy to Total National Energy Consumption*
Burundi*	94
Ethiopia	86
Kenya	70
Somalia*	87
Sudan	84
Uganda	95

*Civil war and instability likely to increase reliance on biomass

Source: Karekezi and Ranja, 1997

While the biomass sector is often perceived to be a problem area, it also offers attractive opportunities for the power sector. Of particular interest, is the abundant agro-waste resource that is still inefficiently utilised. In certain cases, countries incur significant cost in disposing of potentially valuable energy resources such as agricultural and urban waste. Available estimates indicate that 85 per cent of current electricity generation in 16 Eastern and Southern African countries could be provided by power produced from bagasse-based co-generation in the region's sugar industry (Karkezi et al, 1995:8).

2) Limited Links Within the Energy Sector

Another important feature of the energy sector in Africa is the limited linkages that exist between the different sectors. The energy sector is characterised by the pre-dominance of the household sector which derives most of its energy needs from biomass energy resources that are threatened by land clearing for farming; raw material demands of wood industries; urban wood energy demand; and, desertification (Karekezi, 1989a; Karekezi and Ranja, 1997). An estimated 80 per cent of inhabitants of sub-Saharan Africa are wholly or largely dependent on biomass fuels (World Bank, 1986; World Bank, 1989 and Karekezi and Ranja, 1997).

Biomass energy resources are largely utilised in the household sector (Karekezi et al 1991; Karekezi and Ranja, 1997) while petroleum is mainly used by the transport sub-sector (Rutabanzibwa, 1989; Karekezi, Ranja and Kimani, 1997). Industry meets its energy needs through electricity which is often sourced from hydro resources or from thermal power stations. In most cases, linkages between these supply and consumption sub-sectors are limited which constrains opportunities for substitution (World Bank, 1982; Davidson and Karekezi, 1992).

At the institutional level, a parallel absence of links can also be discerned. The policy making machinery in most African Ministries of Energy is pre-occupied with conventional energy questions, namely power and oil. Much of the guidance for action in the conventional energy sector is provided by the World Bank, bilateral donors in collaboration with Western energy private sector firm involved in either actual energy investment or consultancies. The African energy research community has been largely pre-occupied with renewables with special emphasis on biomass and more recently on solar PV. The link between the African policy making community and its research community has generally been weak with detrimental effect on the priorities that the region pursues in the energy sector.

One initiative that has attempted to address this gap is the African Energy Policy Research Network (AFREPREN). The key objective of the African Energy Policy Research Network (AFREPREN) is to strengthen local research capacity and to harness it in the service of energy policy making and planning. Initiated in 1989, AFREPREN is a collective regional response to the wide-spread concern over the link between energy research and the formulation and implementation of energy policy in Africa.

The African Energy Policy and Research Network, AFREPREN brings together about 100 African energy researchers and policy-makers from Africa who have a long-term interest in energy research and the attendant policy-making process. AFREPREN has undertaken policy research studies in 16 African countries, namely: Angola, Botswana, Burundi, Ethiopia, Kenya, Lesotho, Mauritius, Mozambique, Rwanda, Seychelles, Somalia, Sudan, Tanzania, Uganda, Zambia and Zimbabwe. AFREPREN also maintains close collaborative links with energy researchers and policy makers from Cote D'Ivoire, Ghana, Nigeria, Sierra Leone and Senegal.

Since its initiation in 1989, AFREPREN has successfully implemented over 60 research projects involving about 100 African researchers and policy makers from 16 countries of Eastern and Southern Africa and forged close collaborative links with several West African energy researchers and policy makers. Findings of the research undertaken by AFREPREN have been published in 11 major publications as well as numerous journal articles, research reports, working papers, dossiers, and newsletters. Until 1994, the AFREPREN Research Programme has been grouped around six major energy subject areas, namely:

1. Institutions and planning
2. Renewable energy technologies
3. Biomass energy
4. Electricity
5. Oil and gas
6. Coal and gasification

After wide consultations with the African energy research community and senior African energy policy makers, AFREPREN launched a new research programme in early 1995 (which is still ongoing) with the following key focal themes:

1. *Institutions*: Restructuring of energy institutions and impact of privatisation initiatives

2. *Financing*: Financing energy investments in sub-Saharan Africa-sources, institutions and mechanisms.
3. *Capacity building*: The challenge of capacity building for effective energy policy formulation, analysis and implementation.
4. *Management and efficiency*: Management and maintenance in the power sector.
5. *Local and Regional Environment Impact of Energy*: Existing and projected impact of the energy sector on local ecosystems and regional environment.
6. *Climate and Energy*: Policy implications of the Framework Convention on Climate Change in Sub-Saharan Africa.

3) Energy Insecurity

Although Africa continues to register as one of lowest consumer of modern energy on a per capita basis, it has not been able to generate sufficient inflows of convertible currency to adequately finance its import of energy. With the exception of a few countries, notably Nigeria, Algeria, Angola, Cameroon, Congo and Libya, most African countries import petroleum either in the form of crude or refined products. For these countries, their petroleum import bill can account for as high as 50 per cent of a country's export earnings (World Bank, 1989; Rutabanzibwa, 1989; Ogunlade and Karekezi, 1992). The heavy energy import bills often leave precious little surplus for the implementation of environmentally-sound sustainable development initiatives. Consequently, most investments are almost wholly dependent on the vagaries of the generosity of external financiers, donor agencies and a few selected multi-nationals.

The proportion of energy imports to total export earnings has continued to rise from an average of 7 per cent in the early 1960s to over 20 per cent (excluding Nigeria, the region leading oil exporter) at the height of the second oil crisis in the early 1980. In spite of the decline in oil prices in real terms in the 1980s, energy import bills of 32 low-income sub-Saharan African countries is stubbornly stuck at around 25 per cent of their total annual exports earnings (World Bank, 1990). As shown in the table below, energy imports constitute a heavy burden on economies of many African countries (World Bank, 1994). This has been the case, in limited growth in the total volumes of energy imports.

Country	*Energy imports as a % of merchandise exports*	
	1971	**1992**
Ethiopia	14	47
Tanzania	12	40
Uganda	1	73
Burkina Faso	28	58
Ghana	8	52
Sudan	8	41

Source : World Bank, 1996

Many African countries are now trapped in a vicious circle whereby reduced export earnings arising out of declining prices for exported raw materials and virtual absence of manufactured exports diminishes the ability to import adequate energy resources and to adequately maintain the existing energy infrastructure. This, in turn, leads to poor productivity in industries and diminished volume of raw material exports which further limits availability of convertible currency to finance energy imports.

While the rest of world (particularly the North which appears to be succeeding in de-linking economic growth from increased energy consumption) was able to cope with the energy crisis of the 1980s, African countries have continued to face a simmering commercial energy crisis demonstrated by the frequent petroleum shortages and erratic power supply which have combined to bring about a continuing and, in certain cases, increasing dependence on biomass fuels (Davidson, 1986). The recent Kuwait/Iraq crisis that temporarily affected the world oil market, demonstrated the frailty of Africa's energy sector. For example, in September, 1990, Zambia raised its oil prices by 50 per cent, a month later prices were raised by an unprecedented 70 per cent (Reuters, 1990a). In August, 1990, the country's only refinery at Ndola was forced to shut down due to difficulties in procuring crude oil (African Business, 1990).

While cumulative petrol price increases (in the wake of the Kuwait/Iraq crisis) in many African countries were in certain cases well over 100 per cent (Reuters, 1990b; Kenya Times, 1990; Daily Nation, 1990a; Daily Nation, 1990b; and, Daily

Nation, 1990c), increases of petrol-pump prices in North were relatively modest, ranging from a high of 32 per cent in West Germany to a low of 10 per cent in Japan (Economist, 1990). Due to under-exploitation of local energy resources and heavy dependence on energy imports, Africa's energy sector is susceptible to external shocks (Khalema, 1991 and Mohapeloa, 1992). This external dependence compounds the vulnerability of the region's energy sector and limits its ability to shape energy policy (Karekezi, 1988 and Mohapeloa, 1992).

In effect, many African countries are caught in double bind, Their commercial energy sub-sector is being slowly asphyxiated by the relentless pressure exerted by the "simmering commercial energy crisis" described above while their wood-energy subsector is threatened by the rapid denudation of the existing biomass resource base (Karekezi et al, 1986; Karekezi and Ranja, 1997).

The above deep-seated problems facing the African energy sector is not unfamiliar. Attempts to respond to these difficulties have been made by the African energy community over the last 15 years. The question that needs to be answered is to what extent these responses will be successful in addressing the continent's myriad of energy sector problems? What are the prospects for Africa's energy sector in the new millennium?

This paper will attempt to tackle the last question by first examining one of the best know international assessments of Africa's energy future and then juxtaposition this assessment to AFREPREN's perspective which is based on an innovative differentiated approach that provides a comprehensive framework for assessing the highly diverse national energy sector found on the African continent—from the advanced high-tech power sector of South Africa to the biomass-dominated traditional energy sector of the low-income countries of the Sahel.

THE AFRICAN ENERGY SECTOR—FUTURE PROSPECTS

One of the best known attempts to examine the prospects and future of the African energy sector to the year 2010 has been undertaken by the International Energy Agency (IEA, 1996). The IEA assessment categorizes the African continent into 3 major regions, namely: South Africa, North Africa and sub-Saharan Africa. South Africa is perceived to be a predominantly coal dominated region while North Africa is considered to be largely a region that will continue to be fuelled by oil and gas. Sub-Saharan Africa is perceived to be a biomass-dominated region.

Key drivers for the IEA analysis include important assumptions on the current and projected energy intensities of the aforementioned three regions and population growth projections based on statistics from the United Nations. One the most important assumption is the projected economic growth projections which largely assume that Africa will revert to the relatively high economic growth rates that prevailed in the 1970s. The IEA projections incorporate two different differentiated growth projections, one based on high growth assumption that predicts a 5 per cent annual GDP growth after 1996 for South Africa and 4.9 per cent economic growth rate for the rest of Africa. The second growth projection assumes a lower 2.5 per cent annual GDP growth rate for South Africa with the rest of Africa growing at 3.3 per cent (IEA, 1996).

The above differentiated growth projections are then super-imposed on two scenarios, namely: a *capacity constraints* scenario and an energy savings scenario. The capacity constraints scenario is essentially a business-as-usual scenario that assumes that historical trends continue into the future while the energy savings assumes the implementation of exogenously initiated energy efficiency improvements mainly in the oil and gas sector.

The IEA assessment, as shown below, results in 4 different estimates of total commercial energy consumption for Africa by the year 2010.

Projected Energy Consumption
Africa in Year 2010

	Mtoe
Capacity constraints	423
Energy Savings	378
High growth	+70
Low growth	–46

Source : IEA, 1996

While the IEA assessment has some important strengths and provides useful insights on what the future of the African energy sector, it fails to capture what is arguably the most important parameter, namely, socio-economic stability. Research by the African Energy Policy Research Network (AFREPREN) has repeatedly demonstrated that socio-economic stability is often the single most important factor in determining the energy sector development.

To address the challenge of drawing continent wide trends and the wide diversity found in the African continent, AFREPREN researchers have attempted to classify African countries intc five categories based on their energy resource endowment; and, socio-economic stability, as shown in the table below:

Energy resource-rich and stable economies	*Resource-rich and unstable economies*	*Modest energy resources and energy stable economies*	*Modest resources and rapidly reforming economies*	*Modest energy resources and unstable economies*
Egypt	Algeria	Benin	Eritrea	Burundi
Gabon	Angola	Botswana	Ethiopia	CAR
S. Africa	Congo	Burkina Faso	Ghana	Chad
	Libya	Cape Verde	Uganda	Comoros
	Nigeria	C.D'Ivoire		Equatorial Guinea
	DRC*	Djibouti		Guinea Bissau
		Kenya		Gambia
		Malawi		Guinea
		Mali		Lesotho
		Mauritius		Liberia
		Morocco		Madagascar
		Namibia		Mozambique
		Senegal		Niger
		Seychelles		Rwanda
		Swaziland		Sierra Leone
		Tanzania		Somalia
		Togo		Sudan
		Tunisia		
		Zambia		
		Zimbabwe		

**DRC-Democratic Republic of Congo formerly Zaire*

Source : Compiled by author

The *Energy resource-rich and stable economies* together generate about four fifths of the region's electricity. These countries rely heavily on petroleum or coal based thermal electricity as the most economically viable option. This is partly because their geographic location does not favour them with significant hydropower resources. Instead most of them have considerable oil or coal reserves. Gabon for instance is one of the wealthiest country in sub-Saharan Africa, due to its rich oil reserves and consumes a lot of power in regional terms (Financial Times, Apr. 1997:16).

The *Energy resource-rich and unstable economies* even though well endowed with energy resources, have been unable to exploit their resources due to unstable political conditions. In the case of Democratic Republic of Congo (DRC, formerly known as Zaire), the country has by far the region's highest hydropower potential, which stands at about 92,500 MW (SAD–Elec report, 1996:218).

The *Modest energy resource and stable economies* account for an important share of the continent's energy consumption. The energy sectors of most countries in this category are somewhat similar. These countries have low but steady rates of economic growth and relatively well established primary industries, as a base for industrial growth. In some cases, such as for Zambia and Zimbabwe, coal acts as a substitute for electricity especially for heat generation. In general, the energy sectors of these countries have experienced slow but steady growth, since independence and in most cases, have been able to provide sufficient energy supplies (local and imported) to meet growing consumer demand.

The *Modest Energy resource and rapidly reforming economies* include countries such as Uganda, Ghana, Ethiopia and Eritrea. They have in the past been on the brink of complete economic collapse, but recently adopted far-reaching political and economic reforms that have given them a new lease of life and consequently a more promising future. Reform, high economic growth rates and recovery have paved the way for rapid rehabilitation and growth of their respective energy sector.

For the *Modest energy and unstable economies*, countries in this category are Somalia, Rwanda, Burundi and Liberia. The energy sectors of these categories is characterised by general disarray and almost complete collapse of the power sector. Industry and the modern transport sectors have all but died down and prospects for future energy sector development are dim. The prospects for the energy sector in this category of countries is dim and fraught with difficulties.

It is important to note that a number of countries have initiated important socio-economic reforms that could result in improved prospects for the energy sector. Countries that fall under this category include Madagascar and Mozambique that could conceivably change the prospects of their respective energy sector in the near term. Similarly, a number of countries that are currently grouped in the stable socio-economic category could slip into anarchy in the near future. Countries such as Kenya, Senegal, Zambia and Zimbabwe are currently facing serious socio-economic and political instability and could become unstable in the near term.

The additional categories introduced in this paper underline the importance of socio-economic stability. Instead of the three categories (South Africa, North Africa and sub-Saharan Africa) that the IEA uses in its assessment, the author would like to argue that the proposed five categories that incorporate a socio-economic stability dimension are likely to represent a more accurate picture of the Africa's energy sector.

What is required, therefore, is the development of energy projections for the five categories based on national energy sector assessments of each country in the respective category. Category projections can then be aggregated to produce continent-wide estimates. This level of detailed assessment is, however, hampered by the absence of national level energy statistics. A major and continuous effort to track the energy sector status of most African countries is sorely needed. While there has been some limited efforts, notably by the African Development Bank (ADB), to track the energy sector of African countries, there is still no consistent and coherent compilation of the energy sector of African countries available.

The African Energy Policy Research Network (AFREPREN) has initiated a modest in-house effort to track the energy sector of selected eastern and southern African countries but a much more comprehensive and detailed effort is required. The AFREPREN country energy sector database in conjunction with energy data from the African Development Bank as well as a recently completed energy statistics collected by an EU financed project could provide the basis for such an initiative.

AFREPREN could provide an ideal setting for such an

exercise because of its network of researchers found throughout Africa which could be used to not only collect but update the proposed African energy data base. Initially, the African energy data base could be a modest initiative that simple compiles existing energy data and include a brief assessment of the energy institutional set-up.

To ensure early interest, the energy database initiative can concentrate on commercially attractive energy resources, namely coal, oil, gas and hydro. With some initial seed support, it is possible to make the initial effort financially self-sufficient. At a later stage the less commercially attractive energy resources such as biomass, solar and wind can be included. The same principle can be applied to the non-conventional energy resources by placing initial emphasis on resources that could be commercially attractive in the medium term. Examples include biomass waste from agro-processing industries that could provide the basis for the establishment of a cogeneration industry.

Initially investigations by AFREPREN indicate that there would be substantial demand for such a service with a possibility of attracting substantial subscription revenues. The emphasis on early financial sustainability is important in light of the results of past efforts that have largely remained as academic exercises with detailed but largely useless country energy balances for a few selected African countries. Because of limited commercial interest, support for these initiative quickly waned resulting in no updates and rapid obsolescence of the collected energy data.

The enduring commercial interest in the energy sector is demonstrated by the success of a rather mediocre bio-monthly energy newsletter entitled *African Energy and Mining* (Indigo Publications, 1998) edited and published in France with an annual subscription of close to US $ 770 per year. More recently, The Financial Times launched a monthly newsletter in April, 1998 entitled *African Energy* (Financial Times Energy, 1998) with a subscription of just under US$ 1,000 per year. None of these newsletters provide significant amount of energy data.

The aforementioned successful examples of very expensive African energy newsletters indicate a high-level of commercial interest in African energy information and data. This is also supported by the results of AFREPREN research work which indicate a major need for African energy information. It is the opinion of the author that the time is ripe for the provision of a time-bound seed support for the initiation of high quality African energy data base and information initiative that has a clear and unequivocal self-financing target.

REFERENCES

1) AFREPREN. 1990. "African Energy Policies: Issues in Planning and Practice. African Energy Policy Research Network". London: Zed Press.
2) African Business. 1990. "African Feels the Effects of the Middle-East Oil Crisis". Feature in African Business monthly magazine of October, 1990. London: IC Publications.
3) Baguant, J. and Mengistu Teferra. 1991. "Efficiency Use of Petroleum in the Transportation Sector in Ethiopia and Power Sector in Mauritius". AFREPREN Progress Report. Gaborone: AFREPREN.
4) Bachou, Salim. 1990. "Key Factors for the Dissemination of Renewable Energy Technologies—Cookstoves and Briquettes in Uganda: Draft Report". Nairobi and Gaborone: AFREPREN.
5) Bhagavan, M.R. 1990. "Technological Advance in the Third World—Strategies and Prospects". London: ZED Books Ltd.
6) Bhagavan, M.R. and Karekezi, S. (editors). 1992. "Energy for Rural Development". London: AFREPREN and ZED Books.
7) Bhagavan, M.R. and Karekezi, S. (editors). 1992. "Energy Management in Africa". London: ZED Books.
8) Daily Nation. 1990a. "Oil Crisis Starts to Hurt Zimbabwe". Feature in the Daily Nation newspaper of Monday 8 October, 1990. Nairobi: Nation Newspapers Ltd.
9) Daily Nation. 1990b. "Fuel Prices Up". Feature in the Daily Nation newspaper of Tuesday 18 September, 1990. Nairobi: Nation Newspapers Ltd.
10) Daily Nation. 1990c. "Dar Raise Oil Prices". Feature in the Daily Nation newspaper of Tuesday 11 September, 1990. Nairobi: Nation Newspapers Ltd.
11) Datta, Ansu. 1989. "Research for Development and the Development of Research". Gaborone: National Institute of Development, Research and Documentation (NIR), Uni-

versity of Botswana.
12) Davidson, R. Ogunlade. 1986. "Energy Decisions in Developing Countries in Africa—A Case Study of Sierra Leone". IDRC Manuscript Report 192e. Ottawa: International Development Research Centre (IDRC).
13) Davidson, Ogunlade and Stephen Karekezi. 1991. "A New, Environmentally-Sound Energy Strategy for the Development of Sub-Saharan Africa". Nairobi: AFREPREN.
14) Davidson, Ogunlade and Stephen Karekezi. 1992. "A New, Environmentally-Sound Energy Strategy for the Devlopment of Sub-Saharan Africa—Final Statement of the African Energy Experts Meeting." Nairobi: AFREPREN.
15) Davidson, Ogunlade and Stephen Karekezi. eds. 1992. "Environmentally-Sound Energy Options for Africa. Nairobi: UNEP.
16) Diphaha, J. 1991a. "Substitution of Coal for other Fuels in Households and Institutional Applications—A Critical Analysis of the Technological, Socio-economic Pre-requisites and Environmental Impacts: Draft Final Report". Gaborone: AFREPREN.
17) Diphaha, J. 1991b. Personal communication.
18) Eberhard, Anton and H. Trollip. 1994. Background on the South African Energy System. Energy for Development Research Centre. Cape Town: University of Cape Town.
19) Economist, 1990. "Business This Week". Feature in The Economist weekly magazine of 29 September, 1990. London: The Economist Newspaper Ltd.
20) Elgizouli, Ismail. 1991. "AFREPREN Renewable Energy Technologies (RETS) Research Project—Improved Cookstoves and Briquetting in the Sudan: Final draft report". Nairobi & Gaborone: AFREPREN.
21) Energy Research Group. 1986. "Energy Research—Directions and Issues for Developing Countries". Report No. IDRC–250e. Ottawa: IDRC.
22) Financial Times Energy. 1998. "African Energy". London: Financial Times Business Ltd.
23) Indigo Publications. 1998. "African Energy and Mining". Paris: Indigo Publications.
24) International Energy Agency. 1996. "Energy Policies of South Africa". Paris: OECD/IEA.

25) International Energy Agency. 1995. "World Energy Outlook: 1995". Paris: OECD/IEA.

26) Iwayemi, Akin. "Deregulating Public Utilities—An Analysis of the Electricity Industry in Nigeria". Utilities Policy Journal, January, 1994. London: Butterworth-Heinemann.

27) Karekezi, Stephen, et al (editors). 1986. "Development and Dissemination of Wood Energy Technologies in Eastern Africa—Summary Report of the KENGO Regional Workshop on Improved Wood-stoves held Sept.–Oct., 1986". Nairobi: KENGO.

28) Karekezi, Stephen. 1988. "Surviving the Wood Energy Crisis in Lesotho". Nairobi: Kenya Energy and Environment Organisations (KENGO).

29) Karekezi, Stephen. 1989a. "Responding to a Global Need for Local Action in Domestic Energy". Chapter in Proceedings of a Nordic Seminar on Domestic Energy in Developing Countries edited by Elisabet Viklund. Lund: Lund University.

30) Karekezi, Stephen. 1989b. "Review of Mature Renewable Energy Technologies (RETs) in sub-Sahara Africa". AFREPREN research report. Gaborone: AFREPREN.

31) Karekezi, Stephen. 1990. "Review of Mature Renewable Energy Technologies in sub-Sahara Africa". Chapter in "African Energy Policies: Issues in Planning and Practice". London: AFREPREN and ZED Books.

32) Karekezi, Stephen et al. 1991. "Doing More With Less—Sustainable Development of the Wood Energy Sector in Uganda". Nairobi: KENGO/Regional Wood Energy Programme for Africa.

33) Karekezi, Stephen. 1992. "Energy Technology Options for Rural and Agricultural Development: The Major Issues". Chapter in "Energy for Rural Development" edited by M.R. Bhagavan and S. Karekezi. London: AFREPREN and ZED Book.

34) Karekezi, Stephen and Gordon Mackenzie. 1993. "Energy Options for Africa—Environmentally-Sustainable Alternatives". London: ZED Books & Nairobi: AFREPREN.

35) Karekezi, Stephen, L. Majoro, P. Gathu. 1996. Restructuring and Privatisation of the Power sector in Africa. Nairobi: AFREPREN/FWD.

36) Karekezi, Stephen, L. Majoro, A. Tibwitta, J. Muthui and F. Theuri. 1997a. Management and Efficiency of Power Utilities in Africa-Draft III. Nairobi: AFREPREN/FWD.
37) Karekezi, Stephen, T. Ranja, J. Kimani, M. Muthoni and B. Macharia. 1997b. "Pricing and Investment Policy Issues in the Downstream Petroleum Sector in Africa". Nairobi: AFREPREN/FWD.
38) Karekezi, Stephen and Timothy Ranja. 1997. "Renewable Energy Technologies in Africa". London: ZED Books & Nairobi: AFREPREN.
39) Karekezi, Stephen, Timothy Ranja and J. Kimani. 1997. "Petroleum Pricing and Investment in Africa: Policy Issues in the Downstream Sector". Nairobi, AFREPREN.
40) Karenzi, Pierre-Clever. 1989. "Rural and Poor Urban Energy Situation in Rwanda". Gaborone: AFREPREN.
41) Karenzi, Pierre-Claver. 1990. Personal Communication.
42) Karenzi, Pierre. 1991. "Use of Biomass Energy Balance to Determine the Woodfuel Problem in Rwanda". Gaborone: AFREPREN.
43) Katihabwa, Joseph. 1990. "Success and Failure Factors of Biogas Technology Implementation in Burundi". AFREPREN Interim Research Report. Gaborone: AFREPREN.
44) Katihabwa, Joseph. 1991. "La Technologie du Solaire Photovoltaique au Burundi: Ses Facteurs de Succès et D'échec". AFREPREN Interim Report. Gaborone: AFREPREN.
45) Kenya Times. 1990. "Ghana Hikes Fuel Prices". Feature in the Kenya Times newspaper of Tuesday 4 September 1990. Nairobi: Kenya Times Media Trust Ltd.
46) Khalema, Lucy. 1991. "Rural Electrification in Lesotho". AFREPREN research report. Gaborone:AFREPREN.
47) Kubursi, A. and T. Naylor. 1985. Co-operation and Development in the Energy Sector". In Proc. of Symp. on the Energy Sector. May, 1984. Canada: MacMaster.
48) Mohapeloa, L. 1992. "Energy Management in the Transport Sector—The Case of Lesotho". Chapter in "Energy Management in Africa" edited by M.R. Bhagavan and S. Karekezi. London: ZED Books.
49) Nyoike, P. 1993. Personal communication.
50) Davidson O. and Karekezi, S. (editors). 1992. "Environmentally-Sound Energy Options for Africa". Nairobi: AFREPREN.

51) Davidson O. and Karekzi, S. 1992. “A New, Environmentally-Sound Energy Strategy for the Development of sub-Saharan Africa”. Nairobi: AFREPREN.
52) Okot-Uma, W.O. Rogers. 1989. “African Energy Programme, a Perspective of a Regional Network”. Series Number CSC(89)ENP-28.
53) Project Document 272. London: Commonwealth Science Council.
54) Pasztor, Janos and Lars A. Kristoferson (eds). 1990. “Bioenergy and the Environment”. Boulder: Westview Press.
55) Ranganathan, V. (ed.). 1992. “Rural Electrification in Africa”. London: ZED Books.
56) Reuters News Agency. 1990a. “Zambia Hikes Fuel Prices by 70%”. Reuters Feature in Kenya Times newspaper of Tuesday 2 October, 1990. Nairobi: Kenya Times Media Trust Ltd.
57) Reuters News Agency. 1990b. “Uganda in Sharp Fuel Price Rise”. Reuters feature in the Daily Nation newspaper of Friday 21 September, 1990. Nairobi: Nation Newspapers Ltd.
58) Rutabanzibwa, P. 1989. “Tanzania’s Response to the Oil Crisis—Impacts and Lessons”. Nairobi and Gaborone: AFREPREN.
59) US Department of Interior. 1976. “Energy Perspectives”.
60) Wereko-Brobby, Charles, 1990. Private Communication.
61) World Bank. 1982. “Burundi: Issues and Options in the Energy Sector”. Report of the Joint UNDP/World Bank Energy Sector Assessment Program. Report No. 3778-BU. Washington, D.C.: World Bank.
62) World Bank. 1983. “Nigeria: Issues and Options in the Energy Sector”. Washington, D.C.: The World Bank.
63) World Bank. 1984. “World Development Report 1984. Washington, D.C.: The World Bank.
64) World Bank. 1986: “Agricultural Residue Briquetting Pilot Projects for Substitute Household and Industrial Fuels—Ethiopia: Volume I, Technical Report”. Joint UNDP/World Bank ESMAP Activity Completion Report No. 062A/86. Washington, D.C.: The World Bank.
65) World Bank. 1989. “Sub-Saharan Africa—From Crisis to Sustainable Growth, A Long-Term Perspective Study”. Washington, D.C.: The World Bank.

66) World Bank. 1990. “World Development Report—1990: Poverty—World Development Indicators”. Washington, D.C.: the World Bank.
67) World Bank. 1994. “World Development Report—1994: Infrastructure for Development”. Washington, D.C.: the World Bank.
68) World Bank. 1996. “World Development Report 1978–1996 with World Development Indicators 1996—Indexed Omnibus CD-ROM Edition”. Washington D.C.: World Bank.
69) World Bank. 1997. “World Development Report 1997”. Washington D.C.: World Bank.
70) World Energy Conference. 1974. “Survey of Energy Resources”. New York: World Energy Conference.
71) World Resource Institute. 1986. “World Resources 1986”. New York: World Resource Institute.

10

Energy Security—A Venezuelan Perspective

E. PICO-PONTE
Venezuela

Venezuela has been blessed with abundant natural energy resources, which have provided a very important element for the country's relationship with the outside world. It has been an important and reliable international oil supplier during the past eighty years, and has become heavily dependent on oil exports for maintaining economic activity and obtaining fiscal revenue. Located at the northern tip of South America—very near to the centre of the American continent—it is very conveniently close to important geographical markets. With a population of approximately 22 million, the country covers slightly over 900,000 square kilometers.

As to energy resources, it has 452 billion barriers of oil-equivalent of fossil fuel reserves, of which 106 bboe are proven: 75 bboe of oil, 25 bboe of gas, and 6 bboe of coal. There is also a hydroelectric potential of 55,000 megawatts, of which 13,000 MW have generating capacity installed. Hydroelectricity accounts for over 2/3 of the country's total electric generation. Petroleos de Venezuela is a strong national corporation—covering all phases of the oil and gas business—which represents an important asset that makes possible national and international activities in these fields.

The conditions briefly described above provide the basis for the perspective that gives origin to the following ideas on energy security.

In looking ahead towards the third millennium, it is important to remember certain essential facts referred to the actual state of the world in what to energy respects:

a) 40 per cent of its population does not have access to commercial energy, and thus find themselves severely limited as to their potential to participate in economic activity, fact which in turn strongly conditions their level of income and quality of life.
b) The infrastructure in place in today's world is the result of a choice that favoured concentrated energy sources (essentially hydro, coal, oil and gas), made at the advent of the industrial revolution and which, at the same time, made the latter possible. This choice still determines at present a very large fraction of the energy mix used worldwide. Transport, a determining factor in today's national and international economic activity, has developed a specific dependence on oil.
c) These facts, combined with the abundance of resources and the ease of transportation and use of cil and gas, has made strategic commodities out of both of these.

As to the basic elements that help define the direction of Venezuela's international energy activity, the following must be mentioned. The country views its natural energy resources, which greatly surpass its needs, as a privileged platform on which to develop its international relations. These resources, as all other natural resources, are considered neutral in character—their good or bad use does not depend on their essence but on factors of human origin. Poverty and lack of education are considered as the main polluting agents in the developing world. Energy security is considered to be an essential element for the stable development of all countries in the world. Energy is thus seen as a fundamental element for international cooperation and development. Energy, together with the economy and the environment are seen to constitute an integral system, whose parts are not susceptible of being realistically treated in isolation from one another. Oil is the main source of the revenue needed to achieve the country's development and overcome the poverty of many of its citizens.

The above mentioned basic elements constitute guiding principles for Venezuela's insertion in the international community, from the point of view of energy. These principles have translated in practice into the country's active participation as a founder and member of OPEC, and as a promoter of the

Producers–Consumer Dialogue process and of energy cooperation in general, at world level. It is also strongly dedicated to fostering the Hemispheric Energy Initiative, which seeks energy cooperation and integration at the level of the Americas and the Caribbean basin. Venezuela is a supplier, jointly with Mexico, of the San José Accord—now in its 19th year of activity—through which the country has financed projects for a total amount of about 1.5 billion US dollars in Central America and certain Caribbean countries. Of these funds, 65 per cent have been devoted to build electric generation and distribution capacity, highly valuable in a region in which large fractions of the population do not have access to electricity. At the same time, the Accord has guaranteed security of oil supply to these countries—especially valuable in times of market supply upheaval. The country is also a supplier of electric power to neighbouring nations.

Energy security is of paramount importance to the healthy functioning of the world economy and the promotion of human well-being. This, in view of the strategic character of energy to the large number of countries in the world that need to import it to satisfy their needs; as well as of those countries which export it to obtain funds for their development. This concept of energy security in the world market implies security for the consumer-importer (security of supply), as well as for the producer-exporter (security of demand), both of which are necessary and complementary elements of the former. In general terms, it is the producer-exporter who is called to provide security of supply, and the consumer-importer who is called to provide security of demand; the international energy world cannot function in a stable manner without the cooperation of both. At the same time, the viability of both of these aspects of energy security is strongly conditioned by factors of economic, political and environmental nature, which in many cases are determined outside of the domain of energy.

The significance of energy security to Venezuela, a producer-exporter, can be better understood when the 1996 contributions of oil to GDP (27%), fiscal income (49%) and foreign exchange income (46%) are considered. We cannot but conclude that oil activity is essential to the country's functioning, development and progress. Especially when, given its actual socio-economic condition, an annual economic growth rate of 6 per cent would be required for a whole decade if it were to overcome its underdeveloped state.

Venezuela is thus planning to increase its offer of oil, gas, coal, Orimulsion® and hydroelectricity to the outside world. Both, to obtain from these sources the financial resources required for its development, and to help satisfy the growing energy needs being forecast for the world in years to come, specially in developing countries. In the specific case of oil, plans call for an increase of production capacity from 3.3 million barrels per day in 1997 to slightly over 6 mill. bpd in 2006, with very significant participation from the private sector in this effort, channeled through the opening process (apertura) of the national oil and gas industry.

As a result of the elements presented above, we view the following as the main challenges to be overcome in the domain of energy at world level, in the short, medium and long term:

- Providing access to commercial energy to those that don't have it—which should in turn give them access to active participation in economic activity, thus contributing to the eradication of poverty—main polluting agent in the developing world—and the overall world wealth.
- Developing technology for the more harmonious use of energy with respect to the environment.
- Obtaining better scientific understanding of global climate phenomena, and possible ways to mitigate them—among which the sequestering and profitable use of CO_2 could play a very important role.
- Developing consumer awareness as to the use, and specially the waste, of energy—both commercial and non-commercial; including the implicit energy content of products they use.
- Making energy saving a profitable business.
- Deploying energy efficient technology in a massive manner.

The role of technology, main driving force of the exponential changes that have taken human activity to new levels of performance in the last half century, could never be overemphasized as an essential promoter of an ever more harmonious energy-economy-environment world system.

11

Energy Future in the Third Countries: Asia

H. IBRAHIM
Indonesia

INTRODUCTION

The fast growing economy emerging in Asia shall drive for more steady increase in the use of energy in the future. The world's total primary energy demand is expected to grow at an annual average of 1.6 per cent from 8,000 Mtoe in 1995 to 12,600 in 2010. Within the same period oil demand is forecast to increase at an annual average growth of 0.75 per cent, with the extra consumption mainly taken up by the increase of use in the transportation sector. Oil would still retain the lion's share of the global economy, however, falling from 37 per cent share today to 21 per cent share in 2020. Natural gas is therefore the number one substitute taking over the dependency of oil in the power generation sector as well as the industrial sector, with coal playing the next important role, and the biomass and renewables also contributing a smaller but nevertheless very useful share.

While the global trend indicates a steady increase of energy growth per year during the last two decades and projected into the next two decades, further analysis reveals a more fragmented development distribution between region to region or continent to continent. The OECD (Organisation for Economic and Co-operation Development) countries which have very established energy development and utilisation pattern, on one side are observed to have very stable energy growth and as a matter of fact decreasing share of global percentage, whereas the Asian region, the main focus of this paper, is observed to have the other extreme of energy growth pattern.

This paper begins with a preliminary focus on the current energy situation and forecast in Asia as a whole before focusing on the ASEAN Member Countries.

GENERAL ENERGY OUTLOOK IN ASIA

China, East Asia and South Asia are sub-regions in Asia that have demonstrated strong economic performance in the last decade and this situation is predicted to continue into the next one to two decades. Their total energy requirement therefore is of high significance on the energy demand at the global level. These regions enjoyed a 23 per cent share of the world's GDP in 1993. In the decade between 1985 to 1995, average annual economic growth was 7.5 per cent, as compared with the rest of the world (these regions excluded) at 2.4 per cent.

Expecting this rapid growth to continue, the GDP of these regions will increase to 36 per cent in 2010. Primary energy demands of these regions will increase from the 1995 demand of 1,400 Mtoe in 1995 to 3,000 Mtoe in 2010, which is a quarter of the world's energy demand then. Oil dependence in 2010 shall still be dominant with 40 per cent of the increase coming from these regions.

Table 11.1 shows the dramatic increase in the energy resources in two decades from 1973 to 1993, and the energy forecast figures until 2010.

Table 11.1 : Shares of the Dynamic Asian Region in the World (%)

	1973	*1993*	*2010*
GDP	13	23	36
Population	52	53	53
Primary Energy Demand	8	18	26
Solid Fuel	17	34	47
Oil	6	15	23
Gas	1	5	11
Nuclear	1	5	11
Hydro/Others	1	3	7
Electricity Output	6	15	23
CO_2 Emissions	10	22	31
Net Oil Dependence (%)	12	37	65

*Source:*IEA, World Energy Outlook, 1996 Edition

1) Oil: Continuing Increase of Demand

The dynamic Asian regions will account for 40 per cent of the increase in world oil demand between 1993 to 2010. Their net oil import dependence could therefore increase from 40 per cent to 65 per cent in 2010. Currently the Middle East supplies approximately 76 per cent of the regions' total crude oil imports and the dependence on Middle East crude is likely to increase to 79 per cent in 1997 and 84 per cent in 2000. By 2005, close to 92 per cent of all crude imports is expected to come from the Middle East, unless alternative sources of petroleum supplies is favoured, especially natural gas.

By 2015 only Brunei, Malaysia, Laos and Myanmar are projected to be net energy exporters in the region. Current or past giant exporters such as China and Indonesia will move more and more into energy deficit as their huge and increasing populations grow increasingly affluent, and consumes more energy formerly set aside for export. Asia's reliance on external sources of oil to be close to 70 per cent of its demand in 2010.

Oil use has continued increasing in reflection to the strong economic growth, while regional crude oil production has been virtually flattening out. Asian economies, except for certain countries that are currently affected by the financial crises the effect of which is short term anyway, the Asian economies would keep growing at the present pace. Asia's energy demand will increase to 2.38 billion tons oil equivalent (Btoe) in 2000, and 3.55 Btoe in 2010. Out of these figures China alone would need 1.04 Btoe in 2000, and 1.6 Btoe in 2010. With the energy demand in 1995 at 2.08 Btoe, the future demand reflects an average growth rate of 3.6 per year.

In terms of reserves, Asia's energy resources are limited. Excluding Central Asia, Asia's proven reserves account as little as 5 per cent of the world's figure. R/P ratio, obtained by dividing the proven reserves by the production level amounts to 20 years. China, one of the two major oil exporters (with Indonesia) has transformed to a net oil importer in 1993 due to soaring petroleum products demand and falling crude production at home. Indonesia too is expected to become a net importer of oil by the year 2008, unless new significant reserves are found.

2) Natural Gas Use in Asia: Current Situation and Prospects

Until recently gas was seen to have played a relatively minor role in meeting the energy demand of Asian countries. However, recent trend has indicated that the scenario is going to change—more natural gas will be used domestically in the Asian countries over the next two decades.

This optimistic outlook is derived from a number of factors, among others: increase of populations, rapidly expanding economies, shrinking state participation in the energy sector, market deregulation, energy security concern and diversification of fuel source, and higher environmental awareness of fuel burning.

The demand for energy in the next one to two decades of the dynamic regions in Asia is so huge that the countries concerned will be forced to make a shift from over-dependence of oil to more increase of natural gas and coal to some extent. With oil production set to deplete in some countries in the near future, energy security supplies have become a growing concern among the developing countries—and more dialogues with the developed countries are continuously being initiated on the global energy security issues. The need for diversification has been incorporated in the energy policies of many countries, partly in response to concerns on energy security, but also partly, especially for the oil producing countries, to conserve the oil reserves in order to main foreign revenues for as long as possible.

The environmental impact of energy utilisation is also another major reason for the Asian countries to make a shift to use more environment-friendly fuel source. Natural gas, in addition to its lower price per MBTU as compared with oil, has this competitive edge. As shall be discussed further in this paper, more gas infrastructure are now in place and planned that will encourage further domestic utilisation of natural gas.

With the efficiency of power generation now vastly improved with the advent of combined cycle combustion, natural gas shall see itself mostly consumed in the electricity sector within the next decades. China, Japan, India, Indonesia and Thailand are Asia's biggest producer of electricity at the moment with average annual electricity growth of some of these countries and others reaching double digits.

China's 1996 electricity generation has reached 1,075 billion kwh, making China the world's second largest producer of electricity after the United States. China's electricity demand is projected to grow between 6 to 7 per cent per annum and in the next five years 80,000 MW capacity may be required to power up the country. However, China's electricity sector is heavily dependent on coal and it has no serious plan of increasing its natural gas share in the electricity sector. Certain areas, especially coastal provinces being more progressive will be most likely candidates for gas-fired power plants. China has planned to double its annual gas production from 1 TCF (30 BCM) by the year 2005—most of the natural gas production in China is geared towards export by gas-pipelines and by LNG, and domestic industrial use.

3) Coal: Another Alternative Substitute

Coal will continue to become a dominant source of energy for many countries in Asia. While the high share for the use of coal is advantageous in supply stability due to its lower price, its impact on the environment is creating higher concern. With coal use on the increase, there would be more demand for clean coal technologies in this region.

China holds a crucial position to Asia's energy supply and demand. It is estimated that China's total primary energy demand at 1.47–1.50 billion tons coal equivalent (1.02–1.05 Btoe) for the year 2000, and 1.95–2.05 billion tons coal equivalent (1.365–1.435 Btoe) in the year 2010. Of this total primary energy supply coal accounts for 75 per cent, and this trend is likely to last long ahead. In 1995, by sector, 76 per cent of generated output, 75 per cent of industrial energy, and 60 per cent of residential and commercial energy has been covered by coal.

In terms of reserves, Asia's proven coal reserves account 22 per cent that of the world's total reserves. Reserve to production ratio, at 1995 production rate indicate 129 years.

ENERGY FUTURE IN THE ASEAN REGION

ASEAN (Association of South East Asian Nations) is a grouping of nine countries in Southeast Asia, namely: Brunei, Indonesia, Laos, Malaysia, Myanmar, Philippines, Singapore, Thailand and Vietnam. ASEAN has been considered to be the fastest growing region in the world over the past decades. Thailand, for example, has an average GDP growth of 11 per cent over the 1987–1993 period. Together with Malaysia and Singapore, these three countries have been the so-called dynamic ASEAN economies (DAES) due to their rapid economic growth over the past two decades. The current financial turmoil badly hits Indonesia, Thailand and Malaysia, forcing these countries to reschedule their energy infrastructure plans. However, efforts have already been made by the affected countries to make economic reforms in order to bounce back their economies—the crisis is very much regarded as a short-term phenomena. In line with the high economic growth, the energy consumption of the region has also shown a rapid increase. The average annual growth rate of primary energy demand was around P/o per year, with oil having the largest share in the total consumption. Although the relationship is not necessarily absolute, energy is a necessary input for economic activity and trade. Since the nature of growth in the region is very energy intensive, it is expected that energy demand will continue to increase as a result of the industrialization process and the increase in economic specialisation and its concomitant demands for transport.

ASEAN's current share in the world's energy production and consumption amounts to 3.4 per cent and 1.4 per cent, respectively. These figures are apparently insignificant. However, the strong economic growth of most of the ASEAN member countries have indicated an average annual energy growth of 7–8 per cent and this trend is expected to concave until the year 2010. Comparing this growth figure with the average energy consumption growth of the OECD countries ranging at 1.3 per cent from 1990 to 2010, Asia's current and future high energy demand is a topic that has therefore deserved earnest attention in international and global energy security supplies and demand issues.

For oil resources, the region's production increases from 160 Mtoe in 1985 to 298 Mtoe in 1996, corresponding to an average increase of almost 6 per cent per year. The main oil producer in the region is Indonesia. Its production to reserve ratio indicates that oil resource in the country will be depleted in the next 8 to 10 years if there are no new major fields being discovered. Due to this reason, the Indonesian Government has aggressively pursued towards non-oil use especially in the industrial and power sector and towards efficient use of energy especially in the industrial and transportation sectors. Exploration programme has also been intensified in order to found new oil fields. New and more attractive incentives on project sharing contracts have recently been set in place to attract the international players in joint-venture efforts with the national oil company PERTAMINA in exploration and production activities. The main problem in finding new fields is that the location will be further inland and in remote places. This will put additional cost in the development of these resources. New mining techniques will also be expensive and thus incur more funds in the development of oil resources.

Natural gas reserves are large in the region and have been considered the most abundant fossil fuel supply of the South East Asia region. Current recoverable reserves are in the vicinity of 190 TCF or 39 billion barrels of oil equivalent (MMBEq). Barring further discoveries, current gas reserves in the region can provide sufficient fuel to generate as much as 127,000 MW of electricity for the next 26 years.

Natural gas development requires huge amounts of up front investments in the production and distribution infrastructure. Construction of dedicated transport and distribution facilities are a necessity for assurance that natural gas supply will meet the demand at the right time and at a reasonable cost. As regional market for gas matures, a pipeline from a single source to a single demand centre will be expanded into an interconnected pipeline system linking multiple sources and demand centres.

Vast amounts of coal are found in Indonesia, Thailand and Vietnam. Coal reserves are estimated to be around 18 billion barrels of oil equivalent (MMBeq) in the South East Asia region. Production of coal has been increasing rapidly over the ten-years period from 3 Mtoe in 1985 to almost 28 Mtoe in 1996. It is expected that coal will play a significant role in the future especially for industries and power generation. The problem in the development of coal resources is the substantial investment required in the production, transportation and distribution of the resources. For coal to be environmentally friendly, abatement technologies to reduce emission or new clean technologies need to be installed for the future. Without these technologies, the use of coal in large amount will not be warranted since it will have harmful effect to the ecosystem and human health. These technologies of course, will be expensive; thus, putting more pressure in the financing of coal development.

Potential for hydro and geothermal are high in the ASEAN

region especially in Indonesia and the Philippines for geothermal resources and for Malaysia, Indonesia and Thailand for hydro resources. Although the potential is big, both of these resources are under utilised. This is due to the high cost in the development of these resources making it not too economical as compared with oil or coal. For hydro, the area for the dams is extensive and this requires cost of moving the population existing on the appropriate sites. Although these resources cost highly to be developed, but they are clean fuels in the sense that the pollutant emitted to the air is not so hazardous as fossil fuels. The prospect is bright for these resources to be further utilised if there exists in the near future a very stringent regulation that limits the emission of pollutants, especially CO_2.

1) Future Demand of Energy in ASEAN

As mentioned previously, the region is expected to continue experiencing a rapid economic growth in the next few decades. Consequently the demand of energy will also increase as population increases, income level rise and industrialisation grow rapidly.

From a study of AEEMTRC, the ASEAN 2020 programme, it was estimated that under the existing programmes or short term plans of the individual member states (Business as Usual scenario), the primary energy demand of the region, will reach a level of 640 Mtoe by 2020 (2). If biomass is excluded, the primary energy demand would be 600 Mtoe, about 4 times the 1995 level (Figure 1). This corresponds to an average growth of 5 per cent per year over that time period.

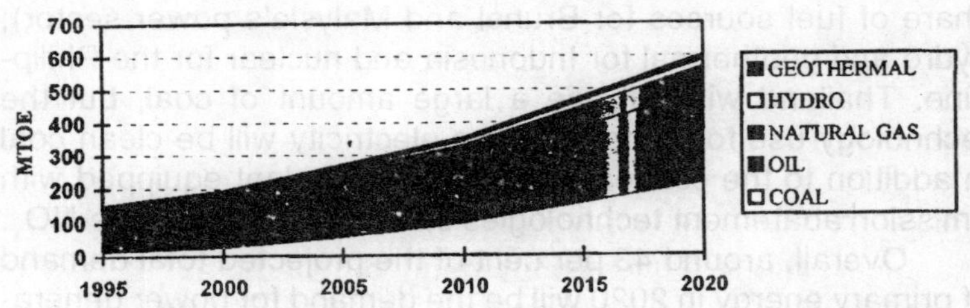

Fig. 1 — **Future Primary Energy Demand of ASEAN**

Oil will still constitute the largest part of the region's total primary energy demand, but its share will decrease to 45 per cent. The main user will still be the transport sector, constituting almost 70 per cent of the region's demand for oil in 2020. The power sector will only contribute around 2 per cent, while the remainder would that of the industry and residential/commercial sectors.

As a substitute for oil in the power sector, coal will be the main choice for countries like Indonesia, Thailand and Philippine. Singapore also chooses coal as its main electricity generating fuel, but through using clean coal technology. From the total of 220 Mtoe of the region's coal demand in 2020, around 90% will be the demand for the power sector. Compared to its consumption level in 1995, the demand in 2020 will result in an average growth of 12 per cent per year over that time period.

Natural gas will also have an increasing role in the future, but its growth will not be as rapid as coal. In 2020, natural gas demand in the region will be 74 Mtoe. Of this amount, 52 per cent will be the demand for power generation, mainly in Malaysia and Brunei.

Hydro and geothermal demand will also increase over the planning period, reaching 21 Mtoe in 2020 for hydro and 16 Mtoe for geothermal. This is an increase by a factor of 4 for hydro and 8 for geothermal, corresponding to an average annual growth of 5 per cent and 8 per cent respectively.

The above findings were that of the "Business as Usual" scenario. If the region imposes a strict regulation on the emission of air pollutant such as SO_2, NO_x, and CO_2, the situation will be different. In this case, coal will not become the main fuel for power generation anymore. Instead natural gas will be the main choice for Singapore (natural gas has already dominated the share of fuel sources for Brunei and Malysia's power sector); hydro and geothermal for Indonesia and nuclear for the Philippine. Thailand will still use a large amount of coal, but the technology use for generating the electricity will be clean coal in addition to the conventional coal power plant equipped with emission abatement technologies such as de-SO_x and de-NO_x.

Overall, around 43 per cent of the projected total demand of primary energy in 2020 will be the demand for power generation, 32 per cent for transport, 20 per cent for industry and the remaining 5 per cent for residential/commercial sectors. Thus, the increase of the region's primary energy demand is contributed mainly from the increase of fuel inputs for power generation as a result of the rapid increase of electricity demand.

The total electricity demand of the ASEAN member countries will reach 1,100 TWh in 2020, corresponding to an average increase of around 5 per cent per year over the 1993–2020 period (Figure 2). Approximately 1/3 of this demand come from Indonesia (400 TWh), and 1/5 from Thailand. By sectors, the major user will be the residential/commercial sector (52%), while the industrial sector constitutes almost 48 per cent and the transport sector almost negligible.

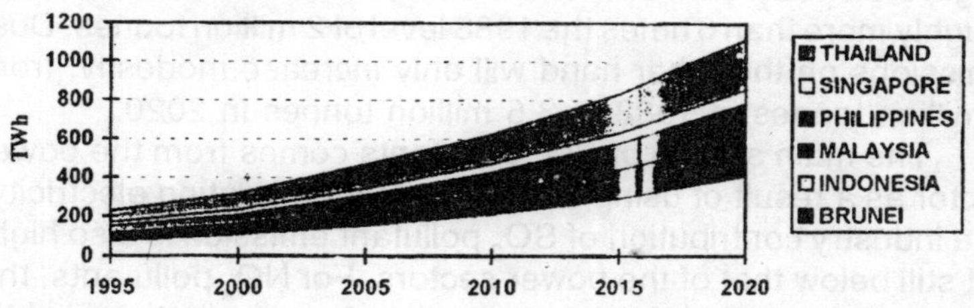

Fig. 2 : **Future Electricity Demand in the ASEAN Region**

2) Implication on Indigenous Supply

The indigenous primary energy production of the region will in total amount to only 360 Mtoe by 2020. Compared to the total demand of 600 Mtoe, this means that the region will rely heavily on energy imports, especially oil. This is based on the findings that oil will still constitute nearly 50 per cent of the total primary energy demand of the ,region. Since Indonesia, the biggest producer in the region, will become an oil importing country during the next decade, oil discoveries should be intensified further in the region so as to anticipate the inadequate oil supply in the torture.

The region's indigenous coal production in 2020 will also not be enough to supply the regions demand of 220 Mtoe (almost 450 million tonnes) in the Business as Usual scenario, thus, contributing to the region's status as a net energy importer.

Natural gas, on the other hand, is produced twice as much as the region's demand in 2020. Therefore, if the region does not impose stringent regulation of pollutant emission in the future there will still be enough natural gas for exports outside the region, Natural gas development and utilisation should be more encouraged, if the region tends to go for the strict emission policy while keeping its position as the largest LNG exporter in the world.

3. Implication on the Environment

The excessive use of oil and coal in the region in the future will increase enormously the emission of air pollutants such as SO_2, NO_x, dust and CO_2. The total SO_2 emission will be around 7 million tonnes in 2020, almost 4 times the 1988 level in 2020 (Figure 3). NO_x emission will reach almost 13 million tonnes, roughly more than 6 times the 1988 level of 2 million tonnes. Dust emissions on the other hand will only increase modestly, from 2 million tonnes in 1988 to 3.5 million tonnes in 2020.

The main source of SO_2 pollutants comes from the power sector as a result of using coal as fuel for generating electricity. The industry contribution of SO_2 pollutant emission is also high, but still below that of the power sectors. For NO_x pollutants, the main source would be the transport sector contributing slightly more than 40% of the region's total NO_x emission for 2020. Dust pollutants come mainly from biomass use especially in the household sector.

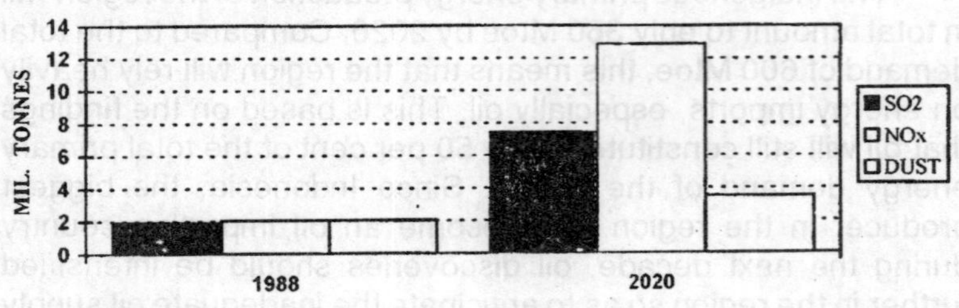

Fig. 3 : **Emission of SO_2, NO_x and Dust Pollutants in the ASEAN Region**

The total CO_2 emission in the region increases from almost 270 million tonnes in 1988 to 1,700 million in 2020, corresponding to an average increase of around 6 per cent per year (Fig. 4). This high growth of CO_2 emission is due to the rapid increase of coal in the region, especially for power generation.

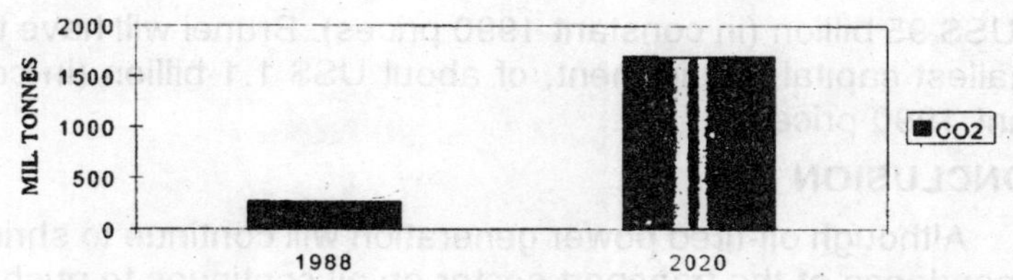

Fig. 4 : **CO_2 Pollutant Emission in the ASEAN Region**

4) Implication on Investment Requirement

The increasing demand of the ASEAN region will impose development of the energy resources in the member countries. This in turn will have impacts on the capital required to proceed with the development.

As mentioned previously, the largest share of the region's primary energy demand comes from the power sector as a result of the rapid increase of electricity demand. Consequently, increment of capacities in both power generation and transmission and distribution of electricity requires vast amount of capital in a short period of time.

The total annual investment requirement of the power sector will be around US$ 400 billion (in constant 1990 prices) over the 1988–2020 period (Figure 5). The bulk of this total annual investment cost is related to the construction of power plants to meet the additional capacity's requirement. The remainder is that of the transmission and distribution system expansion.

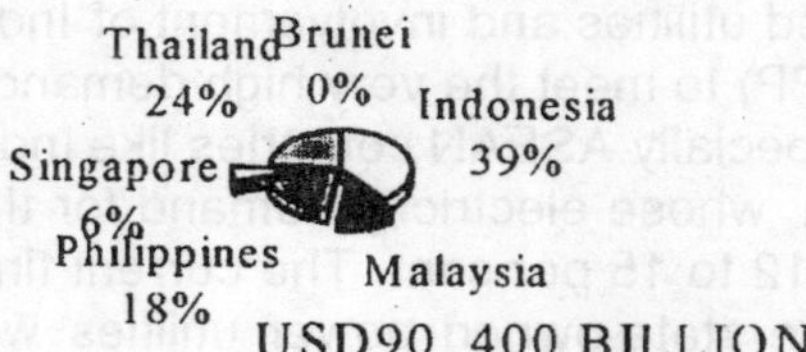

Fig. 5 : **Total Power Sector Investment Requirement over the 1988–2020 Period**

By country, Indonesia's total investment requirement will be the highest, amounting to around US$ 157 billion (in constant 1990 prices). Thailand will constitute the second largest, amounting to US$ 95 billion (in constant 1990 prices). Brunei will have the smallest capital requirement, of about US$ 1.1 billion (in constant 1990 prices).

CONCLUSION

Although oil-fired power generation will continue to shrink, dependence of the transport sector on oil continues to push oil demand every year. The increase in import in Asian Countries on oil and the declining oil reserves have prompted the countries to restructure their energy policies. Energy security and diversification are key strategies pursued by oil exporting countries as measures to preserve longer their oil reserves and to maintain export capabilities. Internally, substitutes such as natural gas, coal and biomass are taking up higher share in the energy mix.

Energy efficiency is also another strategy pursued by a number of Asian countries to reduce their energy demand. In general, especially in developing countries, such programmes are mainly in the form of increasing public awareness. With no strict regulations and by-laws on energy efficiency and savings, the efforts are voluntary in nature and little substantial savings of energy is achieved at the national level. Japan, is one country in Asia that has taken energy efficiency very seriously. Its success, however, is explained by the country being a fully energy importing country.

The private sector has been given more roles to meet the increasing energy demand in most Asian countries. Restructuring and deregulation policies have resulted in the privatisation of state-owned utilities and involvement of Independent Power Producers (IPP) to meet the very high demand of certain Asian countries, especially ASEAN countries like Indonesia, Thailand and Malaysia, whose electricity demand for the last few years ranged from 12 to 15 per cent. The current financial crisis that are hitting the state-owned power utilities would open wider doors for the IPPs to meet the electricity demand each year.

Global concern on environmental aspects of energy generation and utilisation have also created new waves of environmental awareness in Asia. Indonesia, for instance, has imposed strict environmental rules on the use of high-sulphur coal, which therefore restricts use of its own coal production. With coal targeted by many countries to reduce high dependence on oil more clean coal plants will be observed in the near future in Asia, to control the SO_x and NO_x emissions.

REFERENCES

AEEMTRC, ASEAN Energy Review, 1997 Edition
AEEMTRC ASEAN 2020 Programme: Country Reports, 1996
AEEMTRC, Energy in ASEAN—Country Profiles, 1996
IEA, Energy Statistics and Balances of Non-OECD Countries, 1993–1994.
IEA, World Energy Outlook, 1996 Edition.

12

The Newly Palestine Experience

A. MALEK AL JABER
Palestine

INTRODUCTION

This paper examines the potential for a successful Demand Side Management in the newly freed Palestine. The Palestinian electric system suffered a great deal during the Israeli occupation. Palestinians were not allowed to build power plants (some few exceptions) and forced to purchase power from the Israeli Electric Corporation (IEC) at high prices. From the survey that was done by BEC there are 13 per cent of West Bank (WB) population living in 65 communities 223,928 inhabitants are supplied through local diesel generators, still receiving insufficient, inadequate and partial electricity services. In addition to the above these projects also suffered from 20 per cent power losses due to inductive loads connected to the electric supply like motors and fluorescent lamps without capacitors. This creates the low power factor which increases the losses. For that the Demand Side Management is essential for the Palestinian community for many special reasons such as high prices of power and restrictions from IEC on the total consumption.

Palestinian Energy and Environment Research Center (PEC), is working on a comprehensive DSM program in Palestine. A pilot study is under implementation in the Taibeh village. The successful results will be implemented in the rest of Palestine. In this paper, we will explain the procedures and results from the pilot study that was done in Taibeh.

THE OBJECTIVE OF DSM PROGRAM

The Taibeh village was chosen because of the size and the special condition that this village enjoyed. Below are the list of the objectives of this program:

1. The main objective of this project is to improve electricity services and the economical viability of Taibeh electrical

project.

2. Assist Taibeh Cooperation Association to improve and develop their energy project.
3. The joint short term objective of this Cooperative project is to elaborate and test technical guidelines and financial procedures for the dissemination of energy efficient domestic equipment among the inhabitants of Taibeh for their own interest and for the interest of their Cooperative association, who is owner and manager of the electric generation and distribution system.
4. The joint medium term objective is to provide a set of practical references and a base of experience for the joint preparation of a larger program within the province of Jenin for all villages managing their local electricity system.

THE PARTNERS OF THIS PROJECT

There are four partners working on this project, two French Institutes, the Palestinian Energy and Environment Research Center and Taibeh Cooperation Association. The role of each can be summarized as follow.

A) ADEME and EDF are providing financial support in the form of:
   i) Measuring equipment
   ii) An initial stock of 100 efficient fluorescent lamps (high quality types having passed tests in France at an official lab LCIE).
   iii) Expertise on methodological and technical aspects.
   iv) Travels (up to 5) for the selected expert.
   v) An initial stock of energy fridges, expected to be of European Class B for up to 125,000 F.
   vi) A training period for PEC designated staff on the subjects of public lighting, efficient domestic equipment, and CFC recovery from old fridges (Travel + 10 days per diem).

B) The Taibeh Cooperative and PEC will implement the following.
   i) The Taibeh Cooperative Association will circulate to the subscribers information and advises on electricity management and on the energy consumption of domestic appliances.

ii) Taibeh Cooperative Association will facilitate the purchase of efficient equipment by allowing the recovery of terms through adding them to the bills on behalf of the distributors.

iii) Taibeh Cooperative Association will be responsible to collect the credits offered by ADEME and EDF for efficient equipment (lamps and A-B-C classes, fridges...) through the electricity bills.

iv) Taibeh Association must use the recoveries of these credits to improve and develop Taibeh electrical project.

v) PEC and Taibeh Cooperative will arrange for the systematic recuperation of the old fridges (after their substitution by efficient fridges), in order to hand them over to a specialised workshop for the safe recovery of the CFC and for eventual recovery of materials.

vi) PEC will support the village with technical assistance.

vii) PEC staff will collect information, data and conducting field survey to prepare check list of all customers in the village and their equipment consumption.

TAIBEH VILLAGE

Taibeh Village is located on a hill near the green line, 15 km. west of Jenin city. The population of the village is around 1200 residents, living in 238 houses.

The village receives electricity continuously for 24 hours a day by their own Diesel Engines Power Station. This station has three diesel generators.

The electrical project is managed by the Village Electricity Cooperative. The Cooperative consists of 5 persons from the village residents and chaired by an experienced mechanical engineer. Election of the Cooperative is done every year.

The project is feeding 238 residential customers and 6 industrial. The industrial shops are: 4 carpenters shops, 1 blacksmith and 1 aluminium workshop.

The charge rate is 0.5 NIS/KWH (US$ 0.15/KWH), the average monthly consumption of each customer is 150 KWH/month.

The peak load in the village reaches 220 KW between 1700 and 2100 hours in winter.

Due to the strong expansion of the housing sector in the village, there will be an excess of load on the lines and the diesel generators, especially during the peak periods which arise during night hours as mentioned above. The drop of the voltage during the evening period results in the systematic installation of incandescent lamps in each main room, as the fluorescent tubes cannot be started and deteriorate quickly at low voltage (180V), thus accelerating the drop.

It will be vital to manage the electrical consumption of the customers, as most of the customers are residential and as lighting and refrigerators are the main electrical loads at house, the demand side management in the village will be important to keep the sustained electrical service to the customers.

ENERGY SURVEY

The first step that the Palestinian Energy and Environment Research Center and the Cooperative Association of Taibeh is to prepare a check list of all customers in the village and their total consumption per year. A recorded equipment "Ampermetrique" was installed on the major electrical equipment such as refrigerators and washing machines for at least seven days. These devices recorded the total consumption in KWH, Demand KW, Power factor and many other quantities. The information was downloaded by a computer software and the curve of the major equipment consumption was constructed.

The consumers were divided into different groups:

- Customers with less than 110 KWh per month
- Customers with total consumption of 180 KWh per month (five customers).
- Customers with total consumption of 250 KWh per month.

REFRIGERATORS

The main energy consumption in the village were refrigerators. The evaluation of the existing refrigerators was based on the following factors:

1. The brand name.
2. The capacity in liters.
3. The date of purchase.
4. The interior temperature.
5. The total amperes at the start.
6. Total consumption per month.

The EMU MEMO 10 device was used to record the above points.

THE NEW REFRIGERATORS

In Palestine there are no laws and regulation yet for the energy efficient equipment. Wholesale shops are not required to provide information about the total electricity consumption of the refrigerator. We have selected nine different brands available on the market to measure the consumption. Also, the EDF and ADEME, our partners in this joint project, have provided us with a European classification of the refrigerators that are available in Europe. The main points that were considered in our selection of the new refrigerators are:

1. The brand name
2. The country of origin
3. The shape (one door or two doors)
4. The capacity
5. The purchase price

LAMPS

High efficiency lamps are another major electrical consumption in the study that we have done in Taibeh. The compact Fluorescent lamps at capacity of 18 Watt/unit were chosen to replace the incandescent existing lamps.

RESIDENTIAL ELECTRICAL EQUIPMENT

From the survey we conducted in Taibeh village, we found that the main electrical equipment consumption in rural houses are refrigerators and lamps. And most of electricity consumption is from old fridges in the village. Figure 1 shows statistical distribution of monthly consumption in Taibeh village and Figure 2 shows the refrigerators/freezer load per day.

Residential fridges consumed about 70 per cent of total residential electricity use. Using the high energy efficiency of new refrigerators will dramatically reduce electricity used. That is what exactly happened in USA, 62 per cent reduction in refrigerators electrical consumption due using new high efficient refrigerators as shown in Figure 3 and Figure 4.

BENEFITS OF USING HIGH EFFICIENCY APPLIANCES

Using High Efficiency Equipment will provide a wide range of benefits such as.

1. Consumers save money on net, since the reduction in energy consumption and operating cost exceed the in-

crease initial cost, often by a wide margin.

2. The reduction in energy consumption yields environmental benefits in the form of reduced emissions of a variety of pollutants and Greenhouse gases.
3. The benefits to many electric utilities come in the form of reduced need for investment in expensive new plants, transmission lines, and distribution equipment.

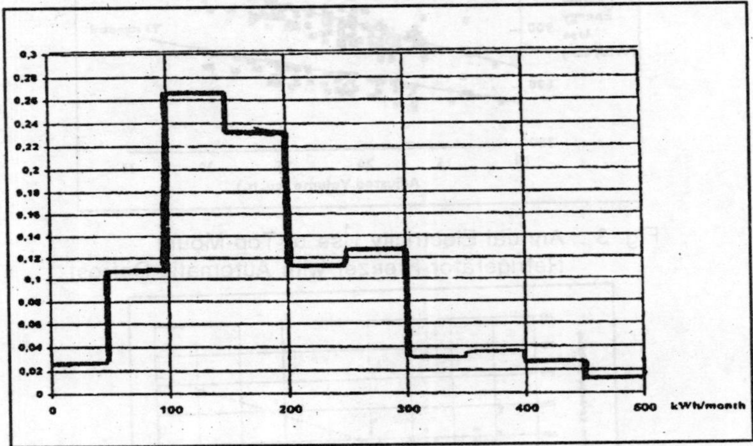

Fig. 1 : Statistical Distribution of Monthly Consumption

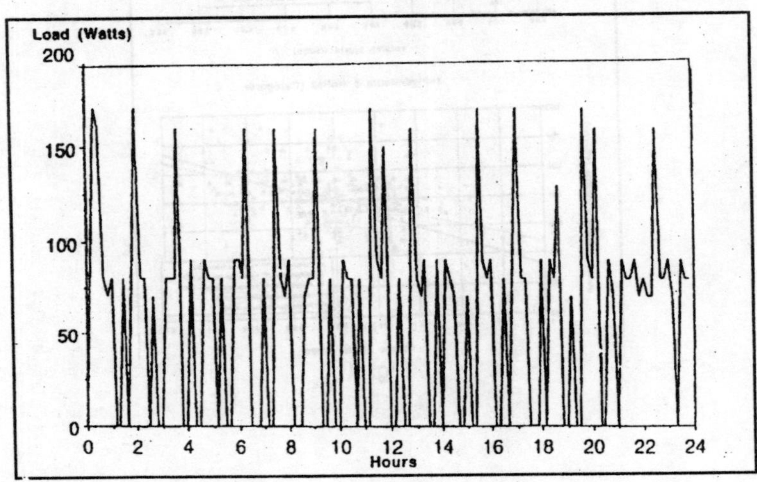

Fig. 2 : Refrigerators/Freezer

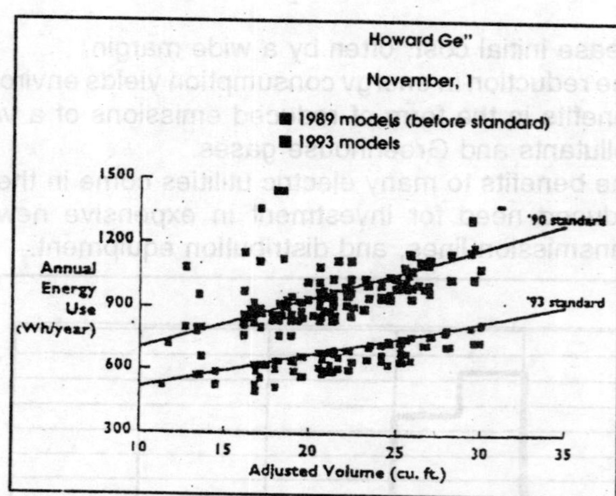

Fig. 3 : Annual Electricity Use of Top-Mount Refrigerator-Freezer with Automatic Defrost

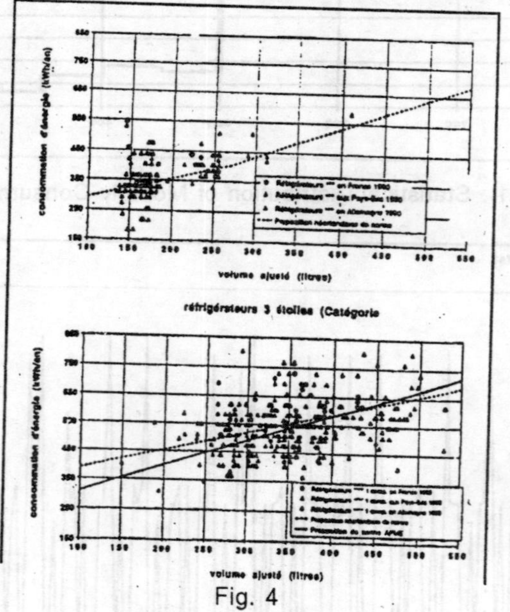

Fig. 4

PART — 4

ENVIRONMENTALLY SECURE ENERGY OPTIONS AND TECHNOLOGIES: RENEWABLE, EFFICIENCY AND COST EFFECTS

13

Sustainable Energy as an Instrument for Development

U. FARINELLI
Italy

INTRODUCTION

The importance of energy for development is certainly not new, but it may have failed to be recognised explicitly in the past. Moreover, the availability of energy influences development not only in a quantitative, but also in a qualitative way. In other words, energy means not only economic growth, it also means social development, improvement in the quality of life, effects on the environment. Such connections have been recently identified[1] and studied[2], and include, among others, the effect of energy on demography and population growth; on gender roles and the promotion of women; on undernutrition and food; on education; on poverty and on distribution of wealth; on job creation and employment; on investment and on foreign exchange; on urbanisation; on water availability; on land degradation, deforestation and desertification; on national security and peace. Each of these connections would deserve a deep analysis. For instance, the availability of energy for water sanitation or to reliably operated refrigerators where vaccines are conserved has a great impact on health. Feeling rural people (generally, mostly women and children) from the daily routine of collecting firewood or carrying drinking water over long distances makes them available for more productive work or for education—both a source of future income, but also a way of reducing demographic growth.

Conversely, the type of energy cycle which is used is not indifferent to the type of economic, social and environmental

1. United Nations Development Programme (UNDP), "Energy as an Instrument of Socio-Economic Development" (1995)
2. United Nations Development Programme (UNDP), "Energy after Rio", (1987).

effects it will produce. It is not sufficient to have "enough" energy, it should also be the right type of energy in that particular context. For instance, alleviating the pressure on utilisation of firewood by introducing more modern forms of energy in the rural environment conserves forests, prevents desertification and land degradation, improves the air quality inside the dwellings (smoke from cooking stoves is a major health hazard especially for women and children).

The importance of energy for development has been fully appreciated for a long time. Without the availability of sufficient energy, the jump in agricultural productivity brought about by the Green Revolution would have been unthinkable, and very little industrial activity could take place.

Consequently, an important part (between 25 per cent and 30 per cent) of all the money devoted to aid to development programmes, both as grants and as loans, by bilateral multilateral or international institutions as well as non-governmental organisation (NGO) go to energy-related projects. It has been mentioned, for instance, that in 1990 for Sub-Saharan African countries[3] 87 per cent of World Bank loans were for the energy sector, with plans of 75 per cent for the power sector alone in the following years; in the same period, the African Development Bank devoted 92 per cent of its loans to the energy sector; UNDP funding also targeted energy, which was its largest single category in 1990.

However, what has been missing until some time ago has been the link between energy, development and sustainability. Most of the energy projects that have been supported have been of the traditional type: first of all, they were supply oriented (i.e. directed to providing energy rather than to using it effectively); second, they were mostly directed to large-scale plants based on traditional technologies (fossil fuels or large hydro-electric plants); finally, environmental assessments were carried out for most projects, at least in the recent times, but a general evaluation and prioritisation of projects in terms of sustainable development has generally been missing. It has been calculated[4] that between 1980 and 1991 only about 2.5 per cent of official

3. O. R. Davidson, "Energy Issues in Sub-Saharan Africa" in Ann. Rev. Energy Environ. 1992, 17:359–403.
4. K. Kozloff and O. Shobowale, "Rethinking Development Assistance for Renewable Electricity," World Resources Institute, Washington D.C., 1994.

development assistance for energy from all donors has gone for renewable energy development.

The new possibilities to obtain energy services which are now available must be fully exploited and introduced in aid-to-development programmes not only because they are more benign to the environment and to global climate, but also because they are best suited to produce development.

New fresh thinking is needed, and developing countries must lead the way: if nothing else, because they are less conditioned by past history, investments, infrastructures, and of course they know better their own situation. It is important that developing countries acquire an independent capability of technology assessment; this process of "capacity building" can be helped by industrialised countries, but the centres of competence thus formed should not be donors-driven.

Sustainable development is based on three components: economic, environmental and social. The process is Western countries has concentrated on one of the components—environment and more specifically global climate, so that sustainability is sometime reduced to the problem of greenhouse gas emissions. On the other hand, in developing countries economic growth is understandably seen as the most important, if not the only, component. It is necessary that all three components are taken up in a consistent, balanced way.

Energy had not been singled out as a separate subject in the discussions of the United Nations Conference on Environment and Development (UNCED) in Rio de Janeiro in 1992, nor was reflected as such in the ensuing Agenda 21, where references to energy were made in several different chapters (most notably in the one on atmospheric pollution). The central position of energy for sustainability was however recognised by the United Nations General Assembly Special Session (UNGASS)[5] devoted to sustainable development in June 1997, which also decided that the 2001 session of the Commission on Sustainable Development will be devoted to energy problems.

This does not mean that energy policies have not been affected by considerations of sustainable development. Great attention has been devoted to energy cycles by the Inter-

5. UNGASS—United Nations General Assembly Special Session—"A programme for the further implementation of Agenda 21" (1997).

governmental Panel on Climate Change (IPCC); the application of the Climate Change Convention, such as deriving from the protocol agreed upon in Kyoto in December, 1997, will greatly influence the energy policies of the signatories. Although developing countries (DC) have not entered any definite engagement for the containment of GHG emissions, their attention to this problem is prominent.

CHANGES IN BOUNDARY CONDITIONS

From 1980 to 1995, energy consumption at the world level[6] has increased from 7021 Mtoe to 8852 Mtoe (+26%). However, this growth is compounded very unevenly: an increase of 18.3 per cent in OECD countries, a decrease of 21.7 per cent in the countries of central and Eastern Europe and of the former Soviet Union (due to the serious economic crisis), an increase 108 per cent in non-OECD Asia (including Middle East countries) and increases of 43 per cent and 61 per cent in Latin America and Africa respectively. It is perhaps even more instructive to look at the per capita energy consumption: this has actually remained constant in Africa at a level of 0.50 toe/person year (against a world average of 1.56) and even decreased in sub-Saharan Africa (from 0.465 to 0.456) due to the massive population growth, while in China energy consumption per capita has grown from 0.54 to 0.83 and in India from 0.21 to 0.33. Differences in rates of development are thus reflected in per capita energy availability.

Energy intensity, defined as the ratio between (commercial) energy consumption and the gross national product in a given country, is a measure of the efficiency of energy utilisation; however it can be taken as a meaningful indicator for the comparison of two countries only when are at similar stages of development. The reason is that in the first phases of development and during the industrialisation process in particular, energy intensity tends to be quite higher than in fully industrialised countries, since production is concentrated on infrastructures and on instrumental goods, which have a high energy content, and in providing durable goods for the first time to the population.

6. Data in this section are taken from: European Commission, Directorate General for Energy DG-XVII, "Energy in Europe, 1997—Annual Energy Review", Special Issue September, 1997.

At the world level, the product has grown more rapidly than energy consumption: correspondingly, energy intensity has decreased from 539 toe/MECU (1985) in 1980 to 477 in 1995 (-11.5%). However, data are very different from one region to another. The energy intensity in the same has decreased by 17.1 per cent from 291 to 241 in the European Union, from 550 to 446 in North America (–18.9%), but it has increased elsewhere, from 1923 to 2349 in the former Soviet Union, where the breakdown of the economy has been greater than that of energy consumption (and where the extremely high values point to a chronic inefficiency); in developing countries, it has dropped from 1469 to 1066 in Asia, but it has increased from 770 to 997 in Africa, and from 463 to 511 in Latin America.

Apart from the evolution of the energy production and consumption patterns, a great shift has occurred in the instruments of energy policies. The main shift is from instruments of the command-and-control, compulsory planning type of market mechanisms.

All over the world, the move toward increased reliance on market mechanisms has progressed during the last years. In the field of energy, it was not only in the so-called Centrally Planned Economies that the governments played a direct and prevailing role. Energy was seen as a strategic sector, crucial to international competitiveness, economic and social development and national security, in which the most important decisions were to be taken by the government. State utilities for electricity, gas and sometimes also coal and oil products were instruments for implementing energy policies in many Western countries, such as the United Kingdom, France and Italy.

This concept was gradually abandoned in many countries, as it was recognised that market mechanisms could perform the same tasks with better efficiency. The introduction of elements of competition even in those sectors which were previously considered as natural monopolies, such as electricity and gas, is required by the directives of the European Union, and is being gradually implemented in the EU countries. Prices of oil products have been deregulated in most industrialised countries. Exploration and exploitation of oil and gas deposits are open to competition. The process of liberalisation in this field has made great progress also in the economies in transition and even in

China.

However, the progress is slower in many of the developing countries. Reservations about foreign interference through investments; preoccupations about the fragility of their economic systems; reluctance to abandon instruments of control; lack of local capital to start energy enterprises have all contributed to delay the transition to free market mechanisms in most developing countries.

Financing institutions consider that the development of sustainable energy would greatly benefit from market and competition. For instance, the World Bank states that "one of the most powerful ways to improve energy supply is to ensure that the energy market is determined by consumers' choices... that means both that the price of energy should reflect its cost and that regulation of energy industries should encourage competition and choice."[7]

The European Union also encourages the introduction of market elements in all fields in its programme of aid to the ACP countries, both in terms of support to structural adjustment (including structural reforms and sectoral policies), and in taking into account in its priorities in allocation of funds the "performance" of the country, including the creation of an environment more favourable to the success of the project"[8].

However, free market is not everything. Effective as market forces are in optimizing the allocation of resources for short-and medium-term objectives, the market is known to be short-sighted, not to respond spontaneously to long-term signals. As the World Bank puts it, "liberalising energy markets, however important may not be the complete answer... private companies have shown little interest in extending electricity supplies to rural areas. {industrial and urban customers are more lucrative}".

Government may provide corrections to the myopia of the market, by introducing long-term signals, correlated with sustainability, that adjust the perspectives of market forces, and allow them to

7. World Bank—"Rural Energy and Development: Improving Energy Supplies for Two Billion People" (1996).
8. European Commission, 1996, "Green Paper on Relations Between the European Union and the ACP Countries on the Eve of the 21st Century—Challenges and Options for a New Partnership," Chapter VI, Brussels, European Commission (also on internet at: http://www.oneworld.org/euforic/greenpap/chap6.htm)

be used in the best way to obtain strategic objectives as well as tactical ones. Accounting for externalities is one of these signals.

The diffusion of sustainable energy forms meets with many barriers, which in a certain sense can be referred to an insufficient or improper functioning of the market. The application of these principles meets a number of barriers, most of which have been identified since quite a long time. One of the most important barrier is the subsidy given by many governments to traditional forms of energy. Once recognised the importance of energy for development, many governments subsidise electricity or various fuels, so that their price to the final consumer is lower than the cost of production and delivery. In many developing countries, energy prices and tariffs are much lower than in industrialised countries, although the cost of producing and delivering energy is by no means lower. This has the double effect of discouraging energy conservation, by making interventions to increase efficiency artificially more expensive than the energy which is saved, and of creating a barrier to the introduction of new forms of energy, renewables in particular, which are not equally subsidized. Moreover, it has been observed that this kind of incentives, originally meant to alleviate poverty, actually favour the richer layers of the population, which can afford consuming much more energy in any case.

Another barrier—which to some extent is common to industrialised countries—is the dispersion of interventions in energy efficiency and distributed energy generation, in particular by renewable sources. The financing of very large scale supply-side projects (a gigawatt-size hydroelectric or coal-powered or nuclear plant) can find capital for investment at much lower interest and longer return times than the corresponding hundred thousand small projects of micro-hydro or wind installation or efficiency improvement. Banks and financing agencies are generally not equipped to deal with a myriad of micro-projects, so that aggregation of demand is a necessity if a "level playing field" is to be established for the different kinds of interventions.

Many more barriers have been identified. Norms and regulations are often not adapted to innovations in the energy field. There is a lack of information to the public of the possibilities offered by alternative energy sources and more efficient devices. Many solutions are not available because equipment is not being distributed by retailers due to the small size of the perceived market, to lack of knowledge or to inertia (for example, compact fluorescent lights).

Another barrier is the lack of skills, not only for installation and maintenance, but also for marketing. Well known are many cases of projects financed by donor agencies which were technically and economically sound, but which were unsuccessful or even counterproductive because they were soon abandoned for lack of appropriate maintenance, unavailability of spare parts, or even ignorance about operational procedures. Even when a project is successful, there is often no capability of replicating and diffusing it.

Finally, as in most industrialised countries, externalities are not reflected in the price of energy. Although there is no clear consensus on the way in which this "internalisation" should occur, and that it would not be fair to expect developing countries to precede advanced countries on this route, it is clear that a gradual introduction of externalities would greatly improve the prospects of increased energy efficiency and of renewable energy sources.

THE CHOICE OF THE RIGHT TECHNOLOGIES FOR DEVELOPING COUNTRIES

Some criticisms have been expressed by developing countries on projects sponsored by donor agencies in the energy field which, they thought, were intended mostly for testing new technologies that had found no application in the countries of origin and which, in many cases, proved to be unreliable or unsuitable or uneconomic. Although very few technologies are independent of context, and most require at least some adaptation to the conditions of the country or the site of application, the criticism does have some element of substance.

Much discussion has gone in the past into the choice of "appropriate technologies" for developing countries; originally, the concept of "appropriate" applied only to very simple technologies developed locally. This concept has been refuted successively for a number of reasons.

Although some energy technologies can be based on simple, locally developed know-how, in most cases they require some advanced and sophisticated technologies that are not universally available. Even mature technologies, such as large-scale hydroelectricity, require a number of accessory know ledge, ranging from organisational methods for large works to maintenance of construction equipment, from training of personnel to quality control and commissioning procedures. Simple, traditional techniques can be vastly improved by the grafting of advanced technology: an electronic control of temperature and humidity can prevent spoiling a crop in a solar or biomass fueled drier; modern computer-aided design methods have been successfully applied to improving the performance of traditional cooking fire-places. Gas turbines cannot be designed or manufactured everywhere; but many developing countries have shown to be able to maintain jet or turbo-propelled planes, and can do the same for power plants employing aero-derivative gas turbines.

The choice of a technology is more successful when it is made in a broader context, taking into due account the general energy policy of the country; the priorities and modalities of economic development; the social and cultural characteristics of the country or region for which it is meant; the geographic and climatic conditions of the site. These factors are now being included in the selection criteria of many donor or lending agencies, including UNDP, the World Bank and the European Commission.

Most energy technologies have been developed and are available only in industrialised countries, and respond to the requirements and specifications of these countries. A process of adaptation of the technology to the particular needs and conditions of the receiving country is necessary, if not sufficient. This process of adaptation should be conducted in large part by the receiving country itself. In general, it is highly desirable that the transfer of a certain technology in a given country corresponds to a well identified need: in other words, that it derives from a "market pull" rather than from a "technology push". Developing countries should strive to set up a capability of comprehensive technology assessment.

A negative aspect in some past experiences of energy technology has been the limitation of the transfer to some intermediate phase of the process, generally manufacturing: this has little scope if it is not followed by system design, system assembly, operation, maintenance, servicing, infrastructure arrangements; and also marketing and financing are key factors for the application and diffusion of a technology. It has generally proved to be more effective to start at the downstream, application end of the technology and proceed upstream rather than vice-versa.

Giving proper attention to locally developed traditional technology has often paid back. These technologies have often paid back. These technologies have often passed through a long process of adaptation to local conditions, and they may respond to specific needs that it is not always easy to identify. They are often ineffective and uncompetitive: but sometime they have shown to have the potential of great improvements. "Blending" new advanced technologies into traditional systems may provide solutions which, while respecting local conditions and adapting to actual needs, are effective, economical and diffusible. Many examples are available, ranging from large earth dams to woodstoves. A further penetration of new technologies may involve the use of advanced materials in traditional ways of producing or using energy. Genetically engineered bacteria can be introduced in a traditional biochemical process (such as the production of alcohol or biogas by fermentation). Electronic control can be inserted to improve performance and facilitate operation and maintenance of a traditional process. Such a component can be seen as a "black box": it is not necessary for the operator to be familiar with its internal mechanism, provided its functions are well understood.

A concept which has been introduced concerning the choice of technology for developing countries is the so-called "leapfrogging", i.e. the direct adoption of an advanced technology without going through the intermediate steps, and even of technologies which have not been generally adopted in industrialised countries. The subject is delicate and has roused some controversy. It is quite obvious that it would not make any sense for a country aiming at producing computers to start building the electronic tube computers of the 1950's or the transistors or the integrated circuits of the following decades become adopting the microchip. On the other hand, looking for very advanced technology to solve a problem is sometime an excuse for not being able to fix a much simpler solution well at hand.

Leapfrogging makes sense when the developing country is not committed to particular solutions by its previous history, and can therefore make a choice less conditioned than an industrialised country. The most common case is the absence of infrastructures. As the prime choice in a sparsely-inhabited country that would set up a telephone system from zero would be to use cellular phones and radio links rather than setting up a wired network, so in the absence of an electricity grid, local generation may well be the most appropriate and economical answer for electrification. A Survey conducted some time ago in Central Africa on the emerging technology perceived as the most useful for the region, saw at the first place the development of Zeppelins for transportation of goods: significantly, this is the transport technology that requires the minimum of infrastructure.

Another element which can possibly favour leapfrogging is the absence of previous investments, in terms of both money and cumulated experience. Certain technologies are used today in industrialised countries with preference to potentially more effective ones because of the great amount of money that has gone into building plants and equipment in former years and of the cumulated know-how. When starting from zero, it may be convenient to look for alternative technologies, more effective and perhaps less capital-intensive, which have become available in the meantime, and which have not been adopted on a large scale because of the reasons just mentioned.

Finally, a possibility of utilising relatively new technology comes from the fact that the market for many goods is expanding rapidly in developing countries, while it is stationary or shrinking in industrialised countries. One case-show example is steel making; essentially no new steel making plant has been commissioned in industrialised countries in the last decade, as steel consumption is slowly decreasing, while several new plants are set up every year in developing countries. New steel-making technologies have been developed up to the pilot scale which are less energy consuming and require less investment, but never applied on full scale yet; this may be on occasion for developing countries (and is actually being considered by China).

14

The World Solar Programme 1996–2005 for Energy Security and Environment

B. BERKOVSKI
France

INTRODUCTION

1) Energy for Sustainable Rural Development

The population of this already crowded planet is growing by almost 250,000 individuals per day—or in one year by ninety million, a figure which approaches the combined population of France and the United Kingdom (11,41). Within the next 35 years, the global population will increase by about half (see Tables 1 and 2). South Asia's population will grow by two-thirds, and that of Sub-Saharan Africa will more than double. Most of this increase occurs in the rural and remote areas of developing countries (see Figure 1). According to Mr. Federico Mayor, Director-General of UNESCO, "The situation is critical, but in no way hopeless. On the contrary, given the will and the resources at both the national and international levels, runaway population growth will be checked and the number of humans on the planet can be stabilised before catastrophe strikes" (69).

Who will provide the incrementally new population with energy? How will these relatively poor masses pay even for the minimum energy services they require? What will be done to relieve the inevitable stress on the environment? These are very difficult questions to answer. What is clear is that population growth represents a colossal challenge for the world energy sector. The latest statistics tell us that average annual world energy consumption per capita in developing countries is of the order of 0.71 tons of oil equivalent (toe). Therefore, a yearly increase of almost 65 million toe will be required to meet the needs of the growing population in the rural and remote areas of developing countries, deprived of a well-developed energy infrastructure. Additional US$ 8 billion will be needed every year to satisfy the elementary energy needs, if conventional energy supply schemes are to be used in the years to come (6–10, 73, 74).

The governments of developing countries put great emphasis on the improvement of living conditions in rural areas while respecting cultural traditions. A better basic energy supply, the creation of small-scale industry to recruit labour and to hinder migration to urban centres are the major concerns. Rural electrification projects have been carried out in many countries in the last 20 years. The experience of these projects has been summarised by the World Bank and USAID as follows:

- The high and increasing cost of investment in the power sector has contributed to be debt burden of developing countries.
- Consumers in rural areas often have very low consumption rates (less than 1kWh/day), and losses due to transmission and theft can be high. This can mean that the supply of electricity is not economical since operating costs are not covered by revenue and that capital repayments have to be subsidised with the result that maintenance may suffer.
- There is often very little growth in production levels following electrification in rural areas.

The above results indicate that, internationally, rural electrification is unlikely to proceed at the same rate as it has in the past and that, where it does take place, there will have to be a clear financial justification for it (34, 79).

It is therefore vital that virtually free primary energy resources be found and exploited. The most promising of these is solar energy. The sun continually emits radiation into space. The Earth intersects approximately 1.7×10^{14} kW, an amount which is 10,000 times bigger than current total world energy consumption. The Earth is actually a huge solar energy collector. The term "solar energy", as used here, refers to all forms of renewable energy, including, but nor limited to, biomass, geo-thermal, hydro, ocean, solar electricity and wind. Solar energy supply is an alternative option energy supply for various energy services, especially in rural areas. Thus in the case of low energy demand in rural areas, solar systems are cost-competitive with traditional solutions; however, they have the advantage of avoiding the environmental problems concerned with fossil energy supply (43).

If, at this point, humankind has not succeeded in harnessing all of these vast reserves of solar energy, it is due chiefly to the fact that this primary energy resource is diffuse rather than concentrated. A surface area of one square meter in the Sahel theoretically receives about 1.0 kW of solar energy; in reality, taking into account atmospheric conditions and the spectral distribution of solar energy radiation, the energy available for use is only a few hundred watts per square meter during daylight hours. This has had an adverse effect on any serious world-wide efforts to develop solar energy technology, despite the fact that research, development and exploitation of renewable energy resources are no more costly than they are for coal, oil and, especially, nuclear energy.

2) The Environmental Imperative

Over the past few decades, scientists and environmentalists have improved their understanding of the impact of conventional energy systems and resources on the environment (42, 67, 68). Formerly, it was thought that the negative effects of the misuse of energy resources and technologies would be felt locally by distinct populations and that these effects would be manageable. In recent years, it has become clear that environmental pollution caused by releases into the atmosphere of radioactive dust and/or greenhouse gases, has no regard for frontiers. Ozone depletion and atmospheric warming have begun to undermine our confidence in our ability to control and remedy their effects. The global impact of traditional energy systems on the environment is fast becoming one of the world's most pressing problems.

The United Nations Conference on Environment and Development (UNICED)—known also as the Earth Summit—which was held in Rio de Janeiro in 1992, was called to focus on the most fundamental and important challenges to the preservation of the natural human environment. At the Earth Summit, it was considered essential to develop environmentally sound technologies and industrial practices.

In Chapter 9 of Agenda 21 of UNCED, it was agreed that "Energy is essential to economic and social development and improved quality of life. Much of the world's energy, however, is currently produced and consumed in ways that could not be sustained, if technology were to remain constant and if overall quantities were to increase substantially. The need to control atmospheric emissions of greenhouse and other gases and substances will increasingly need to be based on efficiency in energy production, transmission, distribution and consumption, and on growing reliance on environmentally sound energy systems, particularly new and renewable sources of energy" (16, 17).

At the global level, climate change is the overriding concern. Of all the environmental challenges facing us today, global warming is the most persuasive. The Inter-governmental Panel on Climate Change (IPCC) reached consensus conclusions in June 1990 that increased atmospheric concentrations of greenhouse gases may result in an additional warming of the Earth's surface. In order to stabilise atmospheric concentrations at present-day levels, immediate reduction of global man-made emissions of the long-life greenhouse gases carbon dioxide (CO_2) nitrous oxide (N_2O) and the halo carbons—mostly chlorofouorocarbons (CFC_s)—by more than 60 per cent, and methane emissions by 15 to 20 per cent, is necessary (75).

Table 3 illustrates the amount of carbon dioxide emissions in different parts of the world between 1990 and 1995. Over the five-year period, global CO_2 emissions from fossil combustion rose 3%. Only in the Commonwealth of Independent States and the countries of Central and Eastern Europe was there a significant decline in CO_2 emissions, mainly due to the unstable economic situation during this period. For the CIS Republics of the former Soviet Union and the countries of Central and Eastern Europe, CO_2 emissions in 1995 were respectively less than 70 per cent and 75 per cent of the 1990 level (71). The total CO_2 emissions from OECD member countries rose by more than 4 per cent. In North America (Canada and the USA), CO_2 emissions rose nearly 6 per cent. OECD member countries in the Asia/Pacific region (Japan, Australia and New Zealand) experienced an increase of 11 per cent in total, with Japan's emissions increasing by 12 per cent and Australia's by 8 per cent over the entire period. It is evident from the statistics that, in regions where developing countries are dominate, there was a significant increase of CO_2 emissions during the five-year period.

3) The Advantages of Renewable Energy Technologies

Most renewable energy sources—wind, solar radiation, ocean waves, etc.,—are free and constitute an autonomous national resource which would be a vital factor in reducing imported energy costs for many countries. Moreover, much of the technology used to exploit these resources does not require expensive high-tech installations or highly skilled experts to operate them and is, thus, particularly suitable for use by developing countries. On the other hand, these technologies are just as attractive to even the most developed countries: as light-industry with a low pollution rate, they adapt better to the surrounding environment and cause minimal disfigurement of the countryside. The possibility of creating small-scale, decentralised energy installations, which are maintained and run by local people, constitute one of the most valuable by-products of this type of technology for both rich and poor countries.

This decentralisation of the energy supply will also contribute to the growth of small and medium-sized industries and businesses in rural areas, whose populations are often dependent upon subsistence-level income from the land. If properly managed, such small-scale production could form the basis of income for rural populations.

Since the 1970s, the world has experienced extreme instability in the cost of conventional fuel, resulting in disruption to the balance of power between local, national and regional groups dependent upon the import of energy from abroad. The various taxes, quotas and subsidies adopted at the national level to manage this problem have often aggravated an already difficult situation. Not only could the exploitation of renewable energy resources and technologies alleviate much of the friction in the trade relations between some developed nations by solving problems of subsidies and how they are used, but it could, at the same time, place some of the struggling developing nations in a stronger trading position vis-ã-vis their powerful neighbours. The growth of energy crops—biomass for energy production—might also hold considerable promise for developed countries, in particular, in solving the difficult problems created by overproduction of food and the subsequent set-aside policies.

There are several key issues which make renewable energy a major component in international policy making. Indeed, the use of renewable energy: i) introduces environmentally friendly technologies, producing a minimum of waste products; ii) revitalises rural communities by creating local industries and businesses; iii) improves the quality of life for inhabitants living in rural areas; iv) reduces mass migration of population from rural to urban areas; v) reduces a nation's bill for imported primary energy resources, vi) might contribute to solutions of problems in the agriculture sector of some developed countries, and vii) contributes to the reduction of unemployment.

The world-wide market for renewable energy technologies is estimated to have a direct turnover of almost US$ 40 billion each year and is an emerging one. The photovoltaic market, for example, has grown steadily over the last 15 years and will reach annual shipments of almost 70 MW in 1996 (33, 70). A further increase in power production through PV power generation can be achieved by improvement of the performance of PV cells and modules and also by developing mass production techniques. Table 4 shows realistic targets for short and medium term performance improvements. Tables 5 and 6 provide basic information on cost of grid-connected PV power generation and also goals for module and system costs.

More than 1,300 MW of new wind energy capability was installed in 1995, representing a 35 per cent jump in world capacity in comparison to 1994. World-wide installed wind power capacity surged to more than 5,000 MW during the first quarter of 1996. This strong growth in international wind energy markets is expected to continue, according to official projections from the American Wind Energy Association (AWEA), which referred to wind power as "the world's fastest growing electric power technology". According to projections, the total installed wind power capacity will surpass 18,500 MW by the year 2005, representing a market of more than US$ 18 billion. However, there exists an imbalance: while many markets flourished in 1995, some slowe considerably, particularly in the US, Germany and India accounted for almost two-thirds of all new installations last year whereas, in the US, new wind capacity only accounted for 41 MW. AWEA's projections indicated that US wind capacity additions will only slowly until the year 2000. (64).

The total world-wide investment in new hydroelectric capacity is projected at US$ 235 billion. Strong growth is expected in China and Latin America, both of which have large, untapped hydroelectric resources. China is expected to increase its hydroelectric capacity by 74 GW, or 32 per cent of total new capacity. In Latin America, hydroelectric plants generated nearly 75 per cent of all power in 1992 and, with a planned capacity increase of 58 GW between 1995 and 2010, it will continue to be the most important power source in Latin America (64, 65).

4) Constraints of Solar Energy Market Penetration

The energy production levels necessary for a quantitatively significant contribution to the world energy supply are not likely to be constrained by resource availability. A number of other practical considerations, however, do limit the renewable resources that can be used, including the following, (8, 18):

- Lack of objective information and training are important factors hindering further expansion.
- The present supply position of fossil fuels and the level of prices make it difficult for renewable sources of energy to compete in the market where a well-established energy infrastructure exists.
- The failure to internalise external costs in energy prices leads to decisive disadvantages for renewable sources of energy in the market.
- Biomass must be produced sustainably, with none harvested from virgin forests. The biomass supply should come from plantations established on degraded lands or, in industrialised countries, from excess agricultural lands.
- The use of wind equipment will be substantially constrained in some regions by land-use restrictions, particularly where population densities are high. Noise is perceived as an important sitting issue and wind turbines can sometimes effect local TV reception.
- The cost of photovoltaic power is too high for grid-connected applications, and cost reductions of about a factor of five are needed for applications where photovoltaic power is attractive. There are many interesting ideas for cost reductions in the future. Given that cost reductions will take place, the potential is extremely large. The obvious tendency of the average cost decrease for PV roof systems is shown in Figure 2 using a standard case of 3KWp solar roof system. This reduction in PV costs contrasts with a

non-decrease of oil prices shown in Table 7.

- The amount of wind, solar-thermal and photovoltaic power that can be economically integrated into electric generating systems is very sensitive to patterns of electricity demand, as well as to weather conditions.
- There is concern that fish may be injured by hydro schemes. Some types of turbine can oxygenate the river waters. Operation of hydro-electric installations may result in the collection and removal of large amounts of water-borne debris that may also flood large areas. Owing to these constraints, it is assumed that only a fraction of potential sites will be exploited, with most growth occurring in developing countries. World-wide, only one-fourth of the technical potential, as estimated by the World Energy Council, may be exploited by the year 2050.
- Relatively high investment costs are caused mainly by low capacity and energy densities and the common fluctuations in the availability of renewable energies over time, which result in a comparatively high need for space and material and the necessity of establishing parallel systems or storage facilities.

UNESCO'S PAST ACTIVITIES

UNESCO has a remarkably long and distinguished record of involvement in promoting solar energy based research and development. Going back to 1951, UNESCO launched one of the first international research projects on solar energy as an integral component of its Arid Zone Programme. Several pilot projects were implemented under the auspices of this programme, one of which was a regional center for renewable energies in Bamako, Mali. Various other programmes were also implemented including considerable renewable energy training, education and information dissemination activities. These activities included the organisation of numerous seminars and conferences and the subsequent publication of a number of journals for the widespread dissemination of information on renewables.

The first international meeting on solar energy in the post World War II era was the UNESCO–Indian Government sponsored Symposium on Solar Energy and Wind Power held in New Delhi in 1954. The proceedings, which were published in 1956, stressed social and political implications as well as the technical possibilities for the use of solar energy (45, 49).

Following the implementation of the Arid Zone Programme, UNESCO continued to provide substantial technical assistance to UNESCO Member states for the solution of energy problems mainly through its system of Regional Science Cooperation Offices. As noted earlier, UNESCO also continued in publishing and providing solar energy related documentation on a world-wide basis paralleled by its financing of various scientific events of an energy related nature.

Interest and activity in renewable energies were advanced in the 1970's when UNESCO instituted a new stage in its program. During this period close cooperation was established between UNESCO and the International Solar Energy Society. At the height of the energy crisis, in July 1973 UNESCO organized and hosted at UNESCO headquarters in Paris a major International Congress entitled "The Sun in the Service of Mankind". This International Congress was held in cooperation with the International Solar Energy Society (ISES), the Coopération Méditerranéenne pour lénergie solaire and the Association francaise pour l'étude et le développement des applications de l'energie solaire. The Congress drew 1,000 participants from more than sixty countries and attracted world-wide attention to the role of and application of solar energy. Concurrent with the Congress, a group of leading experts met to consider UNESCO's role in the future development of solar energy (31). Among the possibilities envisaged was the establishment of an international solar energy commission. This commission's objective would be the promotion of development as well as education and training in renewable energy.

The demand for pollution-free energy resources, and the rising cost of fossil fuels, took solar energy out of the exclusive realm of scientific speculation and brought it squarely into the market place. After the International Congress, UNESCO activities in the field of solar energy were further developed. Within the framework of the promotion of engineering sciences, the UNESCO Medium-Term Plan for 1973–1978 stressed the importance of developing research and international cooperation in solar energy utilization (37, 38). Among major UNESCO actions in this field were the following:

— "Symposium on Solar Energy", Geneva, 1976 (in cooperation with WMO and ISES) (63);

— Solar Energy Training Courses for Research Personnel from Developing Countries;
— "European Solar Energy Working Group" meeting, Genoa, 1976;
— Publication of a book for the broad educated public devoted to solar energy utilisation.

The Medium-Term Plan for 1977–1982 again stressed the significance of research and international cooperation in renewable energy and outlined the main lines of UNESCO activities (50–54). As a result of fruitful cooperation with governmental and non-governmental organisations such as IAEA, IIASA, the World Energy Conference, the International Center for Heat and Mass Transfer, and the International Center for Mechanical Sciences, several successful seminars and conferences were held over this period in the field of fundamental scientific and technical energy problems with special emphasis given to the prospects of developing new and renewable sources of energy (NRSE).

Priority was given to the effective dissemination of information in the area of new and renewable energy resources; several reports were prepared and widely disseminated on these subjects. Among the activities carried out by UNESCO in NRSE in 1977–1982, special mention should be made of the following:

— preparation of the second "International Forum on Fundamental World Energy Problems" (1979) aimed to contribute to a better understanding of the perspective of two major long-term energy options: solar and thermo-nuclear;
— assistance to the "International Solar Energy Congress", organised by ISES in New Delhi in 1978;
— organisation of the "Asian Solar Energy Workshop" immediately after the ISES Congress;
— organisation of the "European Solar Energy Task Force" meeting (in collaboration with the Bulgarian Academy of Sciences) in Varna in 1978;
— organisation of and financial support to many Post-graduate courses and workshops on solar energy in all regions of the world;
— support of the development of regional and sub-regional cooperation:

Establishment of the regional association for solar energy

in Asia and Solar Energy Society in Africa; an European solar energy network; an international liaison group on MHD; the "International Co-operative Program for Research and Training in Energy Planning and Management";

— during 1980, a major contribution was made to the preparatory work for the "United Nations Conference on NRSE" in the form of participation in technical panels and a survey of facilities and needs in the field of education and training related to NRSE; the Organisation took part in drawing up the Programme of Action adopted in 1981 by the UN Conference on URSE which was held in Nairobi, Kenya, and, after the Conference, in implementing the Programme.

Among the most significant activities carried out in preparation for the Nairobi Conference was the "International Workshop on Non-Technical Obstacles to the Use of NRSE in Developing Countries" organised by UNESCO in May 1981 in Bellagio, Italy in cooperation with other organisations concerned. (32) Twenty-five international specialists met under the aegis of UNESCO to test their views on non-technical aspects of new energy development and to identify a number of areas for priority actions as follow-up to the UN Conference on NRSE. In recognising that the solution of technical problems alone will not be sufficient to forward the development of NRSE, the United Nations opened the floor to discussion of such issues as environmental impact, education and training, information, social and cultural conditions, institutional structures, among others.

The group concluded that education and training in NRSE were required over a broad range, covering not only schools and universities, but also training of policy makers, planners, managers and extension workers. NRSE training was identified as necessary as well for the general public in promoting new energy use, as required for the improvement of information facilities and services, as required to place emphasis on the generation of reliable and relevant information and its dissemination to those who can act upon it and as required for change in the energy mix in all societies and cultures. It is unfortunate that the introduction of NRSE relative to this latter point was not well understood.

The group also concluded that the institutional framework for planning, policy-making and coordination should envisage a national focal point for development of NRSE with respect to ensuring that adequate resources are devoted to such basic functions as resource assessment, research and development, planning and promotion, investment promotion and monitoring evaluation. The group closed with the consensus that dealing effectively with this wide range of problems required concerted efforts across the full range of inter-governmental, governmental and non-governmental organisations working in related fields.

Following the United Nations' Conference on New and Renewable Sources of energy (NRSE), held in Nairobi in 1981, and the "Nairobi Plan of Action", in which Member States urged the United Nations agencies to place special emphasis on the development of alternative sources of energy, many organisations in the UN-institutional family developed their activities in NRSE (55–58). UNESCO's renewable energies programme was essentially the responsibility of the Division of Technological Research and Higher Education (now known as Engineering and Technology Division) which undertook major training and information activities in this field (20–22). However, renewable energy applications were components of programs of several other divisions within the Science Sector including multi-disciplinary programmes such as Man and Biosphere (MAB) or with subject-specific divisions such as the programmes of the Division of Hydrology or the Division of Earth Sciences (54–58, 61, 72).

3) HIGH-LEVEL EXPERT MEETING "THE SUN IN THE SERVICE OF MANKIND"

The High-Level Expert Meeting (the summit of world expertise in the field of renewable energy) was successfully convened at UNESCO Headquarters on 5–9 July 1993—exactly twenty years after the World Congress, "The Sun in the Service of Mankind", also held at UNESCO in 1973. Some 346 top experts from 53 countries prepared 66 in-depth critical written assessments of solar energy and related fields including "Solar Energy and Health", "Solar Energy, A Strategy in Support of Environment and Development" and "Financing Solar Energy Development". These reports were debated in 35 specialiSed round tables. At its seven plenary sessions, the meeting discussed and approved: "Conclusions and Recommendations of the High-level Expert Meeting for the World Solar Summit" and "Resolution Addressed to the Founding Sponsors of the World Solar Summit Process" together with the series of specific recommendations adopted at the Round Tables (7).

UNESCO, the European Commission, the International Energy Agency, EUROSOLAR, the International Solar Energy Society and the French Agency for Environment and Energy Management were the founding members and joint sponsors of the Meeting. In addition to these organisations, some thirty cooperating agencies and institutions, including the United Nations Environment Programme, the United States Department of Energy (DOE), the E7 Network of Expertise for the Global Environment, the Friedrich-Ebert Foundation (Germany), the World Health Organization and the United Nations University were involved in this High-Level Expert Meeting of the WSSP.

This meeting was supported by numerous leading international figures. An *ad hoc* International Executive Committee meeting, comprising representatives of the founding institutions, coordinated preparatory activities for this successful first phase. The UNESCO Engineering and Technology Division of the Science Sector served as the Secretariat for the meeting.

The WSSP and the World Solar Programme 1996–2005 (WSP) must be viewed as a cooperative process involving all interested groups and organisations in the field of renewable energies, whose objective is to promote wide-scale utilisation of renewable energies by constant encouragement, support, guidance and pressure on relevant authorities to keep the goal of sustainable development before the entire world. Involved are the creation of infrastructures, the selection of projects, the market penetration of suitable technologies and, above all, an unfaltering conviction by all to achieve success (44).

Consequently, from the very outset, due consideration was given to creating a true partnership of *support groups* to be organized at national, regional, and international levels, to maintain the momentum created during Phase I of the WSSP.

It was envisaged that these support groups could play a major role in advancing the WSSP, for example, through regional mechanisms such as the Commission of the European Union and the Organisation of African Unity, and at the national level through existing specialised institutions devoted to promoting renewable energies such as ADEME in France, the New and Renewable Energy Authority in Egypt, ENEA (Enter per le Nuove Tecnologie, l'Energia e l'Ambiente) in Italy and DOE in the USA.

The creation of such specialised institution in countries where none existed was also encouraged.

During the preparation of the above meeting, a world-wide network of eminent experts was created (***World Solar Summit Process Network***). These experts represented practically all existing national, regional and international authorities in the field. The network included existing groups such as the International Solar Energy Society, the renewable energy experts of the World Energy Council, the European Solar Council, the New Sunshine Project and also those which were in the process of being created, e.g. the World Renewable Energy Network, the Russian Solar Energy Society, the International Programme of Solar Sports, etc.

The High-Level Expert Meeting also considered several worthwhile, large scale projects which would encourage greater use of renewable energies in modern society to promote social welfare and stability, as well as peace-keeping efforts. Such projects would encourage the developed nations to take the lead in design, manufacture and commercialisation of solar energy technologies and devices. They would aim to help the developing nations in particular to overcome disadvantages which they suffer in the fields of energy and environmental protection. The following six (6) strategic projects of universal value were identified:

- **Solar Energy for Rural Development**
- **Public Information and Education**
- **Solar Energy for the Development of Africa**
- **Solar Energy for Peace**
- **A World Solar Found**
- **An International Solar Convention**

THE WORLD SOLAR SUMMIT PROCESS

Following the conclusions and recommendations of the High-level Expert Meeting "The Sun in the Service of Mankind", UNESCO launched in 1993 an initiative that was referred to as the "World Solar Summit Process" (WSSP). Several important programmes should be mentioned as antecedents to, starting points for, and foundations for the WSSP. These include:

- The UNESCO Arid Zone Programme, 1952–1969.
- Activities of the International Solar Energy Society
- An international congress, The Sun in the Service of

Mankind, 1973.

- Japanese Government programs: The Sunshine Project, The Grand Solar Challenge and The New Sunshine Project.
- Programmes of the European Commission: Thermie, Joule, etc.
- Global Solar Pumping Project; UNDP/World Bank, 1978–1990.
- Large-scale US Government and contractor programmes: photovoltaic, conservation and wind projects.
- The Nairobi Programme of Action, 1981
- UNESCO/ISEEK Energy Database
- Large-scale geothermal projects in Iceland, Italy and New Zealand
- The Wiley-UNESCO Energy Engineering Learning Package

The WSSP initiative taken by UNESCO in broad-based partnership with the UN Secretariat, UN specialised agencies and governmental and non-governmental organisations involved, was viewed as a communication, awareness building and coordination process which aimed to (10):

- enhance understanding of the role that renewable sources of energy could play in the preservation of the environment, in the provision of energy services—particularly in ecologically fragile areas—and in contributing towards a solution to unemployment.
- favour access to, transfer and sharing of knowledge on renewable energies by establishing a global information network system using state-of-the-art communications technology
- promote and harmonise cooperation in education, training and research, as well as in the transfer of research disclosures to industry at the regional, inter-regional and international levels.
- demonstrate how wide use of renewable energy is a cost-effective and rapid way for many developing countries to reduce energy costs, save foreign exchange and stretch the energy supply base without heavy investment
- urge non-governmental organisations to enter into partnership with, and make their knowledge and experience available to, global and regional inter-governmental bodies, as well as to establish innovative programmes for the promo-

tion of the use of renewable energies
- reinforce local industrial capacities
- identify and define selected strategic projects of universal value and high-priority national projects for inclusion in the World Solar Programme 1996–2005 (WSP)—a major developmental initiative, which will trigger a wider use of renewable energy sources and create open competitive and sustainable markets for renewable energy technologies, equipment and goods
- reinforce the involvement of the international community and, in particular, that of the multilateral and bilateral sponsors, as well as the national commitment towards large-scale use of renewable energy
- promote small-scale financing and delivery mechanisms
- launch the World Solar Programme 1996–2005 during the World Solar Summit.

The WSSP was conducted through a series of regional meetings, both at ministerial and expert levels, that reviewed the situation in the respective regions, identified priorities and received the draft documents prepared for the World Solar Summit. In addition, participating countries were invited to submit national renewable energy projects that they considered of high priority and to adopt a declaration reflecting the political and policy views of the region in respect of these energies. It should be emphasised that the preparation, holding of these meeting and the follow-up to them contributed substantially to a greater awareness and better understanding of the potential of renewable energies on the part of both decision makers and the general public.

In the autumn of 1994, the Executive Board of UNESCO approved the creation of a World Solar Commission that, among other functions, would act as the high-level body to oversee and guide the World Solar Summit Process.

The United Nations Committee on New and Renewable Sources of Energy and Energy for Development, which reports to the UN Economic and Social Council, was informed at its first and second sessions (New York, 1994 and 1996) on the launching and implementation of the World Summit Process. The WSSP general chart is presented in Table 8.

Following is a summary description of the major WSSP meetings and their results(78):

1) Regional-Level Expert Consultations

A series of regional consultations were organised within the WSSP in preparation for the World Solar Summit. These events included:

- Presentation of highest priority National Renewable Energy Projects
- Identification of activities of global importance and proposed regional priorities to be included in the WSP
- A specialized session on existing national industrial renewable energy programmes and projects and also on ways and means to promote renewable energy industry.
- A specialised session on financing of renewable energy projects in the concerned region
- Discussion of the drafts of major WSSP documents
- A constitutive meeting of the Regional Solar Council or a meeting of the Board of Governors of the Regional Solar Council already established.

A calendar of 17 WSSP events for the preparation of the **World Solar Summit** can be found in reference (78).

2) Coordination and Cooperation Among Participating Organizations.

Partnership arrangements have been made by WSSP International Organising Committee with the UN specialised agencies and programmes, the European Commission, the World Bank, the European Bank for Reconstruction and Development, the International Energy Agency of the OECD, the E7 Network of Expertise for Global Environment, the Latin American Energy Organisation, the US Department of Energy and the Russian Ministry of Fuel and Energy.

i) The UN System and the World Bank Group

Within the United Nations System, several specialized agencies and programmes are involved in various aspects of energy development. Within the Secretariat of the UN itself, the Department for Development Support and Management Services executes energy projects, and the UN regional commission carry out certain activities in this field. As to the multilateral development banks, only the World Bank and the Asian Development Bank have been active in looking at opportunities for both agency efficiency and the increased use of renewable energy sources.

During the period covered by this Report, several international organisations and programmes of the United Nations System adopted programmes geared towards facilitating *inter alia* the enhanced utilisation of renewable energies in the context of ensuring sustainable development. Examples of the these programs can be summarised as follows:

a) **The Global Environment Facility (GEF),** considered as the interim financial mechanism of the UN Framework Convention on Climate Change and other International Instruments is implemented by the United Nations Development Programme (UNDP), the United Nations Environmental Programme and the World Bank. GEF finances projects and programmes of global environmental value. These projects and programmes are designed, *inter alia*, to promote the development of renewable energies by removing barriers and reducing costs.

b) **The United Nations Development Programme (UNDP)** is the main source of founding for energy activities within the UN System, with part of the funds being channeled through other UN agencies and programmes such as FAO, UNIDO, UNESCO and the World Bank. In addition to its participation in GEF, as indicated above, UNDP launched in 1996 its Initiative for Sustainable Energy (UNISE), designed (a) to harness opportunities in the areas of more efficient energy use and increased use of renewable sources of energy, and (b) to build upon UNDP's existing energy activities to help move the world toward a more sustainable energy strategy by helping the formulation and implementation of national energy programmes.

c) **The United Nations Environmental Programme (UNEP),** which is not an implementing agency but has a coordinating and catalytic mandate, has developed a new energy programme whose main objectives are:
   - ***to establish*** UNEP as a key source of information on reducing the environmental impacts of energy utilisation
   - ***to encourage*** environmental sustainability in governmental and private sector decisions on new energy policies and programmes

- ***to encourage*** the increased use of energy-efficient technologies and solar energy technologies in developing countries
- ***to identify*** barriers against implementation of environmentally sound energy options and to encourage their removal
- ***to assist*** with the preparation of projects for submission for GEF financing, aimed at mitigating or eliminating adverse impacts of climate change

d) **The World Meteorological Organization (WMO)** and its members, represented primarily by national meteorological and hydrological services, are engaged in a variety of activities, including special programmes which have direct or indirect relevance to the development of new and renewable sources of energy. The most important of these fall under the World Climate Programme, which aims to assist countries in applying climate information and knowledge for economic and social benefit, for the achievement of sustainable development and the implementation of Agenda 21 and associated instruments including The Climate Agenda—an integrating framework for international climate-related studies. Through internationally coordinated projects, it promotes the improvement of understanding of climate processes; the monitoring of climate variations and changes; exchange of climate applications and impact assessment methodologies; and the development of options for strategies to respond to climate variability and change. The World Climate Programme, as a whole, provides an authoritative international scientific voice on climate and climate change.

WMO activities related to energy, during the decade 1996–2005, are expected to include (a) the further evaluation of weather and climate implications in energy matters; (b) the assessment of the effects of climate variability and change on the energy sector, and (c) facilitating the practical applications of meteorological and hydrological information and related methodologies in various areas of energy conservation, production and distribution. This will mainly be implemented within the framework of the Climate Information and Prediction Services (CLIPS) Project.

e) **The United Nations Industrial Development Organisation (UNIDO)** has a number of energy activities which address both the supply side, through the provision of energy for industry, and the demand side, by improving industrial energy use in developing countries. A major element of these activities has been a promotional programme for the development and diffusion of new energy technology. This programme was developed from a recommendation by the UNIDO General Conference in 1984 which adopted a resolution calling on UNIDO to promote cooperation between institutions engaged in research and development for new and renewable sources of energy. As a result of this resolution, the UNIDO Consultative Group on Solar Energy Research and Application (COSERA) was established. This Consultative Group was seen as a mechanism for enlarging international cooperation with a view to enhancing the effectiveness of solar energy research in developing countries and the commercialisation of such research for industrial scale production. An important outcome of UNIDO's work on the promotion of solar energy technology has been the establishment in 1994 of the International Center for Application of Solar Energy (CASE) in Perth, Australia, which operates under the auspices of UNIDO and is supported by the Governments of Australia and of the Western Australia region.

f) **The International Atomic Energy Agency (IAEA)** has a number of activities supporting the world's search for sustainable development. It has taken the lead among UN organisations in initiating a number of projects designed to foster a better understanding of comparative advances of the various forms of power production. One particularly relevant activity concerns a comparative health and environment impact assessment, it has produced relevant guidelines and manuals. A report on greenhouse gas emission for various energy production and utilisation chains has been prepared by the IAEA. Under its Technical Cooperation Programme the International Energy Agency is supporting projects using nuclear technologies for the exploitation of geothermal energy in Latin America. The exploitation of geothermal sources requires a precise knowledge of the location and quantity of underground hot water, and radioactive isotope techniques offer economic and precise methods for determining the most favourable drilling locations.

g) **The Food and Agriculture Organization of the United Nations (FAO)** has a number of energy activities aimed at assisting developing countries to meet their energy requirements in agriculture, forestry and fisheries, as a means of achieving sustainable rural development. FAO member countries have, during the last years, requested assistance in implementing the recommendations of Agenda 21 calling for an energy transition to enhance rural and agricultural productivity. An integrated approach has been followed for the assessment, planning and implementation of energy and sustainable rural development, one example being the activities of the Latin American and Caribbean Working Group on Rural Energisation for Sustainable Development. Another example is FAO's Regional Wood-Energy Development Programme for Asia. It should be underlined that FAO considers that the role of solar and other renewable energies is vital in the issue of food security, as energy is required at every stage of the food chain.

h) **The World Health Organisation (WHO),** while recognising that population growth has been characterised by migration from rural to urban areas, underlines that rural inhabitants will nevertheless constitute 50 per cent of the global population by the end of WHO's Ninth General Programme of Work (2001). Aspects of particular importance in rural environmental health include integrated water resource management, environmentally sound community development and application of new. appropriate technology—notably renewable energy technology—to solving environmental problems. While recognising that health depends to a large extent on the environment, WHO emphasises that it also depends on the availability of energy and that effective health care in the home, at the health center and in the hospital also requires a dependable supply of clean and affordable energy. Some of the most important health development work done by WHO is in remote areas of developing countries where electric power is erratic or non-existent and fossil fuels are not available. At present, and as a result of investigations conducted in the 1980s, there are more than 5000 solar refrigeration systems in use for storing vaccines in WHO's immunisation programmes worldwide.

i) **The World Bank** has vitalised and intensified in recent years its activities in the field of renewable energies. These efforts have been assisted by the availability of some GEF grants and the launching in March 1995 of the World Bank Solar Initiative, which is designed to work with member countries of the World Bank and energy industry, research and non-governmental organisation communities to hasten the commercialisation of renewable energy technologies, and to expand their applications significantly in developing countries. The World Bank Solar Initiative is closely connected with on-going efforts of the International Finance Corporation (IFC) to create a private "Renewable Energy Efficiency Fund", the on-going work of the World Bank's Asia Alternative Energy Unit, and the technical assistance effort of the World Bank/UNDP/bilateral Energy Sector Management Assistance Programme (ESMAP). Figure 3 shows that most of the GEF funds were so far allocated to bio-diversity and climate change projects, its contribution to renewable energy projects was rather symbolic.

In June/July 1995, the Director-General of UNESCO, at the invitation of the WSSP/IOC, wrote letters to the UN Secretary-General and to the Heads of other concerned specialised organisations of the UN system, as well as to the World Bank and the International Monetary Fund, informing them of the state of implementation of the WSSP, inviting them to participate in the preparation and convening of the World Solar Summit, and requesting them to designate a staff member as the focal point responsible *inter alia* for participating, on their behalf, in an *ad hoc* Inter-agency Solar Task Force. A meeting was convened at UNESCO Headquarters from 29 to 30 January 1996 in order to prepare a UN system-wide input to the WSP to the World Solar Summit. The following UN Specialised Agencies and Programmes participated:

- International Atomic Energy Agency
- United Nations Development Programme

- United Nations Environment Programme
- United Nations Industrial Development Organisation
- World Health Organisation
- World Meteorological Organisation

The participants agreed on the following:

- Draft Leading Documents for the World Solar Summit would be sent for scrutiny and comments to the focal points in UN specialised agencies and programmes by the WSSP/IOC Secretariat. Their comments were due by 15 March 1996. The endorsement by the UN specialised agencies and programmes of the draft WSS Leading Documents was expected before the Mediterranean Solar Summit, (20–24 May 1996).
- UN specialised agencies and programmes were urged to send representatives to the WSS preparatory events for the ensure proper reflection of their interests and proposals in the draft WSP as well as contribute to the preparation and convening of the WSS, according to their possibilities
- A global information referral system on renewable energy was considered as a joint global project. It would include data on sustainable development, education, health, industrial development, meteorology, social aspects, etc.
- The WSP could not be considered successful unless it contained a strong education and training component. The Strategic Global Solar Education and Training project is seen as a common endeavour and required a contribution from all UN institutions concerned.

UNESCO has also submitted to the Secretary-General's Special Initiative on Africa a proposal of high priority for inclusion of a project entitled "Solar Highway for Sustainable Development of Africa". The Governments of Mali, Senegal, South Africa, United Republic of Tanzania and Zimbabwe, as well as relevant authorities of the Organisation of African Unity, gave their support to the project. A concrete strategy of action, together with a well-identified coordination mechanism and resource requirements, were included in UNESCO's project submission.

Special mention was made of the letter addressed by the

UN Secretary-General to UNESCO's Director-General, dated 7 September 1995, welcoming UNESCO's initiative and its desire to develop and implement the World Solar Summit Process as a joint, system-wide effort. The representative of UNESCO at the July 1995 meeting of the ACC Inter-agency Committee on Sustainable Development presented a progress report on the implementation in the WSSP and on the above-mentioned initiative of UNESCO's Director-General to achieve wide inter-agency cooperation in this matter.

ii) The European Commission

External relations are becoming an increasingly important part of the European Union's brief. The Union is operating a number of large-scale cooperation and assistance programmes in Central and Eastern Europe, Africa, the Mediterranean Basin, Latin America and Asia (44). The Union's SYNERGY programme, covering energy cooperation with developing countries, has as one of its priorities the promotion of renewables in certain regions. For example, the programme has provided support to the renewable energy center in Elblang, Poland, which aimed to promote renewable energy sources in the entire Baltic region.

The European Commission thought its JOULE-THERMIE programme, has actively participated in the World Solar Summit Process from its very beginning in 1993. It supported the Harare Declaration on Solar Energy and Sustainable Development and launching of a World Solar Programme 1996–2005 for the period 1996–2005 (23, 24).

As a consequence of the increasing political priorities for establishing a framework for improved cooperation between the European Union (EU) and the non-EU Mediterranean countries, the contributions that renewables could make to the energy needs of this region are being examined. Generally, renewable sources are plentiful and the potential, in particular for solar and wind energy, is currently not exploited to its maximum. Within the framework of continuing Euro-Mediterranean energy cooperation, appropriate cooperation and assistance instruments, such as the MEDA programme and the Euro-Mediterranean Forum, will address the scope for further cooperation on renewables in the Mediterranean region. Therefore, a clearly defined and ambitious renewable energy strategy will be indispensable and will allow EU industry to compete successfully in the global market place.

Given the important role renewable energy sources can play in the external policies of the European Union, the Commission will, on the basis of the debate on a Green Paper, further reflect on ways in which the community can ensure that the potential for developing renewable energy in developing countries is sufficiently well emphasised within the existing and future EU cooperation instruments.

iii) Other Major WSSP Partners

The following organisations have largely contributed to the success of the World Solar Summit Process by organising their own major events:

- Ministry of Energy and Infrastructure, Israel
- Institute of Silicon Technology, Pakistan
- Ministry of Education/Ministry of Energy, Telecommunications and Posts, Malaysia
- Ministry of Transport and Energy, Zimbabwe
- Sultan Qaboos University, Sultanate of Oman
- Compañia de Fuerza y Luz, Costa Rica
- Ministry for the Environment, Malta
- Ministry of Science and Technology Policy, Russian Federation
- Osaka University/Akita Prefectual Government, Japan

The US Department of Energy (DOE) participated as a "cooperating agency" in the World Solar Summit Process. It has monitored developments of the Process and offered opinions on proposed activities for the WSP. DOE has argued for improving coordination among existing renewable energy programs and activities world-wide, thereby contributing to the Harare Declaration on Solar Energy and Sustainable Development which embodies this concept. DOE expressed the expectation that the WSP could help focus world attention on the potential for renewable energy, so that it is in the best interest of DOE to continue its interaction with the WSP. Moreover, DOE cooperation included the secondment of a consultant who served from 1993 to 1997 to oversee and assist developments in the World Solar Summit Process and to play an integral part in the planning and operation of the regional meetings of the WSSP and the World Solar Summit held in Harare, Zimbabwe. Other cooperative measures by DOE included an exchange of information between representatives of the US renewable energy industry and government and those representatives of various international renewable energy programme associated with the WSSP.

3) International Organizing Committee of the World Solar Summit Process

The World Solar Summit Process (WSSP) was guided and managed by an International Organising Committee (WSSP/IOC) composed of one or two representatives of all major WSSP partners. Those persons nominated by the members of the World Solar Commission to follow the work of the Commission on a day-to-day basis have jointed the WSSP/IOC in 1995.

The WSSP/IOC was a major instrument in the preparation and convening of national and regional preparatory events for the World Solar Summit. In the course of several meetings, it discussed *inter alia* the following subjects:

- high-priority national and strategic projects of the WSP
- the preparation and convening of the World Solar Summit
- national and regional meetings
- leading documents
- fund-raising campaign

The first WSSP/IOC preparatory meeting held in Harare on 26–27 February 1996 gave H.E. Mr. Robert G. Mugabe, President of the Republic of Zimbabwe, Chairman of the World Solar Commission, the opportunity to provide direct guidance to the WSSP/IOC. Three additional WSPP/IOC preparatory meetings were successfully held in 1996 (78).

An inter-ministerial task force called the Zimbabwe Organizing Committee (ZOC) supervised the preparation of the World Solar Summit in Harare. Immediately before the Summit the Government declared it as a state occasion. Thus, the Cabinet Committee on State Occasions, chaired by the Vice-President of the Republic of Zimbabwe, was given complete responsibility for handling the final stages of WSS preparation and convening.

THE WORLD SOLAR COMMISSION

During the first stages of the World Solar Summit Process, it became apparent that a high-level body to oversee and guide it was required. At its 145th session, in November 1994, the Executive Board of UNESCO approved the establishment of a

World Solar Commission (WSC) with the mandate to ensure the successful completion of the WSSP and the holding of the World Solar Summit, as well as to oversee the preparation of the WSP. H.E. Mr. Robert G. Mugabe, President of the Republic of Zimbabwe, having been designated Chairman of the World Solar Commission, the latter was then formally established at the level of Heads of State and Government, following consultations between the Chairman and UNESCO's Director-General. The Commission is now composed of eighteen members, serving in a personal capacity (see Appendix 1).

As indicated above, the Commission has acted as the supreme body overseeing and guiding the WSSP and will also strive to foster regional, international and national cooperation in education, training and research to include the transfer of research disclosures to industry and the market place. In addition, the Commission will seek to identify and define the strategic projects to be included in the WSP.

Initially, the Commission identified a series of key questions examined during its work, the answers to which were the major recommendations it put forward to the Summit of Heads of State for their approval. These questions included perennial issues facing governments, societies and educators, which will continue to be important issues during the coming years. There were also questions arising from new configurations of society and new developments in the physical and social realm. The latter will imply new priorities, new study and new actions. Some may be universal, based on inevitable and indispensable responses to a changing world; others will be regional or nation-specific and will focus on the widely differing economic, cultural and social situations prevailing in different countries.

Expected outputs from the World Solar Commission include the identification and definition of the major technological, economic, social and political issues facing renewable energy technologies over the past years, development of recommendations on agendas for future research and development and applications, and better understanding of the potential role for renewable energy technologies over the next decade.

The Commission, which held its formal session in Harare immediately prior to the World Solar Summit, advised the WSS on the best possible strategies to be pursued for large-scale development and deployment of renewable energies.

1) Draft Leading Documents for the World Solar Summit

With the conclusion of the first phase of the WSSP—the High-Level Expert Meeting held in Paris, in July 1993—five Proposed WSSP Documents were prepared by Working Groups, which included eminent world experts. These included: **The World Solar Programme 1996–2005**, Strategic Projects, The World Solar Charter, The International Solar Convention and The World Solar Fund. A Chairperson was carefully selected to direct the work of each of the Groups and given authority and flexibility to organise the work in the most efficient manner. The documents arising from these Working Groups were then submitted for extensive debate to all of the WSSP Regional High-Level Expert Meetings and Consultations organised within the framework of the World Solar Summit Process.

The review of these draft documents made it apparent that, for a variety of reasons, agreement would not be reached on an international instrument entailing legal obligations for signatory governments, i.e. a charger or a convention, and that the prevailing economic situation did not plead in favour of the creation of a new fund. Consequently, and in the interest of as wide a political consensus as possible, the decision was taken to replace the draft International Solar Convention with a draft Declaration on Solar Energy and Sustainable Development which included, *inter alia* a commitment by governments to provide adequate funding for the wider use of renewable energies and for the full utilisation of existing international funds. In addition, a draft outline of the **World Solar Programme 1996–2005** was prepared. These two draft basic documents were considered and endorsed by the World Solar Commission, then submitted to the World Summit, which approved them by acclamation on 17 September, 1996. The text of the Harare Declaration on Solar Energy and Sustainable Development is attached as Appendix 2, and that of the World Solar Programme 1996–2005: An Outline as Appendix 3 (28, 77).

2) Strategic Projects

Five activities of global importance and universal value described below were identified and developed during the regional solar consultations and high-level expert meetings, as well as during the meetings of the regional solar councils, organized within the framework of the WSSP since 1993 (76).

i) Global Renewable Energy Education and Training Programme

The utilisation of renewable energy sources, as one of the components of the global energy system, requires knowledge of the various technologies and their adaptation to different contexts and fields of application. Therefore, development of such energy projects and their success will only be possible if an effective education and training programme is developed and applied.

The Global Renewable Energy Education and Training Programme should aim to provide the competent human resources required for the effective integration and use of renewable energies world-wide by employing new teaching methods, open and flexible methodologies, use of inexpensive teaching materials and ready access to modern communication channels and multimedia techniques. Furthermore, the Program aims to improve and widen renewable energy education opportunities on a global scale, to enhance North-South/East-West technology transfer and to raise and unify standards world-wide.

At its 150th session the UNESCO Executive Board decided that UNESCO should assume primary responsibility for this project of universal value, including provision of sufficient financial, technical and human resources drawn also from extra-budgetary sources.

ii) International Renewable Energy Information and Communication System

The objective is to build awareness among the public at large of the potential of renewable energy activities by providing the necessary information, as well as professional know-how and technology transfer with regard to renewable energy technologies and systems.

Availability of unrestricted information and its ease of communication will help to build awareness, notably among the rural population regarding their capacity to preserve, maintain and improve their habitat.

Know-how to be disseminated will include the utilisation of renewable energy systems and their maintenance, relevant irrigation methods, new methodologies and equipment etc.

Communication and information distribution will take place on two different but complementary levels. Firstly, at the professional level, so as to ensure information and data transfer to all levels of professionals dealing with renewable energies. Secondly, useful data and information resources will be disseminated to the public at large. Thus,

- the first system will consist of an information *network* using state-of-the-art technology, whenever available, to link professional centers concerned with renewable energy around the world;
- the second system will deal with mass media information distribution.

In the operation of both systems, the expected differences in level of development and local expertise will be taken into account. Data will be exchanged between the centers on a regular basis. The public information system will be based upon existing facilities and will use available international, regional or national establishments to prepare the information to be disseminated.

The International System for Energy Expertise and Knowledge (ISEEK) represents another major project of the WSSP in the establishment of an international renewable energy information and communication system. The exchange of information is the basis of international cooperation and no development of renewable energies of energy conservation can take place at the international level without a multi-communication of scientific and technical information.

How and where is this to be done ? What are the basic outputs? Who provides support? Who assists in developing networks needed to make the exchange as efficient as possible? These are some of the questions to which, initially, anyone concerned with the subject needs the answers.

The first step towards a system, taken as early as 1994, was the establishment of the ISEEK, based on the existing UNESCO Energy Database. This database will be decentralized, which implies the development of national databases, the composite of which will constitute the international database.

The ISEEK will provide world-wide information on national governmental organisations, research centers, information centers, professional/trade associations, networks, training and education activities and facilities, databases/data banks, journals and reference publications and audio-visual aids.

Recommendations for the future provide for the development of a knowledge database and, ultimately, an expert system. Only then will the international community have a true state-of-the art information tool.

iii) Renewable Energy for Rural Electrification

In the remote rural areas of the world, more than 400 million households do not have access to electricity. Renewable energy electrification is possible by taking advantage of the modularity of the solutions mentioned in the general considerations. It would facilitate satisfying the basic needs of these populations (water pumping, light, audio-visual equipment, rural dispensaries, schools etc.) and would create or stimulate revenue-generating rural activities (tourism, handicrafts, improvement of agricultural production, etc.).

The strategy aims to implement the following three electrification activities designed for people of the world's remote rural areas :

- *Water:* electricity for drinking water, designed to provide the electricity necessary for pumping existing water from wells in these areas and rendering it potable.
- *Educational facilities:* designed to provide needed electricity to existing schools.
- *Health:* designed to provide electricity to existing dispensaries and field hospitals so as to ensure appropriate health care services.

iv) Renewable Energy for Water Desalination and Treatment

The scarcity of fresh water resources and the need for potable water will be increasingly important in the future. It is very likely that the water issue will be considered, like fossil energy resources, to be one of the determining factors of world stability (5, 14, 19).

The objective of this activity is to initiate the implementation of a world water desalination programme designed to establish a new way of supplying potable water to rural populations, using renewable energies.

The world-wide availability of solar and wind resources and the availability of mature technologies in this field makes it possible to consider the coupling of desalination plants with renewable energy production processes, in order to ensure the production of water in a sustainable and environmentally friendly scheme for the regions concerned.

This ultimate goal of this project is to supply energy to produce fresh water for one billion human beings living in isolated areas, securing a minimum of five liters per day as required by World Health Organisation standards, during the period of the WSP.

v) Industrial Policy, Market Penetration and Technology Transfer for Renewable Energy

The aim is to provide a basis for increasing the use of renewable energies. Presently, the lack of relevant legal, technical and innovative financial mechanisms are hindering the wider use of these environmentally-friendly energies. Among suggested approaches to overcome existing obstacles are:

- Formulation of regulations for the integration of renewable energies in common planning procedures and project developments. Weakneses of local infrastructures, including local and regional financial institutions, need to be identified and remedied
- Establishment of new channels to disseminate appropriate technologies to target groups such as professional associations and marketing specialists, and reinforcement of communication with international and other financing institutions
- Establishment of mechanisms designed to facilitate and support North-South and South-South technology-sharing initiatives and related actions, in order to develop a strong renewable energy expertise and economic activities at the regional and local levels
- Mobilisation of resources for technical assistance programs and leveraging of resources to optimize total financing available from seed money, with special emphasis on micro-financing possibilities that can be delivered with precision at the retail level. This innovative seed financing should be accompanied by training in both technical and business skills and by familiarizing local financial institutions with lending in connection with renewable energy

- Supporting the market with a more intensified research and development effort; identification and stimulation of innovative concepts of commercial viability using a market-driven approach
- Enhancing the role of the private sector whenever feasible by introducing incentives to stimulate renewable energy technologies and thus encourage participation by all sectors of the market.

iii) High Priority National Projects

Since the landmark High-Level Expert Meeting "The Sun in he Service of Mankind" which was held at UNESCO Headquarters on 5–9 July, 1993, and throughout the ensuing World Solar Summit Process, all of the participating Governments have been invited to submit to the WSSP International Organizing Committee high-priority projects proposals, worthy of strong national and international support, in the various fields of renewable energies.

A total of almost 400 national projects have been identified for 89 countries from Africa, the Americas and the Caribbean, Arab States, Asia and the Pacific, Commonwealth of Independent States, Europe, and the Mediterranean region (67, 68). The HPNP descriptions are presented in an identical format. Each description contains a general overview of the project including its background and justification, objectives, main results to be achieved: cost estimates and/or budgets are presented in local currency and/or US$. Projects range in duration from one to seven years.

These projects are devoted to the utilisation of various renewable energy sources such as solar, hydro, wind, geothermal, biomass, wood and wood-waste, for application in rural electrification schemes, for heating and cooling purposes, for water pumping and desalination plants, as well as for domestic use and small-scale industry. A significant number of projects deal with information, research, education and training issues, and the marketing and commercialisation of renewable energy technologies.

Various funding opportunities are being considered for HP-NP implementation such as national, regional inter-governmental, private, lending institutions, etc. Regional and national consultations have identified a variety of ways and means for funding the proposed projects. It is our sincere hope and intention that these suggestions for funding be realised, given the fact that a number of funding sources and donors have already expressed concrete interest in supporting specific projects. It is obvious that all potential sources of financing will be consulted by the World Solar Commission in order to facilitate the realisation of these project proposals.

THE WORLD SOLAR SUMMIT

Fifteen years after the UN Conference on New and Renewable Energy Sources in Nairobi, Kenya, the World Solar Summit was held in Harare, Zimbabwe, on 16 and 17 September 1996, on the initiative of the United Nations Educational Scientific and Cultural Organization. Heads of State and Government—gathered or officially represented—who participated in this event following the invitation of the Government of Zimbabwe, recognized the significance of the role that solar and other sources of renewable energy such as wind, geothermal, hydro, biomass and ocean, as well as energy efficiency, should pay in the provision of energy services and in the sustainable use of environmental resources for the well-being of humanity.

1) The Event

The World Solar Summit was attended by 10 Heads of State and one Vice-President, seven Prime Ministers, two Deputy Prime Ministers, 34 Ministers, four Vice-Ministers and a number of other high-ranking officials. A total of 104 countries were officially represented at the Summit, along with 12 organisations of the United Nations System and 10 regional inter-governmental organisations including the European Commission and the Organization of African Unity. The Summit was attended by some 1,000 people, 593 of whom were official representatives of their respective countries. Representatives of private industry and non-governmental organisations also participated. An exhibition of the latest renewable energy equipment was prepared for the WSS and an exhibition manual was sent to all countries together with a Summit brochure mailed to countries and organizations by the end of June 1996. The World Solar Summit was declared a State Occasion of the Republic of Zimbabwe. This meant that several government ministries were called upon to contribute to the successful hosting of the Summit. It was considered a success by all concerned and will be seen in the future as a milestone in the enhanced utilisation of renewable energies as a major contribution to sustainable development. Never before had a meeting on renewable energy issues brought together participants of such high rank.

Table 9 provides information on all events organised on the occasion of the World Solar Summit.

In his opening address to the World Solar Summit, the Director-General of UNESCO referred *inter alia* to the innovative approach to preparing the Summit, which constitutes a new modality that combines scientific rigor with the urgency and rapidity of action that important issues require, especially those affecting directly the welfare of human beings.

The World Solar Summit was preceded on 14 September by the first session of the World Solar Commission, composed of 16 Heads of State and Government, which reviewed and endorsed the two basic documents prepared for the Summit: the draft Harare Declaration on Solar Energy and Sustainable Development, and the draft outline of the World Solar Programme 1996–2005. The World Solar Summit considered these draft documents and, with only one amendment to the text of the Declaration, approved them by acclamation.

Special attention should be paid to the operative part of the Declaration, which calls on all nations to join in the development and implementation of the WSP, invites the World Solar Commission to provide high-level leadership and guidance in this respect, and recognizes the leading role that UNESCO has played and should continue to play in the development and implementation of this Programme, in cooperation and partnership with all other organisations involved.

It should be noted that, prior to the Summit, a Round Table on Private/Public Partnership for the Development of Renewable Energies was held; the participants addressed a special message to governments, inviting them to facilitate the involvement of both industry and financial institutions in the large-scale development and deployment of renewable energy technology.

A forum of international non-governmental organisations concerned with the promotion of renewable energies also took place in Harare at the time of the World Solar Summit, and its report was brought to the attention of the Summit.

A workshop on the Global Renewable Energy Education and Training Programme 1996–2005 also took place prior to the Summit, at the UNESCO Harare Office on 13 September 1996 (4, 12, 13). As part of the preparation for the World Solar Summit, the workshop was attended by prominent educators from all over the world. The following regions were represented during the meeting: Africa, America and the Caribbean, Arab States, Asia and Pacific, Commonwealth of Independent States, East and Central Europe and the Mediterranean region. Also represented were NGOs and international institutions. Officially opened by the Zimbabwean Minister of Higher Education, the workshop was co-chaired by the Director of Energy Technology of the European Commission Directorate General for Energy, the Permanent Secretary in the Ministry of Higher Education of the Government of Zimbabwe and the Head of the Swiss Delegation to the UNESCO Executive Board. The meeting was considered to be of utmost importance as it was called by UNESCO to recommend the launching of the Global Renewable Energy Education and Training Programme 1996–2005.

Representatives of all regions were in agreement on the importance of education and training in solar energy. Examples were given from the various regions of how installations broke down and sometimes even became useless within a period of two or three years from lack of maintenance knowledge. The importance of training all individuals involved, from the technicians who install the equipment to those who maintain it locally and the users themselves, was emphasised as the need for education and training in solar energy to permeate all levels of the curriculum. As a result, the Global Renewable Energy Education and Training Programme 1996–2005 was recommended by the workshop participants for approval by the World Solar Summit as one of the Strategic Projects of the WSP (77).

2) The Follow-up

The three main results of the World Solar Summit can be summarised as follows:

- Renewable energies are recognised as being an important component of the energy sector of the 21st century and as being worthy of development and use on a large scale in the coming decade.

- The decision was made to launch a **World Solar Programme 1996–2005** as a common effort of national Governments, specialised agencies and programmes of the United Nations, inter-governmental and non-governmental organisations, university and research institutions and the private sector
- The outline of the **World Solar Programme 1996–2005** was approved and a decision was made to elaborate a fully-fledged WSP document by July 1997.

The World Solar Commission was invited by the Summit to continue to provide high-level leadership and guidance in order to attain the objectives of the Summit. UNESCO was asked to continue to play a leading role in the WSP preparation (76).

Following the meeting of a Task Force on Energy of the relevant institutions of the United Nations system in December 1996, a coordination meeting took place on 20 December, 1996 at UNESCO Headquarters to review the respective roles of the UN institutions in the development and implementation of the WSP. Personal representatives of members of the World Solar Commission met on 27–28 January 1997 in Vienna on the invitation of the Federal Chancellor of the Republic of Austria, member of the World Solar Commission, and also on 19 March 1997 at UNESCO Headquarters, to discuss the WSP structure and contents. A Consultation Workshop of government designated experts on the preparation of the **World Solar Programme 1996–2005** was organised at UNESCO Headquarters on 20–21 March 1997. A final consultation on the draft document entitled "**World Solar Programme 1996–2005**" was held on 9–10 June 1997 at UNESCO Headquarters in Paris, with the participation of Personal Representatives of Members of the World Solar Commission and other partners involved.

A meeting of Personal Representatives of Members of the World Solar Commission was organised on 11 June 1997 at UNESCO Headquarters, and approved a final draft of the WSP document, for submission to the World Solar Commission.

The "Harare Declaration on Solar Energy and Sustainable Development" and the "**World Solar Programme 1996–2005: An Outline**" have been presented as official documents to the Special Session of the United Nations General Assembly to review and appraise the implementation of Agenda 21, which was held in New York, 23–27 June 1997.

THE WORLD SOLAR PROGRAMME 1996–2005

The **World Solar Programme 1996–2005** was approved on 23 June 1997 by the World Solar Commission at its second session, held in New York on the occasion of the above Special Session of the UN General Assembly. UNESCO has contributed in a substantial way to the conceptual design of the World Solar Programme 1996–2005. It has undertaken the required research and conducted a series of consultations, to enable the preparation of the WSP document. The WSP is composed of the following elements:

- the vision of renewable energy development in the future
- strategic projects of global importance and universal value
- regional priorities
- high priority national projects
- major programmes of non-governmental organisations.

A reference document has been prepared giving a two-page summary on High Priority National Projects submitted so far to the World Solar Programme 1996–2005. This document is planned to be submitted to funding sources concerned by the end of 1997 (68).

An information package to permanent Delegations to UNESCO is being prepared to provide them with information on what can be expected from UNESCO, the World Solar Commission and the Regional Solar Councils, and what is expected from national competent authorities as far as the implementation of their High Priority National Projects is concerned as well as their participation in WSP global projects (59, 60).

A detailed outline of the African Solar Programme 1996–2005 is also being prepared by the World Solar Commission Secretariat in close cooperation with the African Solar Council.

An Internet Web Site for the World Solar Programme 1996–2005 is being permanently updated:

WEB SITE:

http://www.unesco.org/general/eng/programmes/science/wssp/index.html

http://www.unesco.org/general/fre/programmes/science/wssp/index.html

A series of meetings on business and investment to enhance the implementation of the World Solar Programme 1996–2005 are being organized by the World Solar Commission:

CARIBBEAN

Conférence Caribéenne sur les Energies Renouvelables
Conseil général de la Guadeloupe
Basse-Terre, 1–3 December 1997

AFRICA

African Solar Programme 1996–2005
Strategies and Financing
Bamako, Mail, 25-27 March 1998

LATIN AMERICA

Business and Investment for Renewable Energy in Latin America
Compania de Fuerza y Luz
San José, Costa Rica, 24–26 June 1998

RUSSIAN FEDERATION

Business and Investment for Renewable Energy in Russia
Intersolarcenter
Moscow, 28-30 September 1998

MEDITERRANEAN REGION

Business and Investment for Renewable Energy in the Mediterranean Region
ENEA-Rome, Italy, 29-31 October 1998

ASIA AND THE PACIFIC

Business and Investment for Renewable Energy in Asia and the Pacific—Agency for the Assessment and Application of Technology—Jakarta, Indonesia, 17–19 December, 1998.

PEOPLE'S REPUBLIC OF CHINA

Business and Investment for Renewable Energy in China
State Science and Technology Commission
Beijing, February 1999

An awareness and fund-raising campaign has been launched jointly by the Assistant Director-General for Science of UNESCO and the UNDP Assistant Administrator for a High Priority National Project from Ecuador, "Solar Energy for Development of the Galapagos Islands".

In parallel with the preparation, negotiation and approval of the **World Solar Programme 1996–2005**, the Engineering and Technology Division, the Secretariat of the World Solar Commission and major partners of the World Solar Summit Process have implemented some High Priority National Projects and also WSP global projects.

For example (78):

- one of the Zimbabwe HPNP, "Solar Electrification of Rural Institutions", has received a US$ 10.5 million funding in the form of a grant from the Italian Government;
- a US$ 20 million soft loan was provided by the World Bank for implementation of one of the nine Indonesian HPNP, "One Million Home Photovoltaic Rural Electrification in Indonesia".
- a US$ 97.5 million soft loan is being negotiated with EBRD for one of the five HPNP of the Russian Federation, "Geothermal Electrical Station of 4 MW Capacity for Kamchatka";
- the Onamunhama Demonstration Solar Village has been built in Namibia and inaugurated by the UNESCO Director-General;
- Umbuji village in Zanzibar has been provided with the necessary solar equipment;
- a human settlement in Niger in the National Park "Wu" has been provided with solar power supply;
- a solar school in mountainous area of the People's Republic of China has been equipped with a solar power system;
- a nine-volume UNESCO/WSP learning package on renewable energy for practicing engineers has been printed and largely disseminated (15, 25–28, 30, 35, 36, 39, 40, 62);
- a CD-ROM "UNESCO/ISEEK Energy Database" was prepared and widely distributed in 1996 and 1997;
- a prestigious annual summer school "Solar Electricity for Rural and Remote Areas" has been held at UNESCO Headquarters since 1990 for French speaking experts with technical visits to the best European solar centers.

References

1. Approved Programme and Budget for 1955–1956, General Conference of UNESCO, 8th session, Montevideo, 1954, project 2208, pp. 74, 88, 96.
2. Approved Programme and Budget for 1975-1976, 18 C/5, Paris, UNESCO, 1975, p. 230, 232.
3. Arid Zone Research, Wind and Solar Energy, Proceedings of the New Delhi Symposium, Paris, UNESCO, 1956.
4. BENCHIKH, O. *Education et formation pour les énergies renouvelables*. Paris: UNESCO, 1996, 30p.
5. BENCHIKH, O. et ABOU RAYAN, M. *Technologie de Dessalement de l'Eau, Paris*: UNESCO, 1995, 308 p.
6. BERKOVSKI, B. *Renewable Energies for Sustainable Development*. Paris: UNESCO, 1990. PP. 9/11 (presented at the World Renewable Energy Congress, Reading, September 1990).
7. BERKOVSKI, B. *Round-up Report*/High-level Expert Meeting: The Sun in the Service of Mankind, UNESCO, Paris, 1993. (WSSP/93/10/09)
8. BERKOVSKI, B. *The Solar Highway Leading to Sustainable Development*. Paris: UNESCO, 1994. (WSSP/94/09/04).
9. BERKOVSKI, B. *The World Solar Fund: Draft Project Document.*, Paris: UNESCO, 1994. (WSSP/MSC/12/16).
10. BERKOVSKI, B. The World Solar Summit Process—The Solar High-way Leading to Development and Environmental Protection. In *Proceedings of the World Energy Council*, Congress, 16th, Tokyo, 1995. London: WEC, 1995. pp 386–397. (PS/SRD 1.4.27).
11. BERKOVSKI, B. *The World Solar Summit Process: Background Information.* Paris: UNESCO, 1994. (WSSP/94/11/21).
12. BERKOVSKI, B., GOTTSCHALK, C.M. *Strengthening Human resources: A Global Renewable Energy Education and Training Programme for the 21st Century*. UNESCO Engineering Education and Training. Paris: UNESCO, 1996.
13. BERKOVSKI, B., KOUZMINOV, V.A. *Renewable Sources of Energy in the Service of Mankind*. Moscow: Nauka, 1987. 280 p.
14. BERKOVSKI, B., MALYAVIN, S. UNESCO *Energy Programme for Development and Environment: Problems, Achievements* and Prospects, 1991, 78p.
15. *Biomass Conversion and Technology*/C.Y. Wereko-Brobby, E.B. Hagen. Chichester: John Wiley & Sons, 1996. ISBN 0–471–96246–5.
16. Committee on New and Renewable Sources of Energy and on Energy for Development. Report on the first session, 7–18 February 1994, New York: United Nations (E/1993/25, E/C.13/

1994/8).
17. Committee on New and Renewable Sources of Energy and on Energy for Development. Report on the second session, 12–23 February 1996, New York: United Nations. (E/1996/24, E/C.13/1996/8).
18. *Energy Planning and Policy*/M. Kleinpeter. Chichester: John Wiley & Sons, 1995. ISBN 0–471–95536–1.
19. Energy Planning, Proceedings of the International Seminar on Methodology and Institutions for Energy Planning, Rio de Janeiro, Brazil, 10–14 September 1984, organized by UNESCO in cooperation with FINEP and COPPE/UFRJ, UNESCO, 1985.
20. Energy Scholarships. Grant possibilities Offered by International, National & Regional Organizations, UNESCO, August 1985.
21. European Working Group on Cooperation in the Field of Solar Energy Applications, Genoa, Italy, 1976. Report, Paris, UNESCO, 1976.
22. Fuel Cells: Trends in Research and Applications. Summary Proceedings of a Workshop Organized by the United Nations Educational, Scientific and Cultural Organization, in cooperation with the Council of Europe and the Commission of the European Communities, Ravello, Italy, 10–14 June, 1985, UNESCO, 1985.
23. FUNDACION CANOVAS DEL CASTILLO. *Annex to the Declaration of Madrid*/An Action Plan for Renewable Energy Sources in Europe, Madrid, 16–18 March 1994. Madrid: Fundación cánovas del Castillo, 1994. Point III. The Benefits.
24. FUNDACION CANOVAS DEL CASTILLO. *The Declaration of Madrid* / an Action Plan for Renewable Energy sources in Europe. Madrid, 16–18 March 1994. Madrid: Fundación Cánovas del Castillo, 1994.
25. *Geothermal Energy*/Edited by M.H. Dickson, M. Fanelli. Chichester: John Wiley & Sons, 1995. ISBN 0–471–95366–0.
26. GOTTSCHALK, C.M. *Education and Training for New and Renewable Forms of Energy*/High-level Expert Meeting: The Sun in the Service of Mankind, Paris: UNESCO, 1993.
27. GOTTSCHALK, C.M. UNESCO'S *21st Century International Information and Expert System for New and Renewable Sources of Energy: Strategic Issue: Energy Information.* / High-level Expert Meeting: The Sun in the Service of Mankind, Paris: UNESCO, 1993.
28. *Guidelines for Authors*/Compiled by the Engineering and Technology Division, UNESCO. Chichester: John Wiley & Sons, 1996.
29. *Harare Declaration on Solar Energy and Sustainable Development*: World Solar Summit. Paris: UNESCO, 1996. SC-96/WS/43.

30. *Industrial Energy Conservation*/Compiled by C.M. Gottschalk. Chichester: John Wiley & Sons, 1996. ISBN 0–471–96008-X.
31. International Congress "The Sun in the Service of Mankind", 1973, working papers 4 vol., Paris UNESCO, 1973.
32. International Workshop on Non-Technical Obstacles to the Use of New Energies in Developing Countries, Bellagio, Italy, 1981. Report, Paris, UNESCO, 1981.
33. JOHANSSON, T., and others. *UNDP Initiative for Sustainable Energy*, Energy and Atmosphere Programme, Bureau for Policy and Program Support, United Nations Development Programme, New York 1996, 85 p.
34. KULSUM, A. *Renewable Energy Technologies: A Review of Status and Cost of Selected Technologies*. World Bank Technical Paper n. 240, Energy series. Washington D.C.: The World Bank, 1994 (Energy Series: Technical Paper, 240).
35. *Magneto-Hydro-Dinamic Electrical Power Generation*/H.K. Messerle. Chichester: John Wiley & Sons, 1995. ISBN 0–471–94252–9.
36. MARKVART, T., BERKOVSKY, B., GOTTSCHLANK, C.M. *Solar Energy and Education*/High-level Expert Meeting: The Sun in the Service of Mankind, Paris: UNESCO, 1993.
37. Medium-Term Plan for 1973–1978, 17 C/4, Paris, UNESCO, 1978, p. 117.
38. Medium–Term Plan for 1977–1982, 19 C/4, Paris, UNESCO, 1977, pp. 143–147.
39. *Mini Hydropower*/T. Kiandong, Z. Naibo, W. Xianhuan, H. Jing, D. Huishen. Chichester: John Wiley & Sons, 1996. ISBN 0–471–96264–3.
40. *Ocean Thermal Energy Conversion*/P. Takahashi, A. Trenka. Chichester: John Wiley & Sons, 1996. ISBN 0–471–96009–8.
41. OECD. OECD *in Figures: Statistics on Member Countries*. Paris: OECD, 1992.
42. Our Common Future, World Commission on Environment and Development, Oxford, 1987, Part II, Section 7: Energy, Economy and Environment.
43. PALZ, W., *Solar Electricity*. Paris: UNESCO; Buttenworths: London and Boston, 1978. 292 p.
44. PAPOUTSIS, C., *Community Policy on Renewable Sources of Energy and the World Solar Summit Process*. Harare, 16–17 September 1996.
45. Report by the Director-General and the Executive Board on the Activities of the Organization during the Year 1953, 8 C/3, p. 91.
46. Report of the Director-General on the Activities of the Organization in 1955, Paris, UNESCO, 1956, pp. 68, 70.
47. Report of the Director-General on the Activities of the Organization in 1959, Paris, UNESCO, 1960, pp. 90.

48. Report of the Director-General on the Activities of the Organization in 1960, Paris, UNESCO, 1961, p. 96.
49. Report of the Director-General on the Activities of the Organization in 1962, Paris, UNESCO, 1963, p .104.
50. Report of the Director-General on the Activities of the Organization in 1973, Paris, UNESCO, 1974, p. 100.
51. Report of the Director-General on the Activities of the Organization in 1975–76, Paris, UNESCO, 1977, p.27.
52. Report of the Director-General on the Activities of the Organization in 1977–78, Paris, UNESCO, 1979, p.23.
53. Report of the Director-General on the Activities of the Organization in 1977–78, Paris, UNESCO, 1979, p. 23–24.
54. Report of the Director-General on the Activities of the Organization in 1979–1980, Paris, UNESCO, 1981, pp. 27–28.
55. Report of the Director-General on the Activities of the Organization in 1981–1983, Paris, UNESCO, 1985, p. 23.
56. Report of the Director-General on the Activities of the Organization in 1984–85, Paris, UNESCO, 1988, p. 28.
57. Report of the Director-General on the Activities of the Organization in 1986–1987, Paris, UNESCO, 1988, p. 28.
58. Report of the Director-General on the Activities of the Organization in 1988–1989, Paris, UNESCO, 1990, p. 32.
59. Secretariat of the World Solar Commission. *High Priority National Projects* / World Solar Summit, Harare, 1996. (WSS/HPNP) [on UNESCO Web on Internet: http://wwv.unesco.org].
60. Secretariat of the World Solar Commission. *High Priority National Projects* / World Solar Programme 1996–2005, November 1997. (HP-NP/11/97).
61. Solar Cooling and Refrigeration, by J.C. Mc Veigh,. A Report sponsored by the United Nations Educational, Sceintific and Cultural Organization, UNESCO, March, 1984.
62. Solar Electricity/Edited by T. Markvart. Chichester: John Willey & Sons, 1994. ISBN 0–471–94161–1.
63. Solar Energy, Geneva, 30 August to 3 September 1976, Proceedings of the UNESCO/WMO Symposium, Geneva, 1977.
64. STRONG, M. *The Global Partnership for Environment and Development*, Geneva: UNCED, 1992, 116 p.
65. *Symposium International Energie et Société*. Paris: UNESCO, 13–17 December 1993.
66. The Future for Renewable Energy, Prospects and Directions. 1996 EUREC Agency. Published by James & James Ltd. ISBN: 1–873936–70–2.
67. The Statistical Analysis of the Member States' and the NGO's interest in UNESCO's Science and Technology activities, as expressed in their replies to the 25 C/4. Questionnaire on UNESCO's Third Medium-Plan, SC/OPS, 11/03/88.

68. Third Medium Plan (1990–1995), 25 C/4, Approved, Paris, UNESCO, 1990, p. 58, para. 102.
69. UNESCO Sources, N°47, May 1993, p. 23. Paris: UNESCO, 1993. ISSN 1014–5494.
70. UNITED NATIONS. Department for Development Support and Management Services, *Trends in Environment Support Assessment of Energy Projects*, New York, 1994, 79 p.
71. VELIKHOV, E. *Strategic Energy Challenges*, International Journal of Global Energy Issues, Vol. 2, N°1, 1990, p. 1.
72. Workshop on Training in Solar Energy, Cadarache, France, 14–17 November 1983, Volume I, Summary Proceedings, UNESCO 1984.
73. WORLD ENERGY COUNCIL. *Renewable Energy Resources: Opportunities and Constraints.* 1990–2020/Congress, 15th, London, 1992. London: WEC, 1992.
74. WORLD ENERGY COUNCIL. *Round Up* / Congress, 15th, London, 1992. London: WEC, 1993. 214 p.
75. WORLD ENERGY COUNCIL. *Round Up: Energy for our Common World—What will the future ask of us*? / Congress, 16th, Tokyo, 1995. London: WEC, 1995, 449 p.
76. *World Solar Programme 1996–2005*. [on UNESCO Web on Internet: http:/ / www.unesco.org].
77. *World Solar Programme 1996–2005: An Outline.* / World Solar Summit. Paris: UNESCO, 1996. SC-96/WS/39.
78. *World Solar Summit: Preparatory Process and Results: Final Report*/Secretariat of the World Solar Commission. Paris: UNESCO, 1997. SC-97/WS/16.
79. *Zimbank Economic Review*, June 1991, Harare, 1991. p. 24.
80. 1995 Demographic Yearbook. Forty-seventh issue, Department for Economic & Social Information & Policy Analysis, United Nations, New York, 1997. ISBN: 92-1-051086-0.

APPENDICES

APPENDIX 1

World Solar Commission

Composition at the World Solar Summit 16–17 September, 1996.

Chairman:

His Excellency Mr. Robert Gabriel Mugabe
President of the Republic of Zimbabwe
Chairman of the World Solar Commission

Members:

His Excellency Mr. Soeharto
President of the Republic of Indonesia

His Majesty Juan Carlos I
King of Spain

His Excellency Mr. Abdou Diouf
President of the Republic of Senegal

His Excellency Mr. Zine El Abidine Ben Ali
President of the Republic of Tunisia

His Excellency Mr. Eduard Shevardnadze
President of Georgia

His Excellency Mr. Jang Zemin
President of the People's Republic of China

His Excellency Mr. Sardar Farooq Ahmad Khan Leghari
President of the Islamic Republic of Pakistan

His Excellency Mr. José Maria Figueres Olsen
President of the Republic of Costa Rica

His Excellency Mr. Nelson Mandela
President of the Republic of South Africa

The Honourable Dato'Seri Dr. Mahathir bin Mohamad
Prime Minister of Malaysia

The Right Honourable Percival James Patterson
Prime Minister and Minister of Defence of Jamaica

His Excellency Mr. H.D. Deve Gowda
Prime Minister of the Republic of India

His Excellency Mr. Franz Vranitzky
Federal Chancellor of the Republic of Austria

His Excellency Mr. Benjamin Netanyahu
Prime Minister of the State of Israel

His Excellency Mr. Yasser Arafat
President of the Palestinian Authority

The following Changes have taken place as from September 1996:

New Members

His Excellency General Ibrahim Barré Mainassara
President of the Republic of the Niger

His Excellency Mr. Romano Prodi
President of the Council of Ministers of Italy

Replacements

His Excellency Mr. Victor Klima replaced His Excellency Mr. Franz Vranitsky as Federal Chancellor of the Republic of Austria and Members of the World Solar Commission

The Honourable Mr. Inder Kumar Gujral replaced His Excellency Mr. H.D. Deve Gowda as Prime Minister of India and Member of the World Solar Commission

Secretary-General:

B. Berkovski
Director
Engineering and Technology Division
UNESCO

APPENDIX 2

World Solar Summit

Harare Declaration on Solar Energy and Sustainable Development

We, the Heads of State and Government, gathered or officially represented in Harare on the occasion of the World Solar Summit, following the invitation by the Government of Zimbabwe and at the initiative of the United Nations Educational, Scientific and Cultural Organisation, in collaboration with international organisations and institutions, in order to launch a program for global solar activities to be known as the **World Solar Programme 1996–2005**,

1. *recognise* the significance of the role that solar and other sources of renewable energy such as wind, geothermal, hydro, biomass and ocean, as well as energy efficiency, should play in the provision of energy services and in the sustainable use of environmental resources for the well-being of humanity;
2. *reiterate* our support for the principles and actions for the promotion of energy systems for sustainable development recommended by the United Nations Conference on Environment and Development, held on 3–14 June 1992 in Rio de Janeiro, Brazil, in particular the Rio Declaration on Environment and Development, Agenda 21, as well as the commitments made in the United Nations Framework Convention on Climate Change;
3. *recognise* that the provision of sufficient energy services at affordable prices and the adoption of energy conservation measures are essential for the progress of all countries, developed and developing alike, to meet current and expanding needs in ways which minimise environmental degradation and risks, as well as to realise the full potential of renewable energy sources;
4. *recognise* that there is a need to increase substantially access to energy in developing countries, and that the provision of adequate energy services can improve living conditions, alleviate poverty, improve health and education, promote small scale enterprises and create other

income-generating activities especially in rural and isolated areas thereby reducing rural to urban migration;

5. *recognise* that the role of solar energy in each country needs to be integrated and specified in its national energy policy;
6. *recognise* that the development, deployment and wide-spread utilisation of solar energy face difficulties, particularly with regard to their management, maintenance and financing as well as to the availability and accessibility of relevant data, information, education, training and techno logy;
7. *recognise* that increased use of solar energy can reduce environmental degradation caused by adverse human activities, such as industrial pollution and deforestation, and that it is the responsibility of governments and all sectors of civil society to work together to find lasting solutions to problems threatening the sustainable development of humanity;
8. *emphasise* that the management of atmospheric emissions of greenhouse and other gases and substances will increasingly require efficient and environmentally sound energy systems, including solar energy ones;
9. *recognise* the important role that relevant non-governmental organizations play in the development and deployment of solar energy;
10. *recognise* that women have an important role to play in the promotion of solar energy, and that they would benefit significantly from its use;
11. *are convinced* that sustainable development and utilisation of solar energy would be enhanced by local and national capacity building, policy reform and technology sharing among nations, with emphasis on coherent efforts towards technology acquisition and development in the developing countries.

We therefore,

12. *commit* ourselves to work towards the wider use of solar energy to enhance the economic and social development of all people; support and promotion of these efforts should be very important goals of our governments, the international community and all sectors of society, especially with respect to people living in isolated and under-developed

rural and island communities;

13. *commit* ourselves to work towards policies and effective mechanisms that will speed up and facilitate the use of solar energy avoiding duplication and administrative delays, and the encouragement of international cooperation, including participation in regional and international bodies, scientific and technical organizations;
14. *commit* ourselves to work towards the greater use of solar energy through the provision of adequate technical assistance and funding, the full utilisation of existing international funds, and the facilitation of increased participation by both public and private sectors.

In pursuit of these objectives we:

15. *call* on all nations to join in the development and implementation of the WSP;
16. *invite* the World Solar Commission to continue to provide high-level leadership and guidance to achieve the objectives of the Summit, and UNESCO to continue to play a leading role in the development of the WSP, in close cooperation with relevant international organizations.
17. *invite* the Secretary-General and Heads of Specialized Agencies and Programs of the United Nations, as well as national governments, inter-governmental and non-governmental organizations, academic and research institutions and the private sector, to join in the implementation of the WSP.

Harare, 17 September, 1996.

APPENDIX 3

World Solar Summit

World Solar Programme 1996–2005: An Outline

1. Introduction

This outline of the World Solar Programme 1996–2005 (WSP) to be launched at the World Solar Summit on 16–17 September 1996 in Harare, Zimbabwe is an attempt to briefly state the most salient features of the WSP and its determining activity, the World Solar Summit Process.

The WSP is an open-ended attempt through broad partnership and cooperation of governments and organisations to promote the adoption and wider utilisation of renewable energy sources. This will be achieved through the setting of agreed targets, appropriate standards, cooperation mechanisms, incentives and pooling of resources. The major benefits to be derived from the WSP include the enhancement of the quality of life of large numbers of people, particularly in the rural areas, and the creation of additional employment through the development of new enterprises. The WSP is a response to the challenge posed by the Earth Summit which requires all countries to institute appropriate measures for reducing pollution by introducing clean technology—hence, the WSP will set up off on the 'solar highway' leading to sustainable development.

The WSP will evolve over the decade 1996–2005 through the development and implementation of high-priority national, regional, and global projects, under the political leadership and guidance of the World Solar Commission.

2. Background

Rapid political, social and economic changes are taking place everywhere, bringing both hope and despair to individuals in different parts of the world. As far as energy is concerned, the world today seems to have been polarized. The 75 per cent of humanity living in the developing countries account for only 25 per cent of global energy consumption. More than one billion people in the developing countries have no source of energy available to them other than the traditional ones (wood and other forms of biomass). More than two billion people in these countries have no access to electricity. It was in its endeavour to take up the challenge of providing these people with their basic minimum energy needs that UNESCO, in a broad-based partnership with the United Nations and its Specialised Agencies, international organisations, launched in 1993 a global initiative: a three-year long communication process known as the World Solar Summit Process (WSSP). The objectives of the WSSP are to enhance the global understanding of the role that renewable energies can play in providing clean energies to billions of people in far and remote areas of the developing countries of the world, creating new employment opportunities, improving health services, contributing to the preservation of the international cooperation through inter-regional communication and exchange of expertise and technologies. The World Solar Programme 1996–2005 (WSP) is envisaged as an instrument to attain the objectives of the World Solar Summit Process.

3. The Programme and its Nature

The WSP finds its roots in the high-level regional consultations that were undertaken throughout the world during the last three years. The Programme is conceived as a necessary global commitment and a major coordinated effort of the various national and international actors to develop and implement 300-odd top-priority renewable energy projects of national, regional and international value within a 10-year period in order to demonstrate the technical feasibility, economic viability and social and political acceptability of solar energy. This Program should be seen as a challenging joint effort involving governments, major international organisations and funding institutions, non-governmental organisations, industries, scientific and research institutions, universities, etc. to promote the development and deployment of renewable energy.

4. The Aims and Objectives of the Programme

The aims and objectives of the WSP are to:

- enhance the understanding of the role that renewable energy sources can play in the preservation of the environment, provision of energy services—particularly for rural and remote areas, creation of employment and improvement in the socio-economic conditions of the rural people, particularly of women in the developing countries of the world, and the fostering of increased energy independence;

- develop a favourable political, social and economic climate in favour or renewable energies by demonstrating the economic viability and social acceptability of such projects;
- promote and harmonise cooperation in education, training and research as well as in the transfer of research disclosures to industry at the regional, inter-regional and international levels;
- strengthen the commitment from the international community, especially multilateral and bilateral donors as well as the national commitment from each country; and
- reinforce local technological and entrepreneurial capacity-building and encourage the creation of small-scale financing and delivery mechanisms.

5. Scope of the Programme

The Programme includes projects of varying degrees of geographical limits and accruing values. There are projects which have been identified by various national governments themselves as their highest priority national projects. Such projects, known as the High-priority National Projects, will not only be of benefit to the concerned countries but will also provide valuable information to other countries.

Other projects of regional value have been identified during the various regional consultations that took place in different parts of the world during the last three years. These projects have been termed High-priority Regional Projects, the benefits of which will accrue to the entire region.

Finally, there are global projects of universal value, identified and endorsed during the regional ministerial-level and high-level expert meetings, as well as during the meetings of the Regional Solar Councils, organised since 1993 within the framework of the World Solar Summit Process. A short summary of each of these projects is presented hereunder:

- *Rural Electrification.* In the rural and/or remote areas of the world, more than 400 million households do not have access to electricity.
- *Global Education and Training Programme.* Its objective is to effectively disseminate the information and provide

appropriate training for engineers, technicians and users of solar energy technology as well as decision makers, businessmen and industrialists.

- *Water Desalination and Purification.* The objective of this project is to initiate the implementation of a world water desalination program designed to establish new ways of supplying drinkable water to rural areas, using renewable energies.
- *Information and Communication.* The objectives are: to provide the necessary information, both to the decision-maker and to the public at large, in order to sensitise them on the potential of renewable energies; and to provide professional know-how and expertise sharing with regard to renewable energy technologies and systems.
- *Industrial Policy, Market Penetration and Technology Transfer.* The aims are: to set regulations for the integration of renewable energies in common planning procedures and project development; to establish new information channels to disseminate the appropriate technologies to target groups of professionals and marketers; to set up mechanisms to facilitate North-South and South-South technology-sharing initiatives at the regional and local levels; and to support the market with strong research and development efforts to investigate innovative concepts and simulate the take-up of new technologies in the market based on the feedback from the market's end users.

6. Strategy to Achieve the Programme's Objectives

The WSP envisages the accomplishment of its objectives by mobilising a concerted global effort with the continued cooperation and commitment of various governments, international organizations, bilateral and multilateral funding institutions, non-governmental organizations, the private sector, research organizations and universities etc. As the programme aims at implementing the high priority projects of national and regional importance submitted by various governments, it attaches a strategic importance to continued inter and intra-governmental support and to the mobilisation of the private sector.

7. Funding for the Programme

Various regional consultations on this issue have revealed

that in some countries today, budgetary provision already exists in the areas of research, education, training and development on renewable energies. However, it has become increasingly evident that for the development of renewable energies, the funding windows from government and private sources as well as financial and developmental institutions should be augmented. Furthermore, innovative new financing opportunities including micro-financing, and the need to attract private capital to supplement the insufficient public resources should be created. It is expected that national governments, international, regional development and finance banks will accord high priority to finance renewable energy projects identified in the WSP.

8. Methodology for the Implementation of the Programme

The detailed methodology for the implementation of the Programme will be finalised after its formally launched at the World Solar Summit. Under the overall guidance of the World Solar Commission, the programme will be given a definite structure within a period of nine months after the Summit through intensive deliberations and appropriate negotiations, it being understood that the Programme is open-ended, i.e. that new projects can be included during the decade that it will cover.

The WSP will be implemented having recourse as much as possible to partnership arrangements and extensively using the existing structures. At the international level, the various organisations and institutions should be responsible for the execution of projects within their respective spheres of competence. The same will be true at the regional level, where the regional solar councils created at the regional preparatory meetings for the World Solar Summit Process should play a coordinating and monitoring role.

While UNESCO is prepared to continue to furnish the Secretariat of the World Solar Commission, a substantial increase in the participation of all main actors at the international inter-governmental level will be necessary to ensure the successful completion and implementation of the WSP.

APPENDIX 4

List of Acronyms

ACC : Administrative Committee on Coordination of the United Nations

ADEME: Agence de l'Environnement et de la Maîtrise de l'Energie

AWEA: American Wind Energy Association

CASE: International Centre for Application of Solar Energy

CLIPS: Climate Information and Prediction Services

COSERA: UNIDO Consultative Group on Solar Energy Research and Application

DOE: United States Department of Energy

ESMAP: Energy Sector Management Assistance Program

ENEA: Ente per le Nuove Tecnologie, l' Energia el' Ambiente

EU: European Union

FAO: Food and Agriculture Organisation of the United Nations

GEF: Global Environment Facility

HPNP: High Priority National Projects

IAEA: International Atomic Energy Agency

IFC: International Finance Corporation

IIASA: International Institute for Applied Systems Analysis

IPCC: Inter-governmental Panel on Climate Change

ISEEK: International System for Energy Expertise and Knowledge

ISES: International Solar Energy Society

MAB: Man and Biosphere

MEDA: Mesure d'accompagnement à la réforme des structures économiques et sociales des pays méditerranéens

MHD:	Magnetohydrodynamics
NRSE:	New and Renewable Sources of Energy
OECD:	Organisation for Economic Cooperation and Development
UN:	United Nations
UNCED:	United Nations Conference on Environment and Development
UNCNRSEED:	United Nations Committee on New and Renewable Sources of Energy and Energy for Development
UNDP:	United Nations Development Programme
UNEP:	United Nations Environment Programme
UNESCO:	United Nations Educational, Scientific and Cultural Organisation
UNIDO:	United Nations Industrial Development Organisation
UNISE:	United Nations Initiative for Sustainable Energy
WHO:	World Health Organisation
WMO:	World Meteorological Organisation
WSA:	World Solar Academy
WSC:	World Solar Commission
WSP:	World Solar Programme 1996–2005
WSS:	World Solar Summit
WSSP:	World Solar Summit Process
WSSP/IOC:	World Solar Summit Process/International Organising Committee
ZOC:	Zimbabwe Organising Committee

TABLES

Table 1 : Population Rate between 1980–1995 (in millions)

	1980	*1985*	*1990*	*1995*
World Total	4444	4846	5285	5716
Asia	2642	2904	3186	3457
Africa	476	549	633	728
Europe	693	706	722	726
Latin America	358	398	440	482
North America	252	265	278	292
Oceania	23	25	26	29

Source: 1995 Demographic Yearbook, United Nations, NY 1997.

Table 2 : Rate of Increase in Population (%)

	1985–1990	*1990–1995*
World Total	1.7	1.6
Asia	1.9	1.6
Africa	3.0	2.8
Europe	0.4	0.2
Latin America	2.0	1.8
North America	1.0	1.8
Oceania	1.6	1.5

Source: 1995 Demographic Yearbook, United Nations, NY 1997.

Table 3 : Carbon Dioxide Emissions from Fossil Fuel Combustion (in Gigatones carbon)[3]

	1990	*1991*	*1992*	*1993*	*1994*	*1995*	*%age Change between 1990–1995*
North America[1]	1.618	1.600	1.629	1.661	1.693	1.710	5.69%
Latin America[2]	0.287	0.272	0.278	0.288	0.304	0.310	8.01%
EU 15	0.949	0.949	0.942	0.925	0.923	0.934	–1.58%
CIS/C & E Europe	1.311	1.274	1.170	1.068	0.958	0.910	–30.03%
Middle East	0.177	0.186	0.195	0.216	0.232	0.239	35.03%
Africa	0.183	0.187	0.185	0.189	0.196	0.206	12.57%
Asia/Pacific (total)	1.529	1.561	1.637	1.694	1.836	1.919	25.51%
Asia/Pacific (excl. Japan, Australia & New Zealand)	1.126	1.148	1.216	1.264	1.397	1.471	30.64%
Total OECD	13.035	3.034	3.062	3.155	3.125	3.165	4.28%
Developing countries[2]	1.774	1.793	1.874	1.892	1.128	2.226	25.48%
World	6.119	6.102	6.106	6.117	6.211	6.301	2.97%

([1]) Excluding Mexico, ([2]) Including Mexico, ([3]) Gigatonnes=10^9

Source : World Energy Council Journal, July, 1996

Table 4 : Goals for Performances in the Short and Medium Terms

Item	*Short term 2000*	*Medium term 2010*
Solar cell efficiency		
c-silicon		
laboratory	24%	26%
production (mono-multi crystalline)	16–18%	>20%
thin film (polycrystalline, amorphous)		
laboratory	18%	20%
production	>10%	>15%
advanced devices (tandem, concentrator)		
laboratory	30%	35%
production	>20%	25%
Module lifetime	>20 years	>30 years
maximum degradation during lifetime	<10%	<10%
Inverter efficiency		
100% load	>97%	>98%
10% load	>90%	>95%

Systems. Availabilty on short term (2000) of

- reliable systems for developing countries
- building specific modules satisfying building physical, aesthetical and safety requirements
- reliable concentrators
- lower energy consumption balance-of-systems components

Source : The Future for Renewable Energy: Prospects and Directions 1996 EUREC Agency (p. 74).

Table 5 : Goals for Module and System Cost in the Short and Long Terms

Item	*Short term 2000*	*Medium term 2010*
Module cost	<2.5 ECU/Wp	<1.5 ECU/Wp
of which cell cost	<1.5 ECU/Wp	<1.0 ECU/Wp
System cost		
grid-connected	<0.5 ECU/Wp	<3.0 ECU/Wp

Source : The Future for Renewable Energy: Prospects and Directions 1996 EUREC Agency (p. 74).

Table 6 : Cost of Grid Connected PV

Year	"Turn-key cost"(1) $/Wc	Total cost(2) $/Wc
Power plant		
1993	6,26	12,49
1994	6,20	7,75
1995	5,71	6,62
Home systems		
1993	7,70	8,78
1994	6,23	7,13
1995	5,75	6,60

(1) Including all the system (out of texes) connected to the grid

(2) Including the interconnection, control, site preparation, administration, profile and taxes costs

Source : 1997 UNESCO Summer School "Solar electricity for rural and remote areas"

Table 7 : Oil Spot Prices 1974–1996
(US dollars per barrel)

Year	*Oil Type*			
	Arab Light/Dubai	*Forties/ Brent*	*Nig. L.T.*	*WTI*
1974	10.41			
1975	10.70			
1976	11.63	12.80		12.23
1977	12.38	13.92		14.22
1978	13.03	14.02		14.55
1979	29.75	31.61	32.00	25.08
1980	35.69	36.83	37.18	37.96
1981	34.32	35.93	36.67	36.08
1982	31.80	32.97	33.75	33.65
1983	28.78	29.55	30.01	30.30
1984	28.07	28.66	28.96	29.34
1985	27.53	27.51	27.74	27.99
1986	12.97	14.38	14.60	15.05
1987	16.92	18.43	18.46	19.19
1988	13.22	14.96	15.10	15.98
1989	15.69	18.20	18.50	19.68
1990	20.50	23.81	24.27	24.52
1991	16.56	20.05	20.50	21.54
1992	17.21	19.37	19.92	20.57
1993	14.90	17.07	17.60	18.45
1994	14.76	15.98	16.21	17.21
1995	16.09	17.18	17.35	18.42
1996	18.56	20.81	21.17	22.16

Source : British Petroleum Statistical Review, 1997.

Table 8: The World Solar Summit Process General Chart

WORLD SOLAR PROGRAMME EXECUTION
High-Priority National Projects and global projects of universal value

WORLD SOLAR COMMISSION
Second Session, UN-New York, 23 June, 1997,
approval of the World Solar Programme 1996–2005

CONSULTATIONS OF GOVERNMENT DESIGNATED EXPERTS
Draft World Solar Programme 1996–2005
1997

WORLD SOLAR SUMMIT
Harare, Zimbabwe, 16–17 September 1996

WORLD SOLAR COMMISSION
First Session, Harare, Zimbabwe, 12 September, 1998

LEADING DOCUMENTS

- World Solar Programme 1996–2005 Draft Outline
- Draft Harare Declaration on Solar Energy and Sustainable Development

REGIONAL SOLAR SUMMITS
REGIONAL HIGH-LEVEL EXPERT MEETING
REGIONAL SOLAR COUNCILS
1993–1996

PROPOSED DOCUMENTS

National & Regional Reports	Strategic Project Proposals	World Solar Fund Initiative	Proposals for Regional Solar Councils	International Solar Treaty Initiative	Topical Reports

National Support Groups **NGO's** **IGO's**

High-Level Expert Meeting
"THE SUN IN THE SERVICE OF MANKIND"
UNESCO Headquarters, 5–9 July, 1993 ?

Table 9 : The World Solar Summit at a Glance

Event	*Date*	*Hours*
Meetings of the Zimbabwe Organizing Committee	7 September, 1996 9 September, 1996	15:00–18:00 15:00–18:00
Meeting of the International Organizing Committee of the WSSP	12 September, 1996	8:00–18:00
NGO Forums:		
• Renewable Energy Education and Training	12 September, 1996	9:30 – 18:00
• Better Management of Natural Energy Resources in Rural Areas, Fuelwood, Stoves and Biogas	13 September, 1996	9:30 – 18:00
• How to Implement Energy Efficiency and Renewable Energy in Industrial Economies	13 September, 1996	9:30 – 18:00
• Provision of CO2–Free Energy Services Through Off-Grid Solar Electrification: Organizational Experiences	14 September, 1997	9:30 – 18:30
• The Scope for Sub-Regional Collaboration on Renewable Energy Development in Southern Africa	14 September, 1997	9:30 – 18:00
Open Forum on the World Solar Programme 1996–2005	13 September, 1997	9:00 – 13:00
Round Table on Public/Private Partnership for Renewable Energy Development	13 September, 1997	14:30 – 17:00
Workshop on Global Renewable Energy Education and Training Programme	13 September, 1997	9:00 – 18:00
World Solar Commission, First Session	14 September, 1996	9:30 – 13:30
World Solar Summit	16–17 September, 1996	9:30 – 18:00
World Solar Commission: Post Summit Consultations	18 September, 1996	9:30 – 12:00

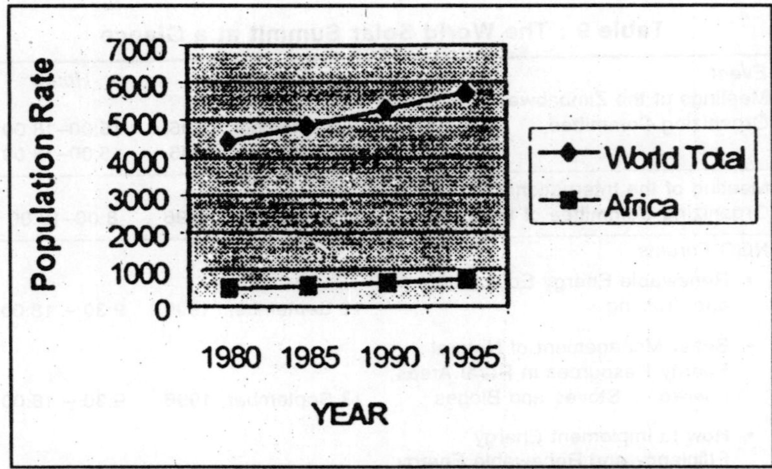

Source: 1995 Demographic Yearbook, United Nations, NY, 1997

Fig. 1 : Population Rate in Africa and the World between 1980–1995 (in millions)

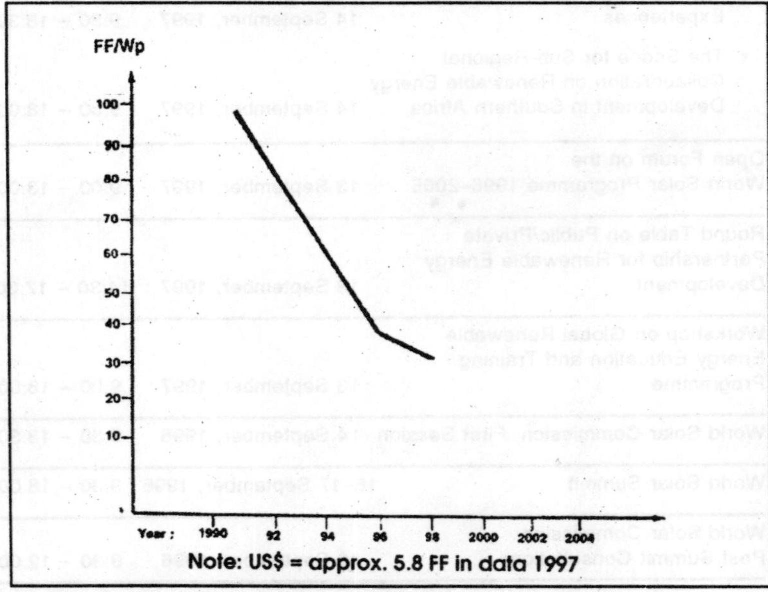

Source: Prof. S. Chabot, ADEME, UNESCO Summer: Sobaco, 1997

Fig. 2 : Average Cost for 3Kp PV Roof System in French Francs/Watt Peak

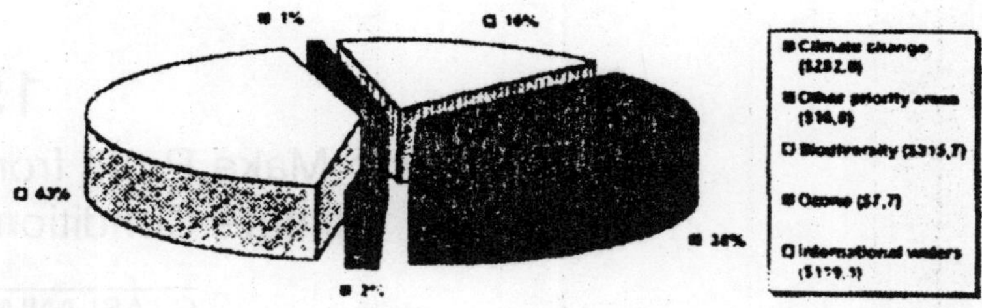

Fig. 3 : Project Distribution by Priority Area for the Year 1993
(Legend figures are given in millions of US$)

15

How to Make Profit from Climate Conditions

G. ASLANIAN
Russian Federation

1. KYOTO PROTOCOL COMMITMENTS

The Kyoto Protocol to the United Nations Framework Convention on Climate Change (UNFCCC) was formally adopted by the third session of the Conference of Parties (COP-3) on 11 December, 1997. The protocol establishes a legally binding obligation on 35 Annex I countries (24 OECD countries plus 11 countries in transition from the former Soviet Union and Eastern Block) to cut total emission of green house gases (GHC) on average by 5.2 per cent below 1990 levels by the years 2008 to 2012. There are no obligations on developing countries under this Protocol.

There is a complex array of alternative approaches that can be adopted by individual countries to achieve their particular emission targets. These policy decisions could require differing contribution by the individual industry sectors as well as individual sources.

The emission reduction targets accepted by developed countries are challenging when compared to projected economic growth and increased energy demand in most of the countries. In reality for the USA the "pain threshold' is about 12 to 15 per cent above 1990 levels. The latest official projection of CO_2 emissions from energy sector is 34 per cent increase by 2010 over 1990 levels. Thus, the USA target of 7 per cent cut require a reduction of 41 per cent within a little more than 10 years provided that energy is developed in "business-as-usual" case and no climate actions are implemented. In Canada with the "pain threshold" about 10 per cent the required reduction of 17-20 per cent should be provided by 2010 (Table 1).

Table 1 : Kyoto Reduction Targets of GHG Emissions for Period 2008 to 2012 over their 1990 Level

Selected Annex B Countries	*Kyoto Reduction Targets of GHG Emissions for the period 2008 to 2012 over their 1990 level, %*	*Real Required Reduction of GHG Emissions by 2010 over their 1990 level in "BAU" case, %*
EU	–8	–16
Canada	–6	–17–20
Russia	0	+8
USA	–7	–41
Japan	–6	–17

The politics of selling such a commitment is difficult in any case, but it is even more difficult when the US commitment is compared to the EU commitment. The EU's 8 per cent cut represents a 16 per cent reduction from the 8 per cent increase officially projected for the EU by 2010. Japan faces the same difficulties: for it the likely "pain threshold" is at least 11 per cent above 1990 levels. Thus the total cut of 17 per cent is required.

To what extent the Kyoto targets can be realistically achieved? What should be done? What cost will be burden by the OECD economies and how to minimize the expenditures? And how environmental commitments should be balanced with economic priorities? A reasonable solution for this problem should be found.

2. ECONOMIC COSTS OF EMISSION ABATEMENT

Current proposals for near-term and middle-terms emissions reductions in the developed countries, which imply curbs on fossil fuel-based energy use, would result in large costs that inhibit economic growth now and would negatively affect trade, investment, competitiveness, employment and life-styles for both developed and developing countries, in all individual nations and regions. Fig. 1 presents estimates of GDP and unemployment losses if 20 per cent reduction and stabilization in CO_2 to 1990 level by 2010 will be committed. As it is seen from Fig. 1 countries more heavily reliant on fossil fuels as Canada, are estimated to experience the largest losses in GDP.

ABARE's (Australian Bureau of Agricultural and Resource Economics) assessment of the economic consequences of reducing CO_2 emission stabilization in OECD countries is pro-

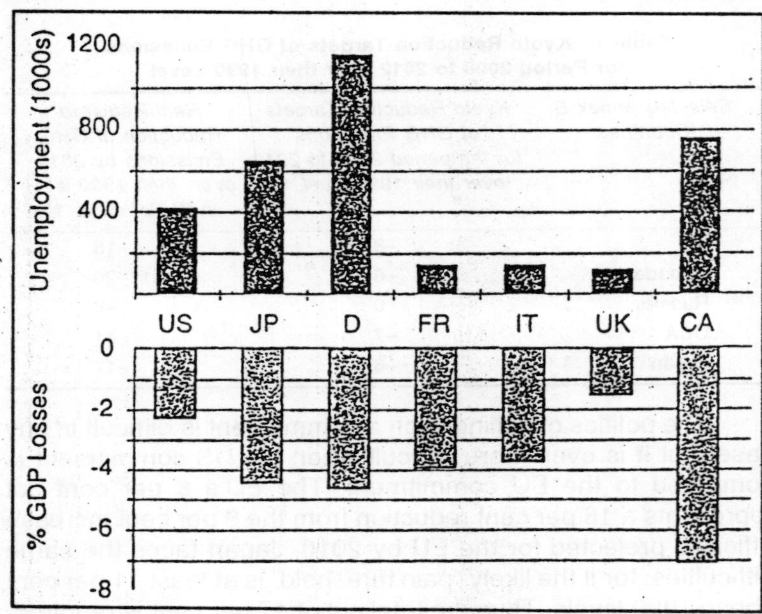

Fig. 1

jected to reduce welfare in all abating countries. Net present value of economic costs due to emission reduction relative to business-as-usual case in selected OECD countries for the period of 2000-2020 is presented in Fig. 2 These costs are

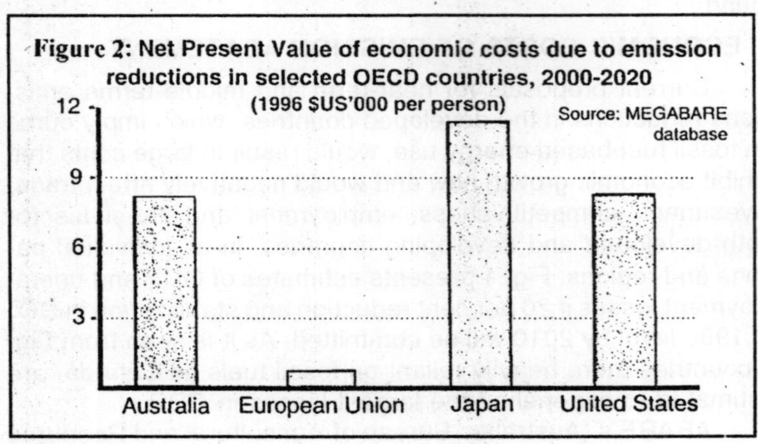

Figure 2: Net Present Value of economic costs due to emission reductions in selected OECD countries, 2000-2020 (1996 $US'000 per person)

presented in terms of net present value per person, USD/Capita; Japan-12000, USA-8000, EU-600. As it is seen from Fig. 2 economic costs are projected to vary significantly among OECD countries: for an average Japanese that is over 20 times that experienced by an average European, and 7 times that experienced by American. For an average Australian the equivalent numbers are 15 times and 5 times.

The costs of the undertaking emission reduction will depend mainly on three factors:

1. the size of the emission reduction required relative to BAS case;
2. the ease of substitutability between fuels;
3. the role of fossil fuels in the trade structure of a particular economy.

Taking into account the estimated welfare losses the benefit of these proposals are highly uncertain and would not be realized for years. It is also unlikely that over the next few decades fuel use will be phased out in the developed world—as this would be necessary to achieve some proposals presented to the Climate Change-Secretariat.

In spite of taken economic costs the effectiveness of abatement measures conducted in the developed country policy is minimal on global emissions level, largely because projected rapid emissions growth in developing countries remains unconstrained. Moreover stabilization policies in OECD countries may lead to a reduction in welfare in some developing countries. Global cooperation is required to ensure an effective and cost minimising strategy to combat global warming. This includes participation of developing countries.

It is clear that many developing nations and countries in transition are not likely to constrain their economic growth to mitigate a potential but unproven problem largely attributed to the developed world. It is obvious that economic and social development and poverty eradication are the first and overriding priorities of the developing countries.

Although developing countries are unlikely to be required to undertake any new emissions abatement commitments, emissions abatement in developed countries would still impact significantly on the welfare of developing countries because OECD emissions abatement will have an impact on the world trade. Stabilisation policies in OECD countries are estimated to lead to an increase in total production for all Rest of World regions relative to business-as-usual scenario. When OECD impose measures to reduce carbon dioxide emissions, Rest of World regions experience an increase in demand for there products because emissions generated in ROW countries during the production of goods and services are neither taxed nor restricted. Hence, production in Rest of World countries is stimulated by OECD country emissions abatement.

It would seem that ROW countries could be expected to gain from OECD stabilisation: given rising economic activity, real incomes could also be expected to increase. However, the available study results project that the ROW regions as a whole would experience a welfare loss, even though they do not abate emissions, because the terms of trade would decline for all Rest of world regions taken together.

In identification of the range of the recommended climate protection measures the priorities must be given to proved and economically attractive no or low regret options, such as economically, justified energy efficiency improvements and carbon sequestration actions. According to available assessments only 29-37 per cent of primary energy ends up as useful energy in the OECD countries (Inter-governmental Panel on Climate Change-29% WEC/IIASA-37%). This fact provides a good ground for energy saving measures implementation, especially taking into account that about '10-20 per cent of energy conservation potential in the OECD are considered as cost effective.

3. WHERE FLEXIBILITY AND WHEN FLEXIBILITY

Analyses show that significant "savings" can be made through flexible collaborative arrangements between countries, by which abatement measures can made less expensively in other locations—referred to as a "*where flexibility*" alternative. For example, many OECD countries could achieve far greater emission reduction in many EIT that they could for the same cost in their home countries—this essentially is the concept behind joint implementation. Looking purely at the economics, it is clearly the way to proceed, but the political and commercial dimension behind this strategy is an extremely complicated issue which should be discussed in the Como forums. Some experts indicated that the "*where flexibility*" alternative could decrease cost to developed countries by 70 per cent from the estimated 2-8 trillion USD cost of stabilizing greenhouse gas emissions by 2000.

Considerable savings can also be achieved through "*when flexibility*"—alternative—allowing flexibility in the timing of policy implementation. The studies show that policy response costs would be far lower and potentially as effective if emissions restrictions where phased in more slowly over time. Allowing reductions to be taken in step with planned turnover of capital stock) rather than to bear the costs of replacing equipment prematurely) and allowing for efficient technology to become available could decrease costs to OECD countries by about 40 per cent.

It is interesting to note that by combining "*where flexibility*" and "*when flexibility*", the economic cost of stabilising greenhouse gas emissions at 1990 levels by 2000 could be reduced by 90 per cent that is saving to the industrialised world of between 2-7 trillion USD. A possible way to combine "*where flexibility*" and "*when flexibility*" is through Activities Implemented Jointly.

4. EMISSION TRADING AND JOINT IMPLEMENTATION

An emission trading mechanism formulated in Article 17 is included in the Kyoto Protocol. Under this mechanism Annex B Parties will have GHG emission quota, either equal to the commitments levels under the Protocol; or the percentage of this. Parties that do no fully use their quota during the first commitment period can sell their surplus to Parties that have exceed the quota.

The concept of "*Joint Implementation*" (JI) is defined in Article 6 of the Protocol, which deals with cooperation among Annex I parties. Contrary to emission trading "*Joint Implementation*" is used to describe a wide range of possible arrangements between interest in two or more countries, leading to the implementation of co-operative development, projects that seeks to reduce or sequester greenhouse gas emissions. JI is based upon classical economic principles—measures to abate greenhouse gases should be taken where they are the cheapest. This means that, for any given reduction target, the most cost-effective alternative to achieve target should be implemented. In this case, countries with high reduction targets may invest in abatement projects in the countries where the returns per unit of investment will be higher than at home. Good prospects for the JIs on energy among the developed countries and EIT are of high interest. We will consider the ways, means and incentives to initiate Activities Implemented Jointly by OECD countries and NIS.

There is a evidence that among the current projects in the EIT the cheapest investment (in terms of USD per t carbon abated) can be found in Russia. The RUSAGAS Fugitive Gas capture project in Palasovka aims at replacing leaking valves on gas pipelines by new ones, thereby reducing the leakage of methane during the transportation of gas. The specific investment cost is less that 1 cent per t of curbed carbon.

5. RUSSIA'S PECULIARITIES

Let us now focus attention at the specific features of the Russian economic situation which in many aspects similar to situation in the other countries of the CIS. First of all we should notice that Russia as the currently second largest emitter and its economy is characterised by higher carbon intensity of unit of GDP, in tons of CO_2 per 1000 USD: Russia - 2.5, USA - 0.85, OECD- 0.597, EU- 0.442) and therefore by high energy intensity. According to official date the technical energy efficiency potential in Russia amounts to 500 million t c.e., which corresponds to carbon emission decreasing potential of 265 Mt, one third of which is in fuel and power complex and industry sectors, 33 per cent, in services and agriculture, 22 per cent—in residential sector and the remaining—in transport.

It is an interesting fact that cost effective energy conservation potential is about 300 Mt c.e., among which about 90 Mt c.e. could be saved at the cost less than 2USD/t c.e., and twice more enerty bulk could be saved at the cost of 20 USD/t.c.e. In this respect it should be noticed that this part of low cost energy efficiency measures practically has been implemented in the OECD economies.

There is a clear perception that ongoing economic restructuring in Russia and in countries with economies in transition (EIT) provides a unique opportunity to foster investments towards the most energy efficient and climate-friendly technologies and practices. While these options are recognised as potential win-win solution for local economies and environment, capital constraints and other barriers have hindered their broad implementation.

Another current Russian specific factor affecting the higher GHG emission results from the structure of economy targeted at production of essential volumes of energy intensive export oriented commodities. The Russian's export is mainly presented by ferrous and non-ferrous metals, fertilisers, chemicals, natural gas, oil and oil-based products. The production and transportation of these products are extremely energy intensive and therefore accompanied by high carbon emission (Table 2).

Table 2 : Carbon Emission in Russia Resulted from of Energy Intensive Export Production

Product	*Export volume, Mt*	*Specific integral emissions, t/t*	*Carbon emission, Mt*
Ferrous metals	20.0	1.13	22.6
Aluminium	2.3	7.5	17.3
Natural gas, billion c.m	196.5	0.08	15.7
Fertilisers	13.3	1.04	13.8
Oil	122.6	0.06	7.4
Oil products	55.5	0.13	7.2
Ammonia	3.4	1.17	4.0
Synthetic Rubber	0.21	8.32	1.75
Coal	28.0	0.06	1.68
Nickel	0.11	12.9	1.45
Others			17.00
Total			110.00

According to available data the total carbon emission linked with export production (in term of money export amount to 50 billion USD) is higher than 110 Mt. which is equal to 25 per cent of total country's emissions. This means that for each dollar income on average about 2 Kg of carbon are emitted into atmosphere. If we assume that tax for a tonne of carbon emission is 10 USD, then payment for emission of 110 Mt of carbon from export products exceeds 1 billion USD which amount 2 per cent of country total foreign currency income.

The question is how this expenditures should be compensated and what part of it finally will come to exporting country.

The real source to cover this expenditures is clear-carbon emission tax. But how will this source be used? There is a sense to propose the targeted allocation of these revenues for Joint Implementation of energy efficiency projects in sectors producing export commodities. This option is more attractive compared to a "hot air trading" system.

Another reason of emission reduction is strong economic decline taking place in Russia and other NIS since the disintegration of the Soviet Union in 1990 as a result: in 1996 the Russian GDP was only 59 per cent of 1990 level. Hence a permanent carbon emission mitigation induced by the economic decline is having place (Fig. 3) resulting in carbon emission reduction of 162 tons below the 1990 level. By forecast for the period 1990-2010 cumulative reduction of carbon emission compared to the 1990 levels would be about two billion tons. Meanwhile according to Kyoto Protocol Russia is committed to return in 2008-2012 to the 1990 level.

Thus Russia as well as the other NIS have "carbon emission credit" which they could use for trading as formulated in Article 17 of the Kyoto Protocol. Under this mechanism Annex B Parties will have GHG emissions Quota, either equal to their commitment levels under the Protocol or a percentage of this.

With respect of quota trading mechanism a speculation issue called "hot air trading" is rising since beginning of 1997. Many countries fear that the given option will result in a capital transfer only without achieving any additional emission abatement. It should be agreed that this concern has a real basis. To

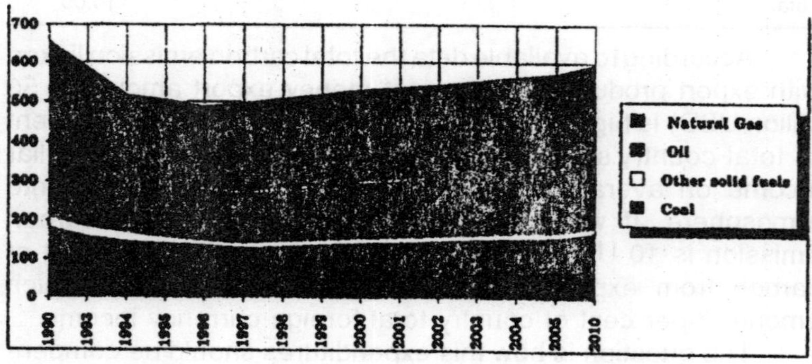

Fig. 3

avoid this speculation and use emission trading as "win-win" mechanism for achieving global benefit the carbon income should be primarily allocated for JI on the project base. The most attractive option to use carbon tax is not as a grant but rather for financing the cost effective energy efficiency projects necessarily on the pay-back ground.

The use of carbon tax for providing incentives for implementation of marginally effective energy efficiency projects is another attractive option. Such incentives could be in form of low interest rate, cover the cost of project's feasibility studies, or be used to provide necessary guaranties. All details of such approaches should be carefully analysed by the joint efforts at the international level.

To provide a legal base for large scale JI actions there is a sense to develop and adopt a legislative document similar to production sharing agreement aimed at the joint exploration of carbon resources in the atmosphere contrary to fossil fuel resources of the Earth. Practically we are proposing a new approach of benefit sharing which combining the options on quota trading and Joint Implementation could create a promising mechanism for climate protection and making profit by both partners involved.

16

Outlook of Fuel and Energy Complex Development in the Commonwealth of Independent States and International Energy Security

E. NADEZHDIN
Switzerland

It is my pleasure to be with you today for the Forum energy security. We are deeply grateful to the government of Italy's invitation and, especially to the UNESCO International School of Science for Peace and Landau Network-Centre Volta for hosting this event.

I believe that this Forum will make a significant contribution to the problems we are going to discuss during these three days. In my presentation I would like to describe the Project "Present Situation and Prospects for the Energy Complexes in the Commonwealth of Independent States" implemented by the ECE Energy Division together with the experts of CIS countries and its interrelationship with International Energy Security.

But before I do it, allow me to convey to you the support and appreciation for the organisation of this Forum of our Executive Secretary Mr. Yves Berthelot, who has expressed his best wishes to this Forum for a successful outcome of our deliberations this week.

Let me start my presentation with consideration of a concept of energy security in general. Energy security is a multifaceted concept and therefore not easy to define. The dictionary defines security as protection against potential harmful incidents or detrimental developments. With this in mind, energy security can be defined as protection against internal and external factors that could potentially disrupt the continuous flow or availability of energy at reasonable costs, resulting in important negative consequences to the economy and people.

There are two time frames of relevance when discussing energy security issues, the short run and the longer term. The first involves issues pertaining to the potential short-term disruption of energy supplies associated with breakdowns, acts of terrorism, natural disasters, social unrest or political action. The second pertains to the longer term development of energy supplies to meet expected demand at reasonable costs. In other words, it concerns the potential for the development of disequilibria between energy demand and supply in the long run associated with inadequate investment, financial constraints or political action.

There is another dimension of particular interest of analysts of energy security questions. This is whether the focus of attention should be on "physical" flows or "economic" flows. In other words, should one concentrate only on assessing the risks and vulnerabilities to the "physical" interruption of energy supplies or should energy security be assessed more broadly in economic terms.

Obviously, the two approaches are interdependent to some degree. The reason sometimes given for restricting oneself to the assessment of physical flows is the difficulty involved in defining what is meant by a reasonable price level and the difficulty in assessing and quantifying the economic consequences or costs associated with energy security issues. Thus, in order to simplify our task we will restrict our future analysis to physical flows of the long term development of energy supply and demand.

Now I would like to draw your attention to the situation in the Commonwealth Independent States which will play an important role in the realisation of an international Energy security concept. The following analysis will be based on the results achieved in the course of the implementation of the above mentioned Project "Present Situation and Prospects for the Energy Complexes in the Commonwealth of Independent States". The master report of the Project was put together from national reports submitted by the experts and the data base at the Russian Academy of Sciences' Energy Research Institute,.

Since the collapse of the Soviet Union and the emergence in its place of independent States, the fuel-energy ties binding the CIS and Baltic countries to Russia and, to some extent. Turkmenistan, have remained largely intact despite steps by a number of CIS States to develop their own energy bases, in part in order to break into the world energy markets and diversify their imports of fuel and energy resources.

The continuous decline in production and output of basic fuel-energy resources in most CIS countries over the past few years is becoming threatening. This is particularly clear in view of current trends and moves to stabilise and revive the economies in CIS countries, for growth could lead to a situation where a shortage of fuel-energy resources became a brake on social and economic progress.

Already today, while gross domestic product continues to fall (Table 1 and 2) along with overall industrial and agricultural output in the CIS countries (though the rate of decline slowed markedly in 1995), meeting the fuel and energy needs of the economy and population is becoming a strain. And this when the depth and rate of the decline in overall output in industry and agriculture have been significantly greater than in the fuel energy complex and the economy as a whole.

According to expert estimates, the aggregate gross national product of the CIS countries declined by 50 per cent between 1990 and 1995 (see Table 1), and industrial and agricultural output declined by 43 per cent and 30 per cent respectively. It should be noted that by 1995 per capita incomes in the CIS countries had fallen on average by 51 per cent (see Table 2).

Over the same period, output of oil (including condensates) in the CIS countries fell from 571 million to 356 million tons (down 42% over five years), that of natural gas fell from 815 billion to 701 m^3 (down 38%) coal production fell from 703 million to 426 million tons (down 38%) and output of basic petroleum products (petrol, diesel and boiler fuel) from 317 million to 190 million tons (down by 4.2%), while electricity production fell from 1,673 billion to 1,263 billion kWh (down by 24.5%).

Per capita consumption of primary energy products over the same period fell in the CIS countries by 28 per cent and per capita consumption of electricity by 24 per cent (see Fig. 1).

Difficulties with energy supplies in virtually all CIS countries were largely the result of a substantial increase over recent years in their already high specific energy consumption per unit of

Table 1 : GDP's of the CIS Countries at 1990 Prices

(Billions of Roubles)

	1990 ct.	*elative o 1990*	*1991 ct.*	*elative o 1990*	*1992 ct.*	*elative o 1990*	*1993 ct.*	*elative 0 1990*	*1994 ct.*	*elative o 1990*	*1995 ct.*	*elative o 1990*
Azerbaijani Reuplic	14.7	1	14.6	0.99	11.3	0.77	8.7	0.59	6.8	0.46	5.6	0.38
Republic of Armenia	9.7	1	8.8	0.91	4.2	0.44	3.6	0.37	3.8	0.39	4.1	0.42
Republic of Georgia	16.2	1	12.9	0.80	7.7	0.48	4.7	0.29	3.3	0.20	4.1	0.25
Republic of Belarus	42.7	1	42.2	0.99	38.2	0.89	34.1	0.80	28.7	0.67	25.9	0.61
Republic of Kazakstan	46.4	1	40.9	0.88	35.6	0.77	31.0	0.67	23.1	0.50	21.1	0.45
Kyrgyz Republic	8.5	1	7.8	0.92	6.7	0.79	5.7	0.67	4.6	0.54	4.3	0.50
Republic of Moldova	12.7	1	10.5	0.83	7.4	0.59	7.3	0.58	5.1	0.40	4.9	0.39
Russia	644.0	1	611.8	0.95	522.9	0.81	477.8	0.74	417.3	0.65	400.6	0.62
Republic of Tajikistan	7.3	1	6.7	0.91	5.0	0.69	4.2	0.58	3.7	0.51	3.2	0.44
Turkmenistan	7.6	1	6.8	0.89	5.8	0.76	5.4	0.71	5.0	0.66	4.7	0.62
Republic of Uzbekistan	32.4	1	32.2	1.00	28.7	0.89	28.0	0.86	26.8	0.83	26.6	0.82
Ukraine	164.8	1	145.7	0.88	125.7	0.76	107.9	0.66	83.1	0.50	72.1	0.44
Total	1007.0	1	941.0	0.93	799.4	0.79	718.5	0.71	611.2	0.61	577.1	0.57

Table 2 : Per Capita GDP (roubles, at 1990 prices)

	1990 ct.	*elative o 1990*	*1991 ct.*	*elative o 1990*	*1992 ct.*	*elative o 1990*	*1993 ct.*	*elative o 1990*	*1994 ct.*	*elative o 1990*	*1995 ct.*	*elative o 1990*
Azerbaijani Reuplic	2041	1.00	1999	0.98	1533.0	0.75	1169.0	0.57	904.8	0.44	744.7	0.36
Republic of Armenia	2691	1.00	2454	0.91	1139.3	0.42	965.5	0.36	1013.0	0.38	1090.4	0.41
Republic of Georgia	2967	1.00	2370	0.80	1415.3	0.48	862.4	0.29	605.9	0.20	760.7	0.26
Republic of Belarus	4164	1.00	4107	0.99	3688.0	0.89	3290.0	0.79	2778.0	0.67	2505.0	0.60
Republic of Kazakstan	2780	1.00	2359	0.85	2005.0	0.72	1572.0	0.57	1128.0	0.41	1274.4	0.46
Kyrgyz Republic	1948	1.00	1759	0.90	1444.0	0.74	1222.0	0.63	905.0	0.46	939.7	0.48
Republic of Moldova	2906	1.00	2403	0.83	1704.0	0.59	1687.0	0.58	1162.0	0.40	1138.3	0.39
Russia	4336	1.00	3787	0.87	3067.0	0.71	2705.0	0.62	2294.0	0.53	2704.0	0.62
Republic of Tajikistan	1361	1.00	1196	0.88	903.1	0.66	736.8	0.54	639.0	0.47	544.2	0.40
Turkmenistan	2048	1.00	1784	0.87	1364.7	0.67	1238.0	0.60	1123.0	0.55	1028.4	0.50
Republic of Uzbekistan	1565	1.00	1520	0.97	1321.4	0.84	1259.0	0.80	1187.0	0.76	1155.1	0.74
Ukraine	3181	1.00	2801	0.88	2408.9	0.76	2071.0	0.65	1606.0	0.50	1405.5	0.44
Total	3570	1.00	3317	0.93	2803.3	0.79	2516.0	0.70	2142.0	0.60	2022.9	0.57

economic output combined with wasteful uses of energy resources (fig.2.) At present, CIS countries flare over 6 billion m^3 of associated gas, 20 per cent of this resources, every year.

The collapse of the USSR left the energy-exporting CIS countries facing a radically new transport and communications situation.

Russia, for example lost a series of ports and terminals on the Baltic and Black Seas and large segments of many pipelines intended for exports to Europe. Over 95 per cent of the natural gas exported to the West transits over trunk pipelines running through Ukraine, and 3 per cent through Belarus.

The situation facing other CIS energy exporters(Turkmenistan, Azerbaijan and Kazakhastan) is even more complicated. Their dependence on the intermediate CIS counties through which the main routes for their energy exports to Eastern, Central and Western Europe run significantly reduces their returns on such exports.

As for Russia, its transport infrastructure carries natural gas from Turkmenistan and oil from Kazakhastan to consumers in Ukraine and the Transcaucasian countries and to Central, Eastern and Western Europe.

CURRENT STATUS AND DEVELOPMENT PROSPECTS OF INDUSTRIES IN THE FUEL ENERGY COMPLEXES OF THE CIS COUNTRIES

The Oil Industry

Oil output throughout the CIS in 1995 totaled 356 million tons. This is 3.5 per cent less than in 1994 (361.3 million tons, itself 39% less than in 1990). Oil and condensate output in Russia, the largest producer in the CIS, continues to decline, having fallen from 516 million tons in 1990 to 306 million tons in 1995 (down 43% in five years)—see Table 3. A similar picture is presented by other Commonwealth States, Uzbekistan apart. The decline is only partly explained by a shortfall in investment for replacing, rebuilding and developing extraction equipment. The main reason is the influence of new market conditions on the workings of the petroleum industry. The freeing of petroleum prices in mid-1994 and the continuing decline in the economy have led to several crises over raw material supplies (the most acute occurred in 1994-1995).

Table 3 : Oil and Gas Condensate Balances of CIS Countries (mt)

	1990	*1993*	*1995*	*2000-min*	*2000-max*	*2010-min*	*2010-max*
PRODUCTION*							
Azerbaijan	13	11	9	9	14	31	46
Belarus	2	2	2	2	2	1	1
Georgia	0	0	0	1	3	3	3
Kazakstan	26	23	21	24	30	35	55
Kyrgyzstan	0	0	0	0	0	1	1
Russia	516	354	307	284	297	290	360
Turkmenistan	6	5	5	4	5	8	11
Uzbekistan	3	4	8	8	9	10	10
Ukraine	5	4	4	4	6	4	8
CIS	571	403	356	336	365	383	494
EXPORTS*							
Azerbaijan	0	0	0	0	5	17	30
Kazakstan	22	13	9	15	8	25	21
Russia	220	123	118	109	104	111	145
Turkmenistan	0	0	1	0	0	1	3
Uzbekistan	0	0	1	0	0	0	0
CIS	243	136	129	124	118	154	199
IMPORTS							
Azerbaijan	4	1	1	0	0	0	0
Belarus	37	18	11	13	15	20	35
Georgia	2	0	0	1	1	2	2
Kazakstan	14	11	7	6	6	6	6
Russia	19	11	5	5	6	10	13
Turkmenistan	0	1	0	0	0	0	0
Uzbekistan	5	4	0	0	0	0	0
Ukraine	54	19	15	17	36	20	38
CIS	136	64	39	42	64	58	94
CONSUMPTION							
Azerbaijan	16	12	10	9	9	14	16
Belarus	39	20	13	15	17	21	36
Georgia	2	0	0	2	3	5	5
Kazakstan	18	12	19	16	28	16	40
Russia	315	242	194	180	199	189	228
Turkmenistan	6	6	4	4	5	7	8
Uzbekistan	8	7	7	8	8	10	10
Ukraine	59	23	19	21	42	24	46
CIS	463	321	266	254	310	286	390

**Including by foreign companies*

According to expert estimates, aggregate output of petroleum and gas condensate in CIS countries in the year 2010 will range between 383 million and 494 millions tons—27 to 136 millions tons more than in 1995. Russia, after a fairly protracted decline in oil output which will probably continue to the year 2000, expects to produce not less than 290 million tons in the year 2010. Maximum possible output in 2010 is put at 360 million tons. Two other major producers, Azerbaijan and Kazakhistan, hope by 2010 to have raised their oil production substantially, to 31-46 and 25-55 million tons respectively as against 9 million and 21 million tons in 1995. If this happens, Russia's share of total oil output will fall from 86 per cent in 1995 to 65-70 per cent in 2010.

This outlook for net oil exports from CIS countries is for continuing decline to a level of 70-76 million tons by the year 2000, followed by an increase that will restore the commonwealth's export potential to pre-crises levels in the former USSR (93-130 million tons per year by 2010). Exports of petroleum products to non-CIS countries will probably undergo less of a change than crude oil: from 46 million tons in 1995 they may reach 52 million tons by the year 2000.

The Gas Industry

Natural gas is expected to play a leading role in addressing energy and environmental problems in Russia and most other CIS countries. The cleanest form of fuel in environmental terms, and one that can be highly regulated and monitored at all stages of extraction, transport, and use, natural gas has long been established as an irreplaceable fuel on the CIS Energy market.

Natural gas output in CIS countries overall, and Russia in particular, continues to be relatively stable (Table 4), although there is a small downward trend. Output in Russia in 1995 was 595 billion m^3 (85% of total output in the CIS countries). Since 1990 it has fallen in Russia by 46 billion m^3 and throughout the CIS by 114 billion m^3 (in Turkmenistan, by 58 billion m^3).

In 1995, Russia exported around 194 billion m^3 (to the West and CIS countries), practically as much as in 1994 but 22 per cent, less than in 1990. In 1995 Turkmenistan exported 20 billion m^3. Total net exports of gas by the CIS countries in 1995 amounted to 104 billion, m^3, 3 per cent more than in 1994 but 18 per cent less than in 1990.

Table 4 : Natural and Associated Gas Balances of CIS Countries (bn m^3)

	1990	1993	1995	2000-min	2000-max	2010-min	2010-max
PRODUCTION*							
Azerbaijan	10	6	5	6	6	7	15
Kazakstan	7	7	5	6	18	17	31
Russia	641	618	595	670	675	750	840
Turkmenistan	88	65	30	57	80	92	130
Uzbekistan	41	45	49	50	59	52	56
Ukraine	28	19	17	18	22	23	30
CIS	815	761	701	807	860	941	1102
EXPORTS*							
Azerbaijan	0	0	0	0	0	1	3
Kazakstan	3	3	3	6	8	13	17
Russia	249	179	194	208	216	244	309
Turkmenistan	72	52	20	45	65	80	110
Uzbekistan	10	4	7	6	6	6	6
CIS	334	239	224	265	295	344	445
IMPORTS							
Azerbaijan	8	4	1	1	2	5	0
Armenia	5	1	0	1	2	3	5
Belarus	15	16	13	15	16	17	20
Georgia	5	3	3	2	2	2	3
Kazakstan	9	10	7	7	7	9	11
Kyrgyzstan	2	1	1	1	1	1	2
Moldova	5	3	3	3	3	3	3
Russia	70	4	23	12	14	18	20
Tajikistan	2	1	2	2	3	3	4
Turkmenistan	0	0	0	0	0	0	0
Uzbekistan	7	2	0	0	0	0	0
Ukraine	88	80	66	71	65	80	65
CIS	215	126	119	114	114	142	132
CONSUMPTION							
Azerbaijan	18	11	6	7	8	11	12
Armenia	5	1	0	1	2	3	5
Belarus	15	17	14	15	16	17	20
Georgia	5	3	3	2	2	2	3
Kazakstan	15	14	9	7	13	13	26
Kyrgyzstan	2	1	1	1	1	1	2
Moldova	4	3	3	3	3	3	4
Russia	461	442	424	474	468	524	551
Tajikistan	2	1	2	2	3	3	4
Turkmenistan	16	13	10	12	15	12	20
Uzbekistan	38	43	42	44	53	46	50
Ukraine	116	99	83	89	87	103	95
CIS	696	648	598	656	670	739	791

**Including by foreign companies.*

The existence of enormous reserves of natural gas in Russia and other CIS countries gives hope for the development of the industry. Over the coming 15 years, output in CIS will increase substantially, largely thanks to a rapid increase in output in Russia (from 595 billion m^3 in 1995 to 750-840 billion m^3 in 2010), Turkmenistan (from 35.6 billion m^3 in 1995 to 100-130 m^3 in 2010), Turkmenistan (from 35.6 billion m^3 in 1995 to 20-30 billion m^3 in 2010). Overall, gross output of natural gas in CIS will increase by roughly 34–57 per cent by the year 2010.

The Coal Industry

Coal output (gross output of all types) in 1995 continued to decline in all CIS countries except Uzbekistan. It was down by 40 per cent on 1990. There have been problems in supplying power stations designed to use particular grades of coal which used to be delivered from other republics. Ukraine is facing a coal deficit (both cooking coal and steam coal) because of the continuing fall in its output.

Despite the difficult situation in the coal industry, all CIS countries forecast stable growth in output after 1995. The fastest growth is expected in Kazakhistan, which expects to double production by the year 2010 to between 125 and 155 million tons. This will make it the second largest coal producer in the Commonwealth, overtaking Ukraine (105-110 million tons in 2010). Russia will be able to mine between 300 and 350 million tons of hard and brown coal by the end of this period, more than 5 per cent of total output in CIS. The scenarios studied indicate that the CIS countries will be able by the end of the forecast period to meet their coal requirements without resorting to imports from outside the Commonwealth Kazakhstan will continue to play a particularly important role in this regard, affording a solid foundation for the entire CIS coal market. At the same time, the commonwealth countries may revive their coal export potential if new markets for coal appear outside CIS (Table 5),

Electricity

Electricity consumption overall in the CIS countries will undergo stable growth: by 2010, compared with 1995, total electricity consumption will have risen by 430 billion kWh. Russia's share of total consumption will fall from 66 per cent in 1995 to 63 per cent in 2010, Ukraine, the second largest

Table 5 : Coal Balances of CIS Countries (mt)

	1990	*1993*	*1995*	*2000-min*	*2000-max*	*2010-min*	*2010-max*
PRODUCTION							
Georgia	1	0	0	0	0	0	1
Kazakstan	132	112	90	90	103	125	155
Kyrgyzstan	4	2	0	1	1	2	2
Russia	395	306	250	267	282	300	347
Tajikistan	0	0	0	1	1	1	1
Uzbekistan	6	4	3	3	4	5	8
Ukraine	165	116	83	87	95	105	110
CIS	703	540	426	448	486	538	624
CONSUMPTION							
Azerbaijan	0	0	0	0	0	0	0
Armenia	1	0	0	0	0	0	0
Belarus	3	1	1	1	2	2	2
Georgia	1	1	0	0	1	1	1
Kazakstan	90	79	63	75	71	115	114
Kyrgyzstan	5	3	2	2	3	4	5
Moldova	5	2	2	2	2	1	1
Russia	398	307	250	265	280	292	340
Tajikistan	1	1	1	1	1	1	2
Turkmenistan	1	1	1	1	1	1	1
Uzbekistan	9	5	3	3	4	5	8
Ukraine	163	119	88	93	96	104	103
CIS	675	519	411	444	459	526	577
DEFICIT/SURPLUS							
Azerbaijan	–0	–0	–0	–0	–0	–0	–0
Armenia	–1	–0	–0	–0	–0	–0	–0
Belarus	–3	–1	–1	–1	–2	–2	–2
Georgia	–0	–0	–0	–0	–0	–0	–0
Kazakstan	42	33	27	15	33	10	41
Kyrgyztan	–1	–1	–2	–2	–2	–2	–2
Moldova	–5	–2	–2	–2	–2	–1	–1
Russia	–3	–2	0	2	2	8	7
Tajikistan	–1	–0	–0	–0	–0	–0	–1
Turkmenistan	–1	–1	–1	–1	–1	–1	–1
Uzbekistan	–3	–1	0	0	0	0	0
Ukraine	2	–4	–5	–6	–1	1	7
CIS	28	21	16	5	27	12	47

consumer of electricity among the CIS countries, will consume 16 per cent in 2010 as against 15 per cent in 1995. The largest rises in electricity consumption (doubling in 15 years) are expected in Azerbaijan, Georgia and Turkmenistan, and under the maximum scenario, in Russia as well.

An increase in electricity production is expected in all CIS countries once economic difficulties have been overcome. Depending on the development scenario, production levels of between 1,680 and 1,870 billion kWh may be expected by the year 2010. It should be noted that under the likely scenario, production will only slightly exceed the levels attained in 1990 (Table 6).

Table 6 : Electricity Balances of CIS Countries (bn kWh)

	1990	*1993*	*1995*	*2000-min*	*2000-max*	*2010-min*	*2010-max*
PRODUCTION							
Azerbaijan	23	19	17	20	22	27	31
Armenia	10	7	5	10	10	12	13
Belarus	40	33	25	28	29	40	45
Georgia	14	10	6	6	7	12	15
Kazakstan	87	77	65	78	86	100	119
Kyrgyzstan	13	12	11	13	13	15	17
Moldova	16	10	6	9	9	8	11
Russia	1082	957	863	890	945	1115	1235
Tajikistan	18	18	14	18	19	20	23
Turkmenistan	15	13	9	14	17	21	27
Uzbekistan	56	49	48	47	50	57	60
Ukraine	299	230	195	230	240	256	270
CIS	1673	1435	1263	1362	1448	1683	1865
CONSUMPTION							
Azerbaijan	23	19	17	18	19	25	30
Armenia	10	6	5	10	11	12	14
Belarus	49	39	33	34	35	42	45
Georgia	17	11	6	10	11	18	20
Kazakstan	105	88	79	82	86	105	115
Kyrgyzstan	8	10	10	11	12	13	14
Moldova	13	10	9	9	9	9	10
Russia	1077	914	835	869	925	1090	1200
Tajikistan	19	17	15	18	19	20	21
Turkmenistan	10	13	11	14	17	19	27
Uzbekistan	53	49	48	46	48	53	58

Contd...

Table 6 : Electricity Balances of CIS Countries (bn kWh)

	1990	1993	1995	2000-min	2000-max	2010-min	2010-max
Ukraine	270	228	195	210	220	245	271
CIS	1654	1405	1262	1330	1412	1652	1824
DEFICIT/SURPLUS							
Azerbaijan	0	0	0	2	3	2	1
Armenia	0	1	0	–0	–1	–1	–1
Belarus	–10	–6	–8	–6	–6	–2	0
Georgia	–3	–2	0	–4	–4	–6	–6
Kazakstan	–17	–11	–14	–4	–0	–5	4
Kyrgyztan	5	2	1	2	1	2	3
Moldova	3	0	–3	0	0	–2	1
Russia	5	43	28	21	20	25	35
Tajikistan	–1	1	–1	–0	0	0	2
Turkmenistan	5	0	–3	0	0	2	0
Uzbekistan	3	0	0	1	2	4	2
Ukraine	28	2	0	20	20	11	–1
CIS	19	30	1	32	36	31	41

Primary Energy Balance

Total output of primary energy resources in the CIS countries was down by 470 million tons of standard fuel, or 26 per cent, in 1995 from 1990. Only in Uzbekistan did overall output of primary energy resources increase over the period, by 15 per cent. Overall output of energy resources in the CIS countries will not, under the likely scenario, have reached 1990 levels by the year 2010, chiefly because of the moderate growth in output expected in Russia. On the other hand, Kazakhistan is planning to double its output of energy resources over the coming 15 years, Turkmenistan to triple its output, putting it in second place, and Azerbaijan, almost to sextuple is its output (Table 7).

Tables 8 to 10 show the final energy balances for the CIS as a whole in 1995, 2000 and 2010, calculated using the International Energy Agency methodology (abbreviated version) for the maximum case scenario.

As you can see the primary energy balances of the CIS countries (Table 9 and 10) prepared on the basis of the official information which had been provided by the energy experts of these countries show that CIS as a group is going to be net

Table 7 : Primary Energy Balances of the CIS Countries (mt oil equivalent)

	1990	*1993*	*1995*	*2000-min*	*2000-max*	*2010-min*	*2010-max*
PRODUCTION*							
Azerbaijan	29	22	19	19	27	53	83
Armenia			1	2	2	2	2
Belarus	3	3	3	3	3	2	2
Georgia	2	1	1	2	4	6	6
Kazakstan	152	131	107	114	147	171	239
Kyrgyzstan	4	2	1	2	2	4	4
Moldova							
Russia	1855	1527	1382	1457	1498	1594	1844
Tajikistan	3	2	2	3	3	4	4
Turkmenistan	109	82	42	72	99	117	165
Uzbekistan	57	61	71	72	85	79	86
Ukraine	201	150	119	131	144	151	171
CIS	2415	1984	1748	1875	2008	2174	2589
CONSUMPTION							
Azerbaijan	45	29	21	21	22	33	37
Armenia	6	1	2	2	3	4	7
Belarus	76	48	36	39	44	51	76
Georgia	11	6	4	6	8	12	11
Kazakstan	115	96	88	91	111	131	178
Kyrgyzstan	7	5	3	4	3	7	7
Moldova	8	6	5	5	5	4	6
Russia	1363	1166	1025	1037	1097	1137	1295
Tajikistan	5	4	5	5	4	7	5
Turkmenistan	27	24	18	20	25	24	35
Uzbekistan	63	64	62	65	76	72	78
Ukraine	378	271	221	208	269	238	292
CIS	2104	1721	1488	1501	1661	1713	2011
NET EXPORT							
Azerbaijan	−15	−7	−2	−1	5	20	46
Armenia	−6	−1	−1		−1	−2	−5
Belarus	−72	−45	−33	−36	−41	−50	−75
Georgia	−9	−5	−4	−3	−3	−6	−5
Kazakstan	37	35	19	23	36	40	61
Kyrgyzstan	−3	−2	−2	−2	−1	−3	−2
Moldova	−8	−5	−5	−5	−5	−4	−5
Russia	491	361	358	421	400	457	549
Tajikistan	−2	−2	−2	−3	−1	−3	−2
Turkmenistan	82	58	24	51	74	93	130
Uzbekistan	−6	−3	9	7	9	7	8
Ukraine	−177	−122	−101	−78	−125	−87	−121
CIS	311	263	260	374	347	461	578

*Including by foreign companies

Table 8 : Summary Energy Balance of the CIS in 1995 (mt oil equivalent)

	Coal	*Oil*	*Petro-leum products*	*Gas*	*Hydro-power*	*Elec-tricity*	*Total*
Production	298	356		587	65		1306
Imports	1		3	2			6
Exports	−11	−90	−46	−94			−241
Consumption of primary energy sources	288	266	−43	495	65	0	1071
Power stations	−62	−5	−36	−169	−65	109	−228
Oil Refining		−261	245				−17
Other Consumption*		0	166	326	0	109	827

* Including losses during production and transport, stocks and reserves, and non-fuel and raw-material applications.

Table 9 : Summary Energy Balance of the CIS in 2000 (estimate) (mt oil equivalent)

	Coal	*Oil*	*Petro-leum products*	*Gas*	*Hydro-power*	*Elec-tricity*	*Total*
Production	340	365		774	76		1555
Imports	1		6	3			10
Exports	−19	−54	−52	−152		−3	−280
Consumption of primary energy sources	322	311	−46	625	76	−3	1286
Power stations	−78	−3	−28	−199	−76	125	−259
Oil Refining		−308	289				−19
Other Consumption*	244	0	215	426	0	121	1007

* Including losses during production and transport, stocks and reserves, and non-fuel and raw-material applications.

exporter in the years 2000 and 2010. But in order to fulfill these plans it is necessary to overcome significant problems with which fuel and energy complexes of CIS are facing now. The following major problems should be mentioned:

Obsolescence in business' fixed assets is rising catastrophically. Two thirds of the fixed assets in fuel and energy

Table 10 : Summary Energy Balance of the CIS in 2010 (estimate) (mt oil equivalent)

	Coal	*Oil*	*Petro-leum products*	*Gas*	*Hydro-power*	*Elec-tricity*	*Total*
Production	437	494		992	82		2004
Imports	0		2	0			2
Exports	–33	–105	–62	–281		–4	–485
Consumption of primary energy sources	404	390	–60	710	82	–4	1522
Power stations	–80	–2	–18	–306	–82	160	–327
Oil Refining		–388	363				–25
Other Consumption*	324	0	284	404	0	157	1169

*Including losses during production and transport, stocks and reserves, and non-fuel and raw-material applications.

industries in the CIS States have reached the end of their rated working lives. Major overhauls, modernisation and reconstruction are occurring only on an extremely small-scale, and lag far behind requirements. A number of CIS countries have an acute shortage of skilled power workers because many skilled people have emigrated.

In the coal industry the CIS countries, more than one half of all pits have been operating for over 30 years without reconstruction.

In the electricity industry, 40 per cent of the equipment needs requires replacing or upgrading.

Over 30 per cent of gas pipelines and 45 per cent of oil pipelines have been in operation for over 20 years.

For the most part, petroleum-product pipelines were built in the 1950s and 1960s; virtually none have been built in recent years.

Only 25 to 30 per cent of the electricity transmission lines and gas pipelines needed to maintain stable operation are being kept under repair. More than 30 per cent of gas compressor units have reached the end of their rated working lives. Many are operating very inefficiently, so large quantities of gas are consumed in transport.

Owing to the inadequate volume of new electricity generating capacity being brought on line to replace obsolete equipment, the breakdown rate at power stations is rising and unit fuel consumption for the generation of electricity and heat is rising.

An epidemic of non-payment on domestic markets and in intra-CIS trade has disastrously reduced receipts to industries in the various countries' fuel-energy complexes, sometimes pushing them into loss. This has led to a shortfall in investment and, consequently, to a still steeper fall in energy output. Even for countries with energy surpluses, therefore, not only the profitability but the survival of the energy sector is at stake; so too, therefore, is national security.

It is now clear that the CIS countries are to varying extents hostage to one another in the energy aspects of their national security and that without overcoming all these problems and reducing energy intensity of their national economies, this region could become a net importer at the beginning of the next century and considerable amount of primary energy which is delivered to Europe from this region now will be just interrupted.

Could the CIS countries settled these energy problems, themselves by using their own financial resources? Let us make use of estimations done in a World Energy Council Study "Financing the Global Energy Sector". The Study assumes investment requirements to remain at about the same level as in the past, at least untill 2020, that is 3-4 per cent of world GDP or 6-7 per cent of world capital formation. The average annual investments needs of the world energy economy until year 2020 is about US$ 530-660 bill.. The issue is to mobilize the finance under a given circumstance. Mobilizing finance is not a general issue in industrialized countries. It is an issue only if the risk/return ratio of a given energy investment project does not compare favourably with competing projects and if this handicap is not compensated for by public finance and government guarantees. How much domestic capital is available in the economies in transition (this group includes the CIS countries)? The answer depends on capital formation in this region. In accordance with WEC/IIASA study "Global energy Perspectives to 2050 and Beyond" an annual capital formation for economies in transition during 1990-2020 is and will be about 280 bill. $/year (GDP: 1400 bill. $/year and saving, % of GDP-20). Compared to other world regions, the energy systems of the CIS countries would absorb a high (in average during 1990-2020 from 90-140 bill. $ per year or 32-94% of capital formation) proportion of domestic capital, if they were allowed to do so. But it is not possible because of limited domestic capital markets and a necessity to invest capital to some other sectors of their national economies transport, industry, agriculture and etc.) As a result, only an estimated 9-13 per cent of long-term investment "needs" of US$ 12 bill. p.a. is presently financed of which 5 bill. $ is an international capital.

It is evident, that with such investments in the energy sector of the CIS countries they will not be able to overcome themselves of problems of this sector mentioned above. In conclusion I would like to stress that without considerable alterations in the investment policy of Western countries and International Financial institutions related to energy sector of CIS, these countries could become a net importer as a group at the beginning of the next century. This fact will reduce the energy security of Europe and the whole World to a great extent.

17

The Earth Atmosphere and the Aviation Problem and Research Goals

O. FAVORSKY AND A. STARIK
Russian Federation

The production activity of mankind has caused during the last decades a considerable change of the gas and aerosol composition of the atmosphere. One of the sources of direct influence upon the atmosphere is aviation. The absolute quantity of substances emitted into the atmosphere from aircraft engines is 40-50 times smaller than the pollution from surface sources (power generation, transport, industry, agriculture). But due to the fact that those emissions occur directly in the atmospheric layers most sensitive to any kind of disturbances (the upper troposphere and the lower stratosphere), the determination of the real influence of the aviation upon atmospheric processes and subsequently upon the Earth climate is gaining more and more importance. An almost double growth of the flight frequency of civil aircrafts predicted for the next 18-25 years and possible development of the second generation of supersonic commercial aircrafts require a more serious approach to the investigation of a possible influence of the aviation upon the atmosphere.

It should be noticed that already some years ago norms limiting the combustion products emissions and the noise of aircraft engines have been introduced with the aim to reduce the environmental influence of the aviation (especially in airport areas). There were restrictions (depending on the engine pressure ratio) for the content of CO, NOx and soot (smoke) in the exhaust flow reduced to the unit thrust.

Those norms were toughening gradually, and it is quite possible that in the next years another toughening of the emission requirements will take place which will concern in the first line the NOx (nitrogen oxides) content—the emission most difficult to control because of the constant increase of the engine cycle temperature (the turbine inlet temperature) as the main factor providing the improvement of the engine weight, dimension and fuel characteristics.

At the same time a comparatively fast (counted in years of development) improvement of the combustion efficiency up to 99 per cent and higher caused in the first line by the tough (as a rule) requirements to the fuel consumption allowed to considerably reduce the actual emissions of the main products of incomplete combustion— CO and soot. The emission of soot was usually estimated according to the actual visual smoke content in the exhaust. But these were practically the first steps in the environmental control of the aviation.

The recent investigations have shown that the actual range of substances emitted into the atmosphere from jet engines is not only much broader than the one included in the ICAO norms, but it also depends on the engine type, on its performance, dimensions and even on the actual composition of the fuel used (the common name-aircraft kerosene).

Requirements to the aircraft kerosene have been existing for a long time, they are stipulated in corresponding standards. But no deep investigations of the influence of, for example, the actual content of different hydrocarbons groups in fuels upon the emissions have been practically required yet. The introduced restrictions on the sulphur S content are also very relative. It is quite possible—and this is one of the goals of the present work that in the next years it will be necessary to perform a number of appropriate investigations to make sure that the modern requirements to aircraft fuels are sufficient or that any new requirements are necessary.

The problem of analysis of the aviation influence upon the atmospheric processes and the climate has been raised relatively recently and was initiated in the early 70s by the first estimations of the possible influence of supersonic passenger planes upon the ozone layer (Johnston H., Crutzen P.).

The influence of aviation upon the atmosphere manifests itself through a complex of interconnected processes. It is well known that emissions of substances like CO_2, N_2O, NO, CO, SO_2 and soot particles can have a considerable influence upon the dynamics of chemical processes in the atmosphere and its radiation balance. But not only a considerable number of other substances can also turn out to be important, the place of their introduction into the atmosphere can be important as well. For example, an increase of NOx ($NO + NO_2$) concentration in the stratosphere caused by the flights of supersonic aircrafts should lead to a decrease of ozone (O_3) concentration and consequently to an increase of biologically dangerous ultraviolet at the Earth surface. But the same time a deeper penetration of the solar ultraviolet would cause an increase of ozone concentration at low altitudes and thus change its radiation balance and temperature. Recent calculations have shown that the radiation balance of the troposphere is thirty times more sensitive to the NOx emissions of the stratospheric aircrafts compared to the surface sources (Peter T., Bekki S., Weisenstein D.).

Complex research of a possible influence of supersonic passenger aircraft flights upon the ozone layer were conducted in the 70s mainly in the frame of the international programs CIAP (USA), COMESA and COVOS (Europe). An example of conclusions made on the base of this research is the fact that regular flights of 100 supersonic Concorde planes at the altitude of 17 km will cause a global reduction of O_3 concentration by 0.25 per cent a year.

However, those conclusions have been later revised many times, and not only because of new information about the speed constants of physical and chemical processes and because two dimensional photochemical and even three-dimensional general circulation models were used for practical estimations instead of simple one-dimensional models, but as well because of deeper penetration into the physical and chemical processes taking place in the atmosphere and in the jets of aircraft engines.

Besides the results of numerical analysis rich experimental data have been collected in the last year allowing a more complex approach to the problem of aviation influence upon the atmospheric processes. For example, it was first of all discovered that an exceptionally important role was played by the heterogeneous reactions taking place in polar stratospheric clouds and on the surface of sulphate aerosols. It was also demonstrated that the gas composition of the products leaving the combustion chambers undergoes considerable changes both in the downstream engine flowpath and in the exhaust jet. The jet itself can serve as an additional source of aerosol particles with a complicated chemical composition. However many aspects of this difficult problem are still unclear.

The strategy and the main sections of the investigations of the aviation influence upon the atmosphere can be illustrated by the scheme shown in Fig.1

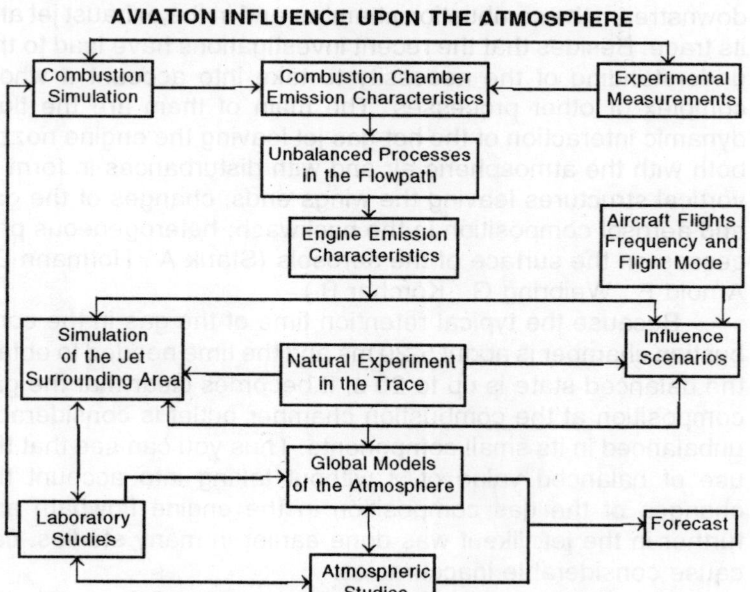

Fig. 1

An essential place in the problem under consideration is taken by the matters concerning the determination of detailed multifactor emission characteristics of engines and changes of the component and phase composition of the exhaust gas in the backwash where the mixture leaving the nozzle is being cooled during mixing with the atmospheric air.

The model calculations made for estimation of the aviation influence upon the atmosphere were earlier based on the assumption that the combustion products emission from jet engines into the atmosphere corresponds to the thermodynamically balanced gas composition at the combustion chamber outlet (Only NO, NO_2, H_2O, CO_2, CO, OH, unburned hydrocarbons and soot were taken into account). Their uniform distribution in a certain atmospheric corridor was assumed, the height of this corridor above the surface of the Earth was determined by the trajectory of a cruise flight. This approach did not take into account for all the actual changes of the gas composition both downstream the combustion chamber and in the exhaust jet and its trace. Besides that the recent investigations have lead to the understanding of the necessity to take into account a whole complex of other processes. The main of them are the fluid dynamic interaction of the hot gas jet leaving the engine nozzle both with the atmospheric air and with disturbances in form of vortical structures leaving the wings ends; changes of the gas and aerosol composition in the backwash; heterogeneous processes on the surface of the aerosols (Starik A., Hofmann D., Arnold F., Weibring G., Körcher B.).

Because the typical retention time of the gas in the combustion chamber is about 5-20 ms and the time needed to obtain the balanced state is up to 20 s, it becomes clear that the gas composition at the combustion chamber outlet is considerably unbalanced in its small components. Thus you can see that the use of balanced value of γi without taking into account the changes of the gas composition in the engine flowpath and further in the jet, like it was done earlier in many studies, can cause considerable inaccuracies.

Due to the chemically unbalanced state of the gas in the outlet cross-section of the nozzle it contains not only the above mentioned H_2O, CO_2, CO, NO and NO_2, but also considerable quantities of HNO_2, N_2O, HNO_3, NO_3 etc. which turned out to be important. The concentration of some of those substances changes considerably along the flowpath. For example, the concentration of NO_3 increases 40 times.

During the combustion of conventional hydrocarbon fuels besides the gas components containing H, N and C the formation as well of noticeable quantities of components containing S (SO_2, SO_3 and H_2SO_4) and soot particles actually occurs which are taking an important part in water steam condensation and in the build up of sulphate aerosols in the jets of aircraft engines (Brown R. Bakirow F.).

As an example, values of T, γi in typical cross-sections of the gas turbine engine RB 211-524B installed for example on the plane Boeing-747 are shown in Table 1. Those values have been calculated by means of combined integration of one-dimensional gas dynamics and chemical kinetics equations (the kinetic scheme included 252 reversible chemical reactions with the participation of OH, H_2, O_2, H_2O, N_2O_2, N_2, N, NO, NO_2, N_2O, NO_3, HNO, HNO_2, HNO_4, HNO_3, NxHy, C, COx, CHy) under the conditions of a cruise flight at the altitude of 10.7 km and the speed of 237, m/s. Table 2 shows that though there are practically no differences concerning the usually discussed components—H_2O, N_2, CO_2, NO_2, C—but for the majority of the components the chemical balance at the nozzle outlet is not achieved. Thus it can be stated that the most important thing for the correct determination of emissions of different components from a jet engine using either hydrocarbon or hydrogen fuels is to take into account the chemical unbalance in all parts of the flowpath (Starik A.).

Table 1 : Radius, Temperature, Gas Composition in Various Cross-Sections of an Engine Jet of the Subsonic Aircraft IL–76 in Cruise Flight Mode
(H=10 km, Mo=0.55)

x, m	0.0	atmosphere	1.0	100	1000
r, m	0.52	——	0.52	1.64	5.74
T, K	450	223.1	44.5	238.9	224.7
H_2O	3.500(−2)	5.768(−5)	3.508(−2)	3.508(−2)	3.535(−4)
O_2	1.618(−1)	2.002(−1)	1.621(−1)	1.963(−1)	1.998(−1)
H_2	5.522(−7)	9.378(−7)	5.538(−7)	8.991(−7)	9.345(−7)
OH	9.478(−5)	1.071(−13)	4.934(−12)	2.890(−13)	2.046(−13)
H	5.179(−8)	2.373(−20)	2.373(−17)	1.240(−19)	5.617(−20)
O	6.054(−6)	1.664(−15)	1.728(−10)	6.890(−12)	4.493(−13)
N2	7.746(−1)	7.995(−1)	7.759(−1)	7.971(−1)	7.993(−1)
HO_2	3.223(−7)	1.406(−11)	4.647(−14)	8.005(−15)	1.138(−15)
H_2O_2	1.346(−8)	8.793(−10)	5.321(−6)	5.365(−7)	4.581(−8)
NO	2.728(−4)	1.003(−11)	2.402(−4)	2.413(−5)	2.034(−6)
NO_2	1.103(−4)	6.027(−12)	1.324(−4)	1.337(−5)	1.111(−6)
HNO_2	1.799(−5)	3.879(−14)	2.593(−5)	2.610(−6)	2.184(−7)
N_2O	4.648(−5)	2.811(−7)	4.656(−5)	4.940(−6)	6.720(−7)
HNO	4.736(−11)	0.0	1.580(−8)	1.5874(−97)	1.332(−10)
O_3	6.463(−9)	7.275(−8)	7.696(−7)	6.268(−8)	7.908(−8)

Contd...

Table 1 : (Contd...)

x, m	0.0	atmosphere	1.0	100	1000
r, m	0.52	—	0.52	1.64	5.74
T, K	450	223.1	44.5	238.9	224.7
HNO_3	2.983(−8)	2.271(−10)	2.561(−5)	2.580(−6)	2.172(−7)
NO_3	2.226(−5)	3.598(−15)	1.053(−10)	2.484(−13)	1.372(−13)
N_2O_5	0.0	1.653(−13)	7.632(−12)	1.074(−9)	8.619(−10)
HNO_4	2.738(−25)	1.061(−10)	1.904(−15)	9.668(−11)	1.058(−10)
CH_3	5.343(−15)	1.855(−21)	2.514(−27)	7.208(−21)	1.817(−21)
CH_4	3.467(−15)	1.496(−6)	3.461(−15)	1.345(−6)	1.483(−6)
CO	1.017(−6)	2.283(−7)	1.008(−6)	3.068(−7)	2.348(−7)
CO_2	2.645(−2)	3.092(−4)	2.649(−2)	2.945(−3)	5.304(−4)
CH_2O	3.792(−15)	1.900(−11)	1.836(−15)	1.710(−11)	1.893(−11)
CH_3O	2.572(−18)	8.917(−18)	1.310(−26)	6.636(−18)	5.766(−18)
CH_3OH	8.950(−18)	0.0	2.894(−15)	2.916(−16)	2.605(−17)
CH_3O_2	1.190(−18)	2.710(−12)	3.036(−25)	3.110(−15)	2.954(−15)
CH_3NO_2	0.0	0.0	2.096(−15)	2.299(−12)	2.478(−12)
CH_3NO_3	0.0	0.0	7.848(−18)	1.501(−13)	2.515(−13)
SO_2	6.91(−6)	0.0	6.742(−6)	6.788(−7)	5.695(−8)
SO_3	0.0	0.0	6.172(−8)	2.066(−13)	1.567(−13)
HSO_3	0.0	0.0	1.993(−16)	1.158(−17)	8.625(−19)
H_2SO_2	0.0	0.0	1.175(−7)	1.805(−8)	1.517(−9)
Cl	0.0	2.316(−16)	9.313(−28)	2.866(−13)	1.054(−13)
ClO	0.0	6.578(−13)	1.745(−28)	5.897(−16)	1.658(−15)
ClO_2	0.0	0.0	0.0	4.811(−14)	4.562(−14)
$ClNO_3$	0.0	3.407(−13)	5.916(−28)	3.614(−13)	5.974(−13)
HCl	0.0	1.934(−10)	3.313(−25)	1.741(−10)	1.920(−10)
HOCl	0.0	4.138(−12)	7.088(−27)	3.721(−12)	4.100(−12)
CH_2Cl	0.0	0.0	0.0	6.061(−17)	3.005(−16)
CH_3Cl	0.0	5.442(−10)	9.323(−25)	4.894(−10)	5.396(−10)
Cl_2	0.0	3.834(−18)	0.0	5.896(−18)	3.108(−17)
CCl_3	0.0	0.0	0.0	7.284(−21)	2.679(−20)
CCl_4	0.0	1.214(−10)	2.080(−25)	1.092(−10)	1.204(−10)
$CFCl_2$	0.0	0.0	0.0	1.610(−22)	2.250(−21)
$CFCl_3$	0.0	1.406(−10)	2.408(−25)	1.264(−10)	1.394(−10)
$CF2Cl_2$	0.0	2.339(−10)	4.006(−25)	2.103(−10)	2.319(−10)

A(−n) corresponds to $A.10^{-n}$

Emissions of H_2SO_4, HNO_3, HNO_2, N_2O_5, H_2O can play an essential part in the changes of the aerosol composition of the atmosphere and in the formation of polar stratospheric clouds. A recent analysis has shown that if 100% of SO_2 contained in the jet leaving an engine of a supersonic plane are conversed into H_2SO_4 than the surface area of the stratospheric aerosol layer can be practically doubled. In fact, the results of stratospheric aerosol layer density measurements are indicative of an increasing aerosol content in the stratosphere at the rate of about 5 per cent a year during the last 15-20 years, that is, approximately corresponding to the increase rate of the fuel mass consumed by aircrafts (Nesenstein D., Hofman D.).

An increase of the H_2O and HNO_3 concentration can cause an increase of the temperature of the polar stratospheric clouds formation. This problem is closely connected with the heterogeneous reactions taking place on the surface of aerosol particles and of the polar stratospheric clouds. It should be noticed that the heterogeneous processes can take place not only in the atmosphere but also in the backwash of a jet engine. That's why they must be already taken into account today (Solomon S. Stolarski R.).

Intensified formation of the polar stratospheric clouds especially in Arctic latitudes can also be indirectly caused by gases emitted from jet engines into the troposphere—CO_2, CO, CH_4, H_2O, N_2O—absorbing the infrared radiation and thus increasing the greenhouse affect. this results in cooling of the stratosphere and in a simultaneous increase of its humidity. And this fact can in its turn cause an expansion of the area of the polar stratospheric clouds formation in lower latitudes if there is a high content of HNO_3 in the atmosphere also due to emissions from the aircraft engines (Hofmann D.).

Water steam emitted from the aircraft engines not only takes part in the formation of polar stratospheric clouds and of the sulphate aerosol layer, but also plays an important part in the radiation balance of the atmosphere. Its influence manifests itself in the direct absorption of the infrared radiation and also indirectly through the ice particles build up in the aircraft trace in the upper troposphere and through the formation of cirrus clouds leading to the intensification of greenhouse effect.

According to (Lion K.) the effect of the several per cent increase of the cirrus clouds area is equal to a double increase of the CO_2 content in the troposphere.

Table 2 shows the calculation results for the gas composition changes in the jet of a subsonic aircraft IL-76 flying at the altitude of 10 km with the speed M0=0,55. The values of T and γi in the nozzle outlet cross-section (x=0) and in the atmosphere are shown. It can be noticed that not only H_2SO_4, N_2O_5, HNO_3, HNO_4, but also chlorine-containing components Cl, ClO, $ClNO_3$, CH_3NO_2, GH_3NO_3 are intensively forming in the jet.

The components of the first group due to the oxidation of SO_2, NO, NO_2 and NO_3 contained in the combustion products, and those of the second group result from reactions with chlorine-containing components entering the jet from the atmosphere. The whole complex of chemical processes taking place in the jet is rather complicated. Modern models include up to 250 reactions. Concentration changes of Cl-containing components and CH_3NO_3 along the jet after a flight of the aircraft IL-76 are shown in Fig. 2.

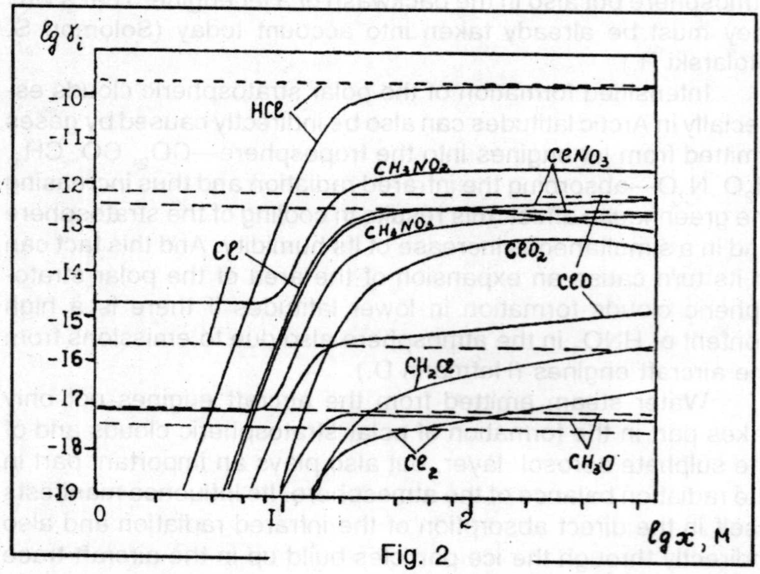

Fig. 2

It is important that if the altitude is higher than 19 km then one must take into account by the numerical simulation of chemical transformations even in a single jet mode not only the chemical reactions but also the photodissociation processes (the last column of Table 2 includes values obtained without taking into account those processes).

Thus it can be stated that the gas composition of the jets of both subsonic and supersonic engines undergoes consider-

Table 2 : Temperature and Gas Composition in Typical Cross-Sections of the Gas Turbine Engine RB 211–524B in cruise flight mode
(H=10.7 km, M_0=0.95)

Parameter	*Combustion Chamber Outlet*	*High Pressure Turbine*	*Transition Piece between Interme-diate Pressure Turbines*	*Interme-diate Pressure Turbine*	*Low 1st Stage*	*Pressure 2nd Stage*	*Turbine 3rd Stage*	*Nozzle Outlet*
T, K	1541	1336	1346	1100	901	736	614	590
H_2O	3.97(–2)	3.99(–2)	3.99(–2)	3.99(–2)	3.99(–2)	4.00(–2)	4.00(–2)	4.00(–2)
O_2	1.54(–1)	1.54(–1)	1.54(–1)	1.54(–1)	1.54(–1)	1.54(–1)	1.54(–1)	1.54(–1)
H_2	6.79(–6)	2.77(–6)	2.28(–6)	1.90(–6)	1.54(–6)	1.48(–6)	1.48(–6)	1.47(–6)
HO	3.68(–4)	1.14(–4)	1.05(–4)	4.26(–5)	1.27(–5)	2.54(–6)	9.00(–7)	2.98(–7)
H	1.58(–6)	3.95(–7)	3.27(–7)	1.63(–7)	2.44(–7)	1.16(–9)	1.94(–10)	3.41(–11)
O	6.36(–5)	1.71(–5)	1.43(–5)	6.87(–6)	1.87(–6)	3.12(–7)	1.12(–7)	1.83(–8)
N_2	7.74(–1)	7.74(–1)	7.74(–1)	7.74(–1)	7.74(–1)	7.74(–1)	7.74(–1)	7.74(–1)
HO_2	8.61(–7)	2.14(–7)	1.99(–7)	1.27(–7)	4.10(–8)	6.05(–9)	2.26(–9)	7.26(–10)
H_2O_2	9.80(–7)	9.04(–7)	7.13(–7)	1.56(–6)	1.20(–6)	1.10(–6)	1.09(–6)	1.08(–6)
N	2.23(–12)	4.98(–14)	4.18(–14)	2.19(–15)	4.58(–17)	6.61(–19)	7.48(–20)	6.51(–21)
NO	7.12(–5)	9.10(–5)	9.13(–5)	1.01(–4)	1.03(–4)	1.04(–4)	1.04(–4)	1.04(–4)
NO_2	5.13(–5)	4.92(–5)	5.19(–5)	4.61(–5)	4.55(–5)	4.49(–5)	4.49(–5)	4.49(–5)
HNO_2	7.20(–5)	5.51(–5)	5.23(–5)	4.72(–5)	4.33(–5)	4.19(–5)	4.17(–5)	4.16(–5)

Contd...

Table 2 : Contd...

Parameter	*Combustion Chamber Outlet*	*High Pressure Turbine*	*Transition Piece between Interme-diate Pressure Turbines*	*Interme-diate Pressure Turbines*	*Low 1st Stage*	*Pressure 2nd Stage*	*Turbine 3rd Stage*	*Nozzle Outlet*
N_2O	3.35(−5)	3.07(−5)	2.88(−5)	2.79(−5)	2.79(−5)	2.78(−5)	2.78(−5)	2.78(−5)
NH	3.17(−11)	3.01(−12)	2.47(−12)	3.38(−13)	1.68(−14)	5.57(−16)	1.52(−16)	2.26(−17)
NH_2	9.30(−13)	1.17(−13)	9.18(−14)	4.69(−14)	6.50(−15)	1.76(−15)	7.17(−16)	1.22(−16)
NH_3	1.47(−12)	3.42(−13)	2.59(−13)	2.14(−13)	1.89(−13)	1.85(−13)	1.84(−13)	1.84(−13)
N_2H	7.26(−11)	8.37(−11)	7.16(−11)	1.72(−11)	1.85(−12)	4.33(−14)	8.26(−15)	2.24(−15)
N_2H_2	7.54(−15)	7.23(−16)	5.01(−16)	3.17(−16)	1.50(−16)	1.27(−16)	1.25(−16)	1.24(−16)
N_2H_3	2.80(−19)	1.07(−19)	6.31(−20)	3.00(−19)	1.79(−19)	1.49(−19)	1.46(−19)	1.45(−19)
N_2H_4	1.14(−21)	4.69(−22)	2.57(−22)	2.05(−22)	9.33(−22)	8.15(−22)	7.69(−22)	7.32(−22)
HNO	5.64(−10)	6.12(−11)	5.23(−11)	1.25(−11)	1.52(−12)	1.16(−13)	2.64(−14)	7.53(−15)
O_3	4.67(−8)	2.74(−8)	2.19(−8)	4.11(−8)	6.05(−8)	6.17(−8)	7.41(−8)	7.73(−8)
HNO_3	9.20(−7)	9.65(−7)	8.82(−7)	1.98(−6)	1.72(−6)	1.04(−6)	3.70(−7)	2.23(−7)
NO_3	1.74(−8)	2.90(−8)	2.21(−8)	3.14(−7)	2.43(−6)	4.88(−6)	5.75(−6)	5.98(−6)
C	3.30(−22)	3.15(−22)	3.09(−22)	3.07(−22)	3.06(−22)	3.06(−22)	3.06(−22)	3.05(−22)
HCO	5.61(−13)	6.08(−14)	2.86(−14)	6.89(−15)	4.36(−16)	8.21(−18)	5.64(−19)	8.13(−20)
CO_2	3.15(−2)	3.15(−2)	3.15(−2)	3.15(−2)	3.15(−2)	3.15(−2)	3.15(−2)	3.15(−2)
CO	1.19(−5)	6.93(−6)	3.87(−6)	3.14(−6)	3.01(−3)	2.99(−6)	2.99(−6)	2.99(−6)
CH_2O	4.48(−15)	2.59(−16)	1.26(−16)	4.08(−17)	1.07(−17)	5.43(−18)	5.05(−18)	4.87(−18)

A(−n) Corresponds to $A.10^{-n}$

able changes, and the engine jet is a source not only of NO and NO_2 as it was assumed earlier, but also of other components of the NOy group—HNO, HNO_2, HNO_3, N_2O, NO_3, and of other groups as well: H_2SO_4, Cl, ClO_2, $ClNO_3$, CH_3NO_2, CH_3NO_3.

Recently it was found out that the exhaustion of the ozone layer due to the presence of aerosols in the atmosphere can take place in the particular according to the following scheme: in a heterogeneous process the inactive compound $ClONO_2$ in its gaseous phase interacts with water inside the aerosol, the result is the formation of Cl_2 inside the aerosol. In spring the solar radiation in the Northern hemisphere causes the evaporation of aerosols of Cl_2 appears in gaseous phase. Having absorbed a solar quantum it transforms to 2Cl. As a result of the following catalytic cycle the ozone layer gets exhausted: $Cl + O_3 \rightarrow Cl + O_2$.

A flight of a subsonic and especially of a supersonic aircrafts calls forth a whole number of complicated fluid dynamic phenomena in the atmosphere. The succession and the importance degree of those phenomena depend on the design of the aircraft (number and arrangement of engines, wingspread etc.) and on the flight mode. But there are typical areas where one or another fluid dynamic phenomenon is dominating.

At the initial stage hot exhaust gases leaving the engine nozzle are mixing with the backwash in the atmospheric air (single jet mode). Exactly here the main changes of the temperature and of the gas composition take place. This mode continues for less than 10s, for subsonic planes that means the distance of about 1 km from the nozzle outlet cross-section.

At the same time with this process formation and transformation of vortical structures leaving the wings ends take place; at a certain distance from the engine nozzle outlet those structures begin to interact with the flow formed by the combustion products mixing with the ambient air ("vortical mode" of the aircraft trace). A typical feature of this mode is the prevailing vertical growth of the trace and its much smaller change in the horizontal direction. The length of the trace forming after a plane is about 20 km. The flow within this area is much more complicated than in the single jet area.

The gas phase chemical reactions taking place in this area are not an intensive as in the single jet area but at the same time an intensive formation of ice particles and of small sulphate aerosol particles (about 10 nm in diameter) occurs just in that very area. On the ice particles the heterogeneous reactions take place, and the small sulphate aerosol particles are condensation centres playing a certain part in the formation of clouds (Brown R., Busen R.).

The next mode (on the time scale) is the mode of "large ring formations" developing in the atmosphere after the disintegration of the aircraft trace due to the fluid dynamic instability. This mode hasn't been studied yet, though exactly here the components formed in the aircraft jet are transferred to the lower atmospheric layers. Only after a detailed analysis of this process the reaction of the atmosphere on the aircraft flights could be estimated more or less exactly.

At the same time with the change of the gas dynamic structure of the aircraft trace changes in the chemical composition of this trace are taking place. This happens not so much because of the combustion products dilution with the atmospheric air as because of gas phase and heterogeneous chemical reactions in the trace. The phase composition of the jets themselves changes too. Let's discuss those processes closer (Miakae- Lye R., Kärcher B., Brown R., Persianseva N.).

Main reasons for the beginning of chemical processes in the jet of a jet engine and for the change of the exhaust gas composition are gas cooling due to the backwash expansion and penetration of small atmospheric components into the hot gas flow. That's why a kinetic model used to describe chemical transformations in the jet should include a number of reactions typical both for the high temperature processes and for the atmospheric chemistry. But until recently the models of gas phase processes in jets of jet engines didn't include reactions with small atmospheric components (Starik A., Weibring G.).

Today the main means for studying the unbalanced chemical processes in jets of aircraft engines is numerical simulation. The reasons for this are the excessive difficulty and the high costs of in-flights experiments. That's why the recently obtained unique experimental data about the measured concentrations of NO, NO_2, HNO_2, HNO_3, SO_2, H_2SO_4, in cross-sections distant from the nozzle outlet (–10 km) are used mainly for testing the results of numerical experiments (Arnold F. Fakey D.) To study the mixing dynamics of a chemically reacting gas a two-dimensional approximation is used—that is, a model of an axially symmetrical single turbulent jet. It allows to identify main factors influencing the speed of chemical reactions. For example, Fig.3 shows the changes of the temperature and of the nitrogen oxides content in the backwash (Anderson M.).

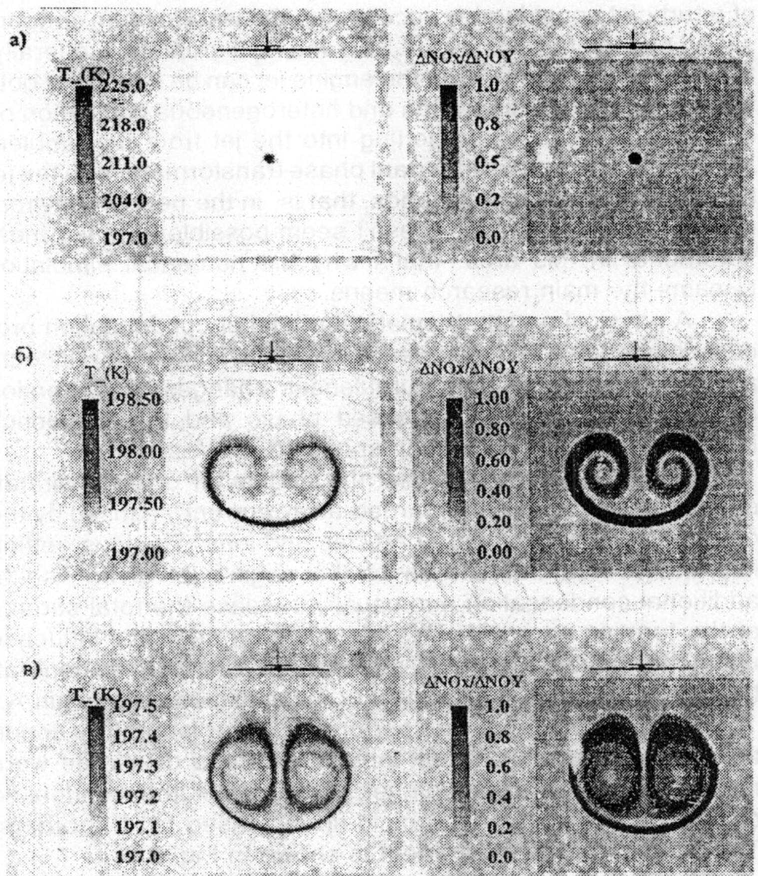

Fig. 3 : Temperature Fields and Concentration Rations NOx/NOy ($NOx=NO + NO_2$; $NOy = NOx + HNO_2 + HNO_3 + NO_3 + N_2O_5$) at various distances downstream at aircraft ER-2 (H = 18.9 km, M = 0.71): a) 0.1 km; δ) 1.7 km; b) 3.2 km.

When the temperature in the jet of a subsonic or a supersonic aircraft using the hydrocarbon or the hydrogen fuel decreases due to the mixing of the hot gases with the atmospheric air then the conditions of water steam oversaturation or even of ice crystalization or sublimation can be reached. The calculations show that the presence of HSO_2, H_2SO_4, and soot particles in the air craft kerosene combustion products leaving the nozzle of a turbojet considerably accelerates the water steam condensation and the formation of H_2SO_4 and H_2O droplets. Generally, the formation of aerosols in the engine jet can be a result of both the homogeneous nucleation and heterogeneous nucleation on soot and other particles getting into the jet from the ambient atmosphere. the most important phase transformations in the jet happen within the first seconds, that is, in the nearest environment of the jet. Today it doesn't seem possible to make measurements in this area. that's why the numerical simulation remains the main research means.

A physical and mathematical model of condensation processes in engine jets was developed which takes into account a three-dimensional flow, turbulences changes of dispersion characteristics of the condensed phase and the turbulence influence upon the nucleation speed (A. B. Vatazhin).

Earlier it was shown that the engine jet of each aircraft carries electrical current ("carried-away current") caused by the charged solid phase micro-particles and ions. The ions getting into the oversaturated area of the engine jet given rise to additional condensation and usually increase the total surface of the forming aerosols. This can have an essential influence upon the characteristics of atmospheric aerosols in general; these effects are also a subject of investigations (Vatazhin A.)

A key experiment was performed under laboratory conditions (A. B. Vatazhin) when ions of a corona discharge were introduced into a steam/air jet. As a result the condensed dispersion phase (aerosol) build-up increased by 1–2 orders. The scheme of the experiment is shown in Fig. 4.

Furthermore, it was shown in a flying laboratory that additional introduction of ions by means of specially designed devices can have essentiall influence upon the condensation in the aircraft trace (A. G. Leut). Fig. 5.

It is till unclear which part of the charged particles at the

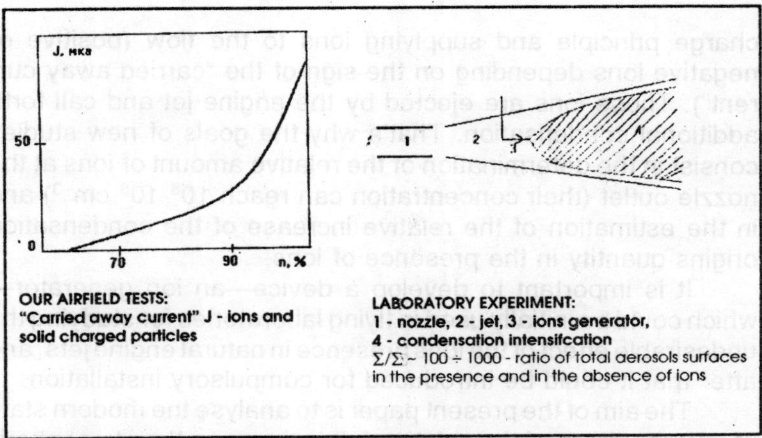

Fig. 4 — Aircraft Aerosols Formation Due to the Presence of Electric Charges in Engine Jets

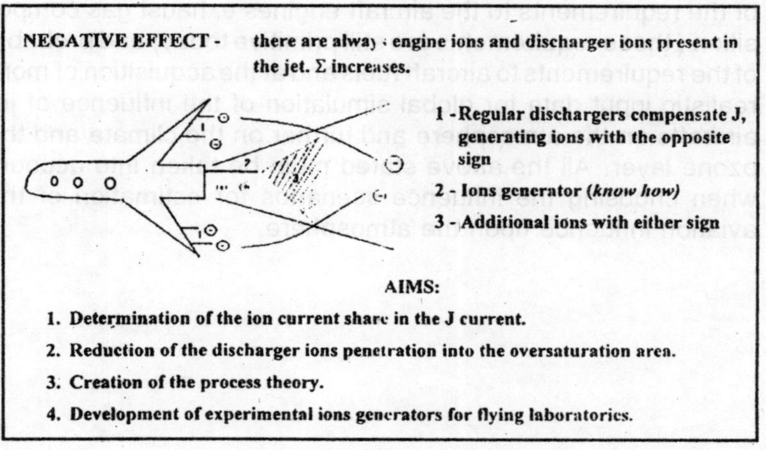

Fig. 5 — Flying Laboratory

nozzle outlet is made up by ions. It depends on the construction and operation mode of the engine. But even if there are no ions at the nozzle outlet and the whole "carried away current" is caused by charged solid micro-particles (unburned fuel particles) the situation remains difficult nevertheless. The "carried away current" is compensated by the current from regular dischargers installed on the aircraft operating on corona discharge principle and supplying ions to the flow (positive or negative ions depending on the sign of the "carried away current"). Those ions are ejected by the engine jet and call forth additional condensation. That's why the goals of new studies consist in the determination of the relative amount of ions at the nozzle outlet (their concentration can reach 10^8-10^8 cm^{-3}) and in the estimation of the relative increase of the condensation origins quantity in the presence of ions.

It is important to develop a device—an ion generator—which could be initially used in flying laboratories for studying the undesirable effect of the ions presence in natural engine jets, and after that it could be introduced for compulsory installation.

The aim of the present paper is to analyse the modern state of investigations of the aviation influence upon the atmospheric processes and to show the main directions of further investigations. Those could be aimed at possible changes (improvement) of the requirements to the aircraft engines exhaust gas composition (those requirements are still primitive today) and probably of the requirements to aircraft fuels and at the acquisition of more realistic input data for global simulation of the influence of jet aircrafts on the atmosphere and further on the climate and the ozone layer. All the above stated must be taken into account when choosing the influence scenarios for estimation of the aviation influence upon the atmosphere.

18

Measurements of the Electricity Consumption of Home Entertainment Appliances: the CEMEDA[1] Collaboration, a French Case Study

J. ROTURIER
France

ABSTRACT

In most homes, electronic appliances have now become commonplace. However alike office equipment in the late 80s, their impact on the electricity demand is now a greater and greater concern. In OECD countries, policies are now implemented in order to optimise both the power load and consumption. This is a key issue, in particular in the framework of the implementation of the international co-operation aiming to limit the greenhouses gas emissions. In the present paper some very recent information regarding these policies are first presented. Then, the CEMEDA collaboration, a French case-study is presented: the main data, segmented by equipment, that are accurately described in a quantitative manner, pay very special attention to the "Electricity leakage" importance. In the conclusion, the most recent news concerning the EU policy and manufacture's partnership are shown.

1. INTRODUCTION

Alike in other OECD countries, French households became systematically equipped, in the 70s and 80s, with the so-called "White Goods", ("wet" and "cold") appliances. Progressively, the "Brown Goods", (television, video recorder, HiFi, etc.) have become commonplace in most homes. However, both markets being saturated, the replacement market is now a priority not

1. CEMEDA: Consommation Électrique des Matériels Électroniques Domestiques en Aquitaine

ignoring that, since early 90s, the home market has been enriched with a multitude of new electronic appliances: cordless telephones, answering machines, clocks, energy controllers... Also, since mid-90s, the high rate growth of multimedia units may not be neglected. For example, in 1997, on the French market 2.7 millions PCs have been sold of which 0.9 millions for home users, a significant 25 per cent more than a year before.

Each of these home electronic appliances is, alike the office equipment, a segment of a world wide market, however showing an important difference, with the "White Goods" market, at least up to now as long as energy efficiency policies are regarded. In such condition, it is necessary to take into account the electricity demand growth, particularly the resulting large amount of "Electricity Leakage" (1) (section 3-f). Then, any measure (labelling programme, standard...) intending to improve the efficient use of the electricity, should better be decided on a world wide basis in a partnership that involves at least both manufactures and public bodies, and preferably electric utilities and users representatives too.

Today, nobody ignores the direct and indirect impact of the office equipment end-uses on the electricity demand demonstrated, in several studies started 10 years ago, by US and French researchers. In the last past 5 years that issue has been commonly analysed in many other regions of the world. In survey reports aiming to provide the European Commission with the most recent data (2), a brief overview is given that shows the huge power load growth of the I&CTs (Information and Communication Technologies) sector (about 2,000 MW annually on the whole EU territory). The "Energy Star Office Equipment Programme" having been started in 1992 by US-EPA, a partnership is now being negotiated particularly by the EU and the US. Although not yet definitely established, such co-operation aims to give the present the international status to be officially recognised by all actors of the computers market. For the same reasons, a similar concern is quickly increasing regarding the home electronic appliances market. In partnership with European National Energy Efficiency Agencies, several international conferences have been recently organised in order to (i) disseminate the main specifications and future trends and (ii) analyse the solutions both from technology, economics and policy point-of-view as shown in Section 2.

2. HOME ELECTRONICS APPLIANCES: A BRIEF INTERNATIONAL SURVEY

In recent Conferences (3), (4) the impact of domestic appliances on the electricity demand-side management programmes has been underlined. The need to pay a greater attention to this end-use, nowadays recognised as an important one, results in a growing interest by various bodies in this area of investigation. In a survey analysis of the Florence Conference (4), J.P. Harris has presented some recent data and trends (5) from countries that are known to push the electronics industry. Facing with such an increase of both electricity consumption and electricity leakage too, these countries are now decided to not remain in a "Laissez-Faire" status.

In Switzerland the best models (about 25%), are identified each year. In terms of energy efficiency, a deep decrease is found from 1990 to the present, i.e., a 60 per cent standby power decrease for PCs. This is a part of a programme for voluntary consumer information of energy-efficient TVs and VCRs, implemented by the Group for Efficient Policy through an annual of the stand-by maximum power load being defined in a yearly label. In Japan (7), 56 types of home electronic appliances were identified in a study involving field measurement of standby consumption: standby power represents about 10 per cent of residential electricity use about 24 TWh/yr when extended to the whole country.

As recently announced, the US-EPA Energy Star programme is extended, since January 1998 to TVs and VCRs with standby power lower than 3W or 4W respectively. In a near future "advanced" functions (such as satellite decoding) and more stringent requirements i.e. lower stand-by power load, are expected to be considered. Such programme has been signed up by manufacturers representing 70 per cent of sales in the US. As again quoted by J.P. Harris (5), "home electronics represents a very important segment of the overall Energy Star labelling programme, both because of their visibility to consumers—during purchase and on a daily basis as users—and due to their significance as a source of growing electricity demand. In the US, a recent study found that "miscellaneous" uses of electricity in the home accounted for all of the net growth in residential electricity use by 2010. EPA and DOE are exploring ways to broaden the Energy Star Programme as a basis for international co-operation, including offering the free use of promotional material from a multimillion dollar public awareness campaign now underway in the US....". J. P. Harris (5) also notes some important questions or comments, presented below, that have to be answered:

- should we expect every appliance and device in the home will be an "information appliance?"
- will the efficiency improvement be fast enough to offset the growing number of products?
- tomorrow's "smart" products alike smart children: should "talk" properly and "listen/and speak a common language and, at the proper time, quickly go to sleep, and wake up promptly.

In order to achieve the present short survey, a reference should necessarily be done to a recent special issue, entirely devoted to Energy Efficiency Standards for Appliances, of "Energy and Buildings" (8) containing a very detailed presentation by most of authors quoted in the present paper of the present situation.

3. CEMEDA: A FRENCH CASE-STUDY

a) Presentation

In 1973, final energy consumption in the French residential sector was 38 Mtoe, which could by broken down into 73 per cent for heating, 15 per cent for hot water and cooking and 11 per cent for other specific electricity uses (9). With the oil crisis, the greater awareness of the risk of shortages and efforts has been resulting in a decrease of the energy bills, especially for heating. Twenty years later, the residential sector accounts for 48 Mtoe, 30 per cent more than in 1973! All efforts to reduce energy consumption have brought the proportion used for specific purposes now represents more than 23 per cent of the total. Over the same period, households have acquired more appliances (10): the proportion of households with televisions, for example, has now risen from 78.5 per cent to 95 per cent and, in the case of telephones, 90% of households have a second phone. Table 1 shows the change of the proportion of households where appliances are found.

Statistics currently available never make mention of the new types of appliances that are now commonly found in French homes, such as microwaves ovens, answering machines, VCRs, clock, radios, digital television decoders etc. For several years, a limited number of studies have shown the major impact of domestic appliances on the electricity bill and hence its present non-negligible impact at national level. One of the most significant work is the CIEL[2] study (11), a comprehensive and detailed survey carried out in South Eastern France that has permitted a precise analysis of the share of common home appliances in the electricity bills.

Table 1: Share of Most Common Household Appliance (1973–93)

% of households with goods	*Washing machine*	*Dish-washer*	*TV*	*Tel.*	*Fridge*	*Fridge/ Freezer*
1973	65	4.5	78	20	85	10
1993	89.5	33	94.5	95.4	98.7	44.5

The present CEMEDA study grew from the need of new data observation with two main goals (i) to measure precisely the impact of new technologies on household power load and consumption (ii) to define an energy pattern, alike an ID power load profile for specific types of homes and end uses. The experimental protocol relies on a methodology and measuring equipment previously tested and implemented in the frame-work of the French case-study, part of both OT3E and MACEBUR[3] collaborations [see Ref. (2)] funded by the EU SAVEPACE Programme and Ademe. The equipment is accurate enough (1W) to measure exactly the electric specifications of appliances often having a power load of a few Watt only. Searching out end-uses where energy savings do exist, the exact power requirement as well as the actual Time-Of-Use were identified for each of them. Subsequently, by knowing the precise role of each piece of equipment and each power level, any possible energy conservation potential may be calculated. Also we give a few data regarding a perfect indicator of the harmonic pollution, a growing factor for electric utilities: the Power Factor (PF), that is, as well known, quite different of the so-called coso, commonly met in

2. CIEL: acronym standing for Controlling the Demand for Electricity, through an In-House Measurement Scheme.
3. OT3E and MACEBUR are acronyms standing for "Energy Efficient Office Technologies in Europe" and (in French) "Maitrise des Consommations d'Electricité de la Bureautique" respectively.

linear electric loads (i.e. motors).

Table 2 : Distribution of Appliances According to Power Levels

Level of Power	*Number of Appliances*	*%*
1	115	68
2	98	58
3	59	35
4	50	30
5	49	29
6	52	31
7	62	37
8	139	83

b) Implementation

Ideally, CEMEDA should be implemented in two steps. In the present first step, to be regarded as a feasibility study, approximately 400 appliances are measured in twenty homes in a partnership between Ademe, MD3E and University of Borodeaux. Briefly introduced here, the main results, represent a validation of the experimental methodology in the residential sector and provide a set of new data. The choice of housing was based on the distribution of different types of housing given in INSEE reports (9), (10). In a second step, measurements are hopefully expected to be performed on a Europe wide enlarged sample, carried out in several EU countries, a more accurate experiment, taking into account for example the differences in user's behaviour should permit a detailed analysis of the impact of energy efficient labels and information campaigns on household consumption.

c) Results

All metered appliances are distributed as listed Figure 1. Monitored in 20 households from June to September, 1997, over a period of three weeks each, they result in 220 sets of data, now analysed. In Table 2, the distribution of appliances listed by power level is also shown, permitting to list the related electricity end-use sectors. The identification consists of specifying each group of equipment or power use, in order to quantify precisely the average annual consumption, the different levels of power load and their respective roles in the operation of each appliance and the related active and apparent energy consumed, that are really found, permit to define the power level of operation. A detailed analysis will bring out typical electricity patterns, as in the case of electronic office equipment, a sector which is beginning to encroach on the domestic sector as shown in the MACEBUR Report (2). In Table 2, the relative power levels found from the analysis, from level 1 the maximum or "Active" power load to level 8, the lowest "Suspend" mode, are shown, well demonstrating that such appliances are operated according to complex power regimes. Column 3 of Table 3 shows the percentage of appliances that are metered with the power level given in column 2. For example 68 per cent of appliances only are found in the active mode meaning that 32 per cent are, at the time of measurement left in stand-by or suspend modes that is confirmed by the fact that Level 8, the lowest power level may be associated with 83 per cent of appliances The main benefits of the present analysis are (i) to remove any possible anomalies and (ii) possibly detecting an opportunity for energy savings.

(d) Summary of Results and Conclusions for Each Type of Equipment

TV Sets

From the measurements, it may be shown in a TV set, the power load is on average 1.8 W per centimeter of the diagonal measurement of the screen, ranging from 1.5 to 2.1 W/cm. The most recent models appear to fall into the lower part of the group. Average annual consumption for all groups in about 140 kWh/year, for an average daily use of 2.5 hours/day[4]. The average power factor i.e. the ratio between the Active and Apparent power, is 0.6 in "ON" mode and 0.5 in "Standby" mode. The standby power level, not depending on the screen size, is about 8 W, in a 5–12 W. The large difference between the power level from "standby" to "ON" suggests that the annual consumption is mainly resulting from the "ON" mode. In such condition, the electricity savings from a standby mode can be exploited through user's awareness campaigns. However, improving the energy

4. Such Time-Of-Use, twice lower than average French one, may be explained by the period of the year (summer) when the occupants may prefer outdoor activities, then, a TV set may remain unused.

efficiency in "ON" mode, may request much R-D-D efforts that possibly need some delay before reaching the marketplace.

In Figures 2 and 3, the range of the average levels of power per hour (Wh/h) is shown for each TV set. Power levels are colour coded: "*white* corresponds to the television being in "Use" use, *black* indicates that the television is switched off (in theory, "Use" power = O W) and *grey* indicates standby. It has been found that TV set # 1 was always on standby when non-used, 3 periods of use being clearly identified: in the morning, 7–8.30 a.m., 12.30–4 p.m. and then in the evening from 6 to about 11. On the contrary TV set # 2 was usually switched off at night. Periods of use can be observed here too. However, in some cases the users may turn the set off using the remote control, and this accounts for the periods of standby (shown in *grey*). Even more energy savings could therefore be made here. Nevertheless, although those viewing TV set # 2 watched about 25 per cent, more TV per day, the extrapolated annual energy consumption should be 21 per cent less than that of TV set # 1.

Table 3 : Influence of User's Behaviour on TV's Energy Demand

TV	*TOU (h/day)*	*Standby mode Consump. (kWh/year)*	*Active mode Consump. (kWh/year)*	*Total Consump. (kWh/year)*	*kWh/TOU ratio kWh/hr*
1.	1.32	83	94	177	0.21
2.	1.64	8	138	146	0.07

The power load is 23 W, 17 W and 9 W in each mode respectively. In terms of annual consumption (i.e. 140 kWh/year), about 90 per cent, is the result of standby mode (see § 3-f "Electricity Leakage"). In Table 3, these two TV sets data are again compared to illustrate the effect of user's behaviour. In the first column, the "active" Time-Of-Use (TOU) is shown while from the last, the kWh/TOU ratio a difference of a factor of 3 between the two TV sets is found: in the first case, 47 per cent of annual consumption is "Electricity Leakage" (see section 3-f). From all TVs data, the annual consumption and daily TOU are, on average, 143 kWh and 4h respectively.

Video Recorders (VCRs)

Their average consumption is 112 kWh/yr, 3 operating

modes being identified:

- "ON" corresponds to watching a cassette or recording a programme;
- "standby" means that there is a cassette in the machine but it is not being used either for viewing or recording;
- "OFF" means that only the time is being shown.

Hi-fi

These appliances represent only a very small proportion of consumption in the households sampled. The average was about 36 kWh/year but for a listening time which represents about 6 per cent of total time.

Audio Video Combinations

Under this heading, appliances such as TV sets, VCRs, satellite, receivers, decoders for coded channels... possibly plugged on the same box, are grouped. The average power of the standby mode being 45 W, the annual consumption for these combinations was approaching 400 kWh, of which only 20 per cent results from the "Active" mode. Only 5 per cent of homes studied had all the equipment for satellite reception (both digital and analogic channels) complete with audio-video amplifier to reproduce "movies" sound. This is therefore, a sector to be monitored closely in future, as consumption will increase as this new technology penetrates the market.

Clock Radios

By far the most common appliances in the sample, 2 or even more clock radios were typically found. Their average permanent power was 2 W, (about 4–5 VA, the average PF being lower than 0.5). Finally, they consume 20 kWh/yr.

Computers

In 40 per cent of households, a PC typically a multimedia home computer was found, equipped with a 14 or 15" screen and a printer. Interestingly, in 40 per cent of cases the computer continued to consume electricity, even when switched off. With a typical "Active" power and Time-Of-Use of about 100 W, and 2.8 hours/day respectively, the average consumption is 106 kWh/year, much lower than the 300 kWh/year of a workstation (1). PCs consumption is not yet taking over from televisions, but

Table 4 : Active and Apparent Power: P.F. Influence

	#[5] Power (W)	*Stand-By Active* Power (VA)	*Stand-By Appar.* Power (W)	*"Use" Active* Power (VA)	*"Use" Appar.*
Answer machine	10	3	6	2.5	6
Hi-Fi	8	5	9	25	33
Microwave	8	2	9	1019	1133
VCRs	8	9	18	21	35
Clock radio	12	1.5	3		
Pay-TV decoder	4	11	15		
TV set	8	8	18	116	177

one computer per person is required ("personal" being the operative word!).

(d) Comments

The current state of the CEMEDA study, and based on results obtained to date, the following points may be highlighted:

- through the present set of data, the real TOU in each operating mode may be calculated for each appliance.
- as an extra benefit, the present results, would facilitate the accurate assessment of the internal heat gains under real operating conditions, a very useful information used in the MTBF[6] calculation.
- The related consumption represents about 30 per cent on average, of the total electricity bill. However, such appliances frequently showing the standby mode, a change in user behaviour and improvements in he energy-efficiency device could produce savings of between 40 and 90 per cent.
- The case of TVs merits particular attention, given the ever--increasing length of time that these are in use. It is certainly necessary to limit as much as possible the amount of time in standby mode, but it would also be wise to look at the power.
- Level of the appliance when it is in its normal operation mode.

5. Number shown in that column represent the number of units having permitted the PF evaluation of the Power Factor, not the total number of equipment found.
6. MTBF: Mean Time Before Failure.

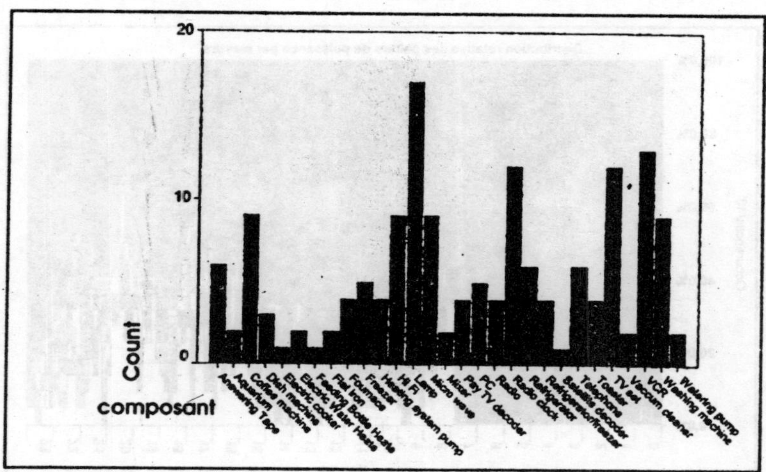

Figure 1 — Distribution of Metered Appliances

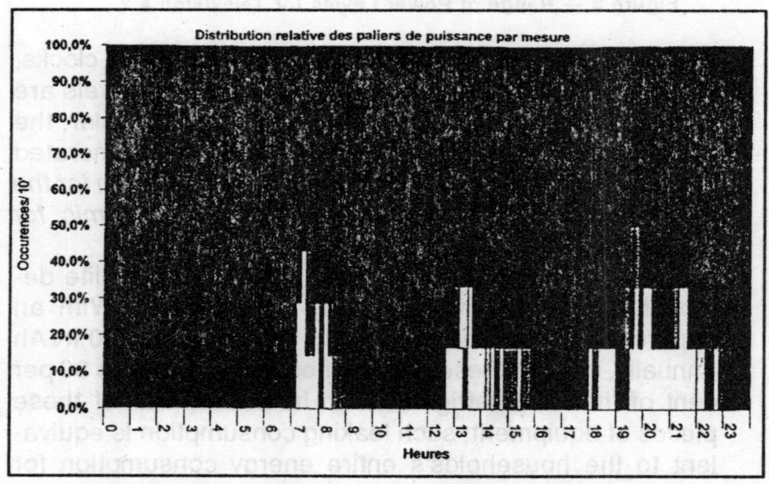

Figure 2 — Range of Power Levels for Television # 1

- In each home, every single appliance connected to the main supply has been measured. In certain cases, the total consumption was also measured directly at the point of delivery; so providing by difference, an analysis of the consumption for every end-use sector.
- As previously told, the 1 W accuracy of the meter has permitted a very precise study of the impact of low power

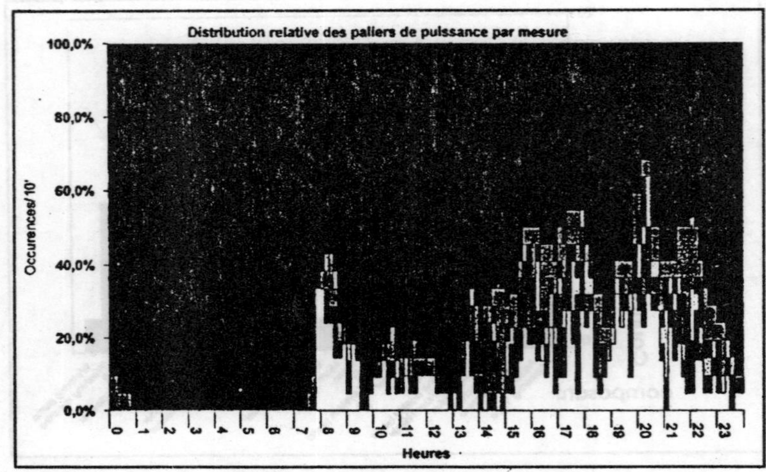

Figure 3 — Range of Power Levels for Television # 2

appliances such as telephones, answering machines, clocks, etc. In Table 4, the active and apparent power levels are given for the main appliances that show, in particular, the effect of the PF on the power that is to be really generated at the busbar, especially in standby mode. *Although for the consumer the standby mode appears to be economic, for the electric utility, it is very much less so.*

- A similar comment applies to TV decoders, satellite demodulators or receivers for pay-TV channels. With an apparent power of the order to 15 VA, a typical 130 kVAh annually, each of these pieces of equipment is 20 to 30 per cent of that of a refrigerator. In homes with all of these pieces of equipment, such leaking consumption is equivalent to the households's entire energy consumption for refrigeration.
- A simple definition of the leaking electricity is given by A. Meier (1), (8): "... it is the energy consumed by appliances when they are switched off".
- A similar comment applies to TV decoders, satellite demodulators or receivers for pay-TV channels. With an apparent power of the order of 15 VA, a typical 120 kVAh annually, each of these pieces of equipment is 20 to 30 per cent of that of refrigerator. In homes with all of these pieces of equipment, such leaking consumption is equivalent to

the household's entire energy consumption for refrigeration.

- A simple definition of the leaking electricity is given by A. Meier (1), (8): "... it is the energy consumed by appliances when they are switched off or not performing their principal function...". Its importance has been underlined by many scientists concerned by the efficient use of electricity, particularly the office equipment (2) and the home electronics appliances (see Ref. 4). Such wasted consumption is so high (12), that it is actually one of the main reasons that is commonly referred by those decision-makers who want to support the energy-efficiency programmes (labels, standards...). A few examples, from the present work are briefly shown here:
- out of 300 appliances that were metered (in 203 plugs) 69 are *never found* with a 0 W power load.
- in *stand-by modes*, the average power load and power factor are 7,6 W and 0,5 respectively then equivalent to produce about *15 VA at the busbar* concerning unused TV sets, *30* per cent *are always left in stand-by*.

4. THE EU ANALYSIS

Firstly shown in the MACEBUR Report (2), a few numbers may help to confirm the importance of both Office Equipment and Information Appliances sectors in the framework of the EU energy policy presented at the Climate Change Conference recently held in Kyoto:

- out of a total of 3,200 MT of CO_2 emissions in 1990, the Buildings sector contributed 654 MT. The EU 2010 proposal forecast (15% decrease) is equivalent to MT of CO_2;
- the annual leakage electricity consumed by any system actually plugged-in but providing no service to the user, is about 250 kWh per workplace for office equipment or per household for all home electronics appliances.[7] With an estimated number of European households of about 100 million, the overall "electricity leakage" is 25 TWh;
- these avoidable 50 TWh are equivalent[8] to 25 MT of CO_2 emissions, 25 per cent of the expected reduction in the

7. These data are simply an order of magnitude. More detailed data are shown in ref. (4) and (8) for example.
8. The EU Equivalent emission is about: 0.5 CO_2/kWh.

Domestic/Tertiary Sector or 3.3% of 800 MT, the whole technical reduction potential.

Finally, the present paper may not ignore the good news i.e. the strong signal now launched by manufacturers both in the US (see section 2) and the EU, where the electronics industry has decided to contribute to the efforts thanks to a strong partnership with the European Commission. As told by P. Bertoldi (13), the EU policy under the SAVE/PACE Programme aims to transform the domestic appliances market. The efforts are now being concentrated on the major end-uses i.e. refrigerators, washing machines, electric water heater and enetertainment electronics. Such policy relies on several instruments such as energy labels, minimum efficiency standards, demand-pull actions. A successful example of such partnership, has been presented by W. Bruens (14): the European Manufacturer Association (EACEM) is now preparing a few simple measures, to start by Jan. 1, 2000, to reduce the electricity leakage of TVs and VCRs:

- "... TVs and VCRs will no anymore have an energy consumption higher than 10 Wh;
- the average of every company will be <6W; later on the stand-by power consumption will be further reduced...".

5. GENERAL CONCLUSION

The present CEMEDA experiment has demonstrated its perfect flexibility from both the technical and scientific points of view, enabling to measure accurately and efficiently the impact of new technologies on energy consumption in the residential sector. Places, where energy savings can be made in the residential sector as well as in the tertiary sector were identified. Aslo, it is revelated that *in terms of total energy consumption of the household, the domestic appliances sector, a sector undergoing rapid expansion, accounts for one equivalent-refrigerator.* Moreover, if the demand for electricity is considered from the point of view of the electric utility, this consumption has to be multiplied by 1.5 to 3.3, due to a poor power factor. It is often the case that 90% of this consumptioin is the result of "electricity leakage" through the standby mode.

In the near future, a massive growth in complete audio/ video kits (TV, VCR, video disc player, satellite receivers and decoders for digitised programmes, audio/video amplifiers, HiFi, etc.) is expected in homes in France and throughout Europe, with each piece of equipment being independent and hence incorporating its own AC/DC conversion system. This will lead to an annual consumption by standby modes of about 560 kVAh per year and per household. In France, on the basis of 22 million households, such total consumption, is equivalent of two 1,000 MVA power plants.

Finally, from the present analysis, and resulting in particular of the importance of the electricity leakage briefly introduced in section 3-f leads, an important question is arising: In their active mode, most domestic electrical appliances when (i.e. when providing a real service) probably require an average power ranging from 10 to a few hundred mW only. However, the actual measured power falls within a 1-10 W range. Would it not be worthwhile to rethink completely the techniques used at present for the AC/DC conversion, a systematic requirement although probably not optimised yet, for the functioning of these appliances ? For all electronic appliances, an improved AC/DC conversion would probably reduce, the power load and the electricity generation of these appliances, by a factor of 7, an interesting and realistic challenge and a very affordable way.

References

1. Meier A., Huber W. "Results from the Investigations on Leaking Electricity in the USA" (ref. 4).
2. Roturier J. OT3E (Contract EC-DG XVII/Univ. of Bordeaux 1 # 4.1031/E/92-01) and MACEBUR (Contract EC-DG XVII/IN2P3 # 4.1301/S/94-87) Final Reports.
3. Energy Savings Home Entertainment Electronics, Copenhagen, 15–16 May, 1995.
4. First International Conference of Energy Efficiency in Household Appliances Florence (Italy), Nov., 1997.
5. Harris J.P. Private Communication
6. Schmitz R. "The GEA Consumer Information Scheme" (ref. 4).
7. Nakagami H. "Home Electronics Appliances in Japan" (ref. 4).
8. "Energy and Buildings" (special issue of Energy Efficiency Standards for Appliances Nov. 1997), ed. by A Meier).
9. "Housing and occupation of the housing stock in 1988 and 1992", INSEE Report 1995.

10. Monteiro S., "Household appliances in 1993" INSEE report no. 408, July, 1995.
11. Sidler O., "Lessons Learned from Measuring Compaigns of Electrical Appliances in the Residential Sector" (ref. 4).
12. Molinder O., "Leaking Electricity: an EU Characterisation Study" (ref. 4).
13. Bertoldi P., "European Union Efforts to Promote More Efficient Appliances" (ref. 4).
14. Bruens W., "The EACEM Commitment for TVs and VCRs" (ref. 4).

19

Strategies of the German Government to Encourage the Use of Renewable Energies

P.G. GUTERMUTH
Germany

ABSTRACT

The German Federal Government aims at promoting energy production from renewables as it provides domestic low-polluting energies, saves finite resources of fuel, and is at the same time widely acceptable to the public. Measures in favour of renewables are included in the list of the Federal government's actions to curb C+ emissions. This policy is supplemented at local level by under and municipal measures. The barriers in the way of faster market penetration by renewable energy has led to a package of political steps in the following fields: support for research, development, and demonstration; improvement of competitiveness; improvement of regulatory, administrative, and institutional conditions; improvement of information and education/training; improvement of international cooperation. Experiences from practical steps in Germany are manifold and can be of interest to other countries, especially as the most important market barrier—i.e. the lack of competitiveness in relation to conventional technologies—is typical for most countries.

1. POLITICAL GOALS

It is difficult for renewable energies (RE) to gain a foothold in the market. As prices of fossil energy sources are declining in real terms, the situation has become even tougher. Apart from some specific applications, they are not yet competitive. Economic incentives are needed to entourage wider use of RE.

Many countries promote the use of renewables. So does Germany, and Germany has been able to gain experience with a large variety of instruments. Perhaps this could be of help to other countries as well.

For a number of years, Germany's policy has been aimed at improving the scope for RE in the supply of energy. It is based on the assumption that RE can and must make a larger contribution to energy supply. In its policy the Federal Government is guided by considerations of energy and environmental policy in addition to considerations of industrial, employment and development policy. As an industrial country, Germany feels particular responsibilities towards developing countries.

In the view of the Federal Government there are the following possibilities, in the medium term, to enhance the use of RE in Germany:

- making greater use of the potential available from small hydropower plants
- making larger wind energy converters in appropriate sites competitive
- efficient use of heat pumps for hot water and space heating
- greater use of geothermal energy as far as this is economic
- support for reducing costs of photovoltaics facilities
- competitive use of solar collectors in water heating installations
- greater use of biofuels in transport
- more extensive use of active and passive solar energy in construction.

If a country is to have a sustainable energy policy, it must accept that there are technical and economic limits to the use of existing energy resources. This includes renewable energies. Their share in primary energy consumption in Germany is about 2 per cent. Experts estimate that technically—if not economically— exploitable additional resources account for about half of energy consumption. With a view to Germany this means: all of the forms of energy we use today will continue to be needed in the foreseeable torture. Nonetheless, by reducing specific energy consumption and shifting the emphasis on the various energy sources, it will still be possible to achieve the necessary alternations to energy supply structures.

Increasingly, one aspect of this is a variety of renewable energy sources which complement one another and for which natural fluctuations in availability are increasingly evened out by output management and storage. In addition to renewable energy sources and efficient use of energy, sustainability also requires changes in life styles.

2. FIELDS OF ACTION

i) The variety of aspects suggesting a more extensive use of RE, the different development stages of the technologies, and the nature of the barriers to RE necessitate a host of different reactions. There is no ideal way, no ready-made solution. Hence the strategy of the Federal Government emphasises three main aspects:

- **To prepare the market**, research, development and demonstration projects in connection with installations and materials for the use of RE are being promoted.
- **To support the market** itself, general conditions for the use of RE in the widest sense are being improved.
- To spread the use of RE by taking them also into account in **development assistance**.

This shows that a number of policy fields are involved (research, economic, finance, environmental, development policies and so forth).

ii) By concentrating on market measures in my remarks, I shall thus confine myself to a few specific aspects of the Federal Government's overall concept.

This is by no means a matter of financial measures alone, even though, seen as a whole, they are not insignificant:

- Total federal expenditure at market level reached approximately DM 165 million in 1994 to 1997. This does not include tax allowances (Home-owner Allowance Act) and tax revenue shortfalls due to special depreciation schemes (in particular Section 82(a) of the Income Tax Law Implementing Ordinance).
- Since 1974, roughly DM 32 billion has been spent on RE research and development,
- more than DM 3 billion on development assistance.

iii) On principle and for practical reasons, the Federal Government has refrained from setting quantitative RE objectives and quotas. It is relying on measures in conformity with the market and on voluntary commitments by those involved. Nonetheless, it announced its goal of a 25 per cent reduction of CO_2 emissions by the year 2005 compared with 1990. One of the stated measures for attaining this objective is the promotion of RE.

iv) Government measures in the market must be directed toward removing barriers to the market. Hence these must be identified. Major barriers are in particular.

- Low world market prices for competing energy resources.
- A lack of full cost-pricing when determining the cost of energy services.
- The high investment costs (due to low energy density and to the fact that the facilities are decentralised and their construction therefore open requires a very individual approach) and the frequently excessive profit anticipation make financing renewable technologies difficult, combined with a lack of financing sources and/or access to appropriate financing systems.
- Specific institutional, administrative and legal framework conditions.
- Lack of general awareness, of access to information, and of education/training.

To remove these barriers, the Federal Government is primarily taking the following steps:

1. Internalisation of external affects as far as this is justifiable in view of uncertainties as to quantities and of existing structures.
2. Market-economy incentives to improve competitiveness, in particular low-interest loans, grants, allowances tax benefits.
3. Regulatory measures to facilitate market access and to remove legal impediments.
4. Promotion of information and advisory services.
5. Improving general awareness and training.

3. THE MEASURES IN DETAIL

i) Internalisation of External Effects

When it comes to internalising external effects, the measures described earlier are inter-linked. In other words, to include the external costs of energy production and consumption (health hazards, damage to property, the environment and the climate) in decision making and cost calculations, both market-based incentives (levies and charges, taxes, emission rights/certificates) and regulatory measures, such as statutory minimum prices for electricity produced from renewable energy, should be considered (Electricity Feed Law, EFL).

Admittedly such measures with largely indirect effects will only have a generalised impact. But they are practicable and may at least correct ecologically wrong signals. More sophisticated steps may not be possible, anyhow, given the methodological problems in defining external effects and quantifying them (Prognos, 1992, 1995).

Such a concrete step would for example be a C per cent tax. In the past few years, the European Commission has submitted proposals on this to the Member States, most recently concerning a tax on energy. Since 1991, the Federal Government has continuously argued for a solution which is co-ordinated at international level, or at least harmonised at EU level and neutral in its effect on revenue. The debate within the EU is in full swing it is impossible at present to forecast any results.

In my view, the oil tax also helps to set the right cast signals, at least for two sources of energy (oil and natural gas). The signalling effect is even more obvious if renewable energy sources, such as fuel produced from biomass, are exempted from this tax, as is the case in Germany.

ii) Financial Incentives

A) Effective financial promotion measures include low-interest loans with long maturity periods. The Deutsche Ausgleichsbank (DtA) grants such loans within the framework of the ERP environmental protection and energy conservation programme as well as the DtA environmental protection programme. More than DM 4 billion has been made available since 1991. The large share spent on wind energy converters stands out in particular, followed by hydropower and biomass systems.

In May 1996, the DtA launched a "50,000 roofs-solar-initiative" which is not only addressed to small and medium-sized enterprises, as the regular loan programme is, but is also directed toward private households. It may be combined with grants.

So far, just under 3,000 projects have been supported within the framework of this special initiative by loans worth more than DM 80 million. For the most part, loans were used to finance solar collectors (72% of approvals) while heat pumps accounted for 15 per cent, photovoltaics facilities for a mere 11 per cent and biomass plants for 2 per cent.

These figures are interesting because they demonstrate quite clearly that demand for loans varies greatly. They are required above all

- in the case of larger investments (which in any case are highly dependent on loan financing) in technologies dose to economical operability (lower risks), and
- where specific circumstances require a flexible combination of support instruments (e.g. in the case of the need for risk assumption or total financing).

B) Grants : The Federal Government has gained experience with four programmes in this context:

a) The Federal Ministry of Economics has provided grants for solar collector systems, wind energy and hydropower plants since 1994, and also for heat pumps, photovoltaics facilities, biomass combustion and biogas plants since 1995. Funds made available in the period 1994 to 1998 total approximately DM 100 million. Grants were extended for the following projects (as of 01.04.98):

766	photovoltaics facilities
112	hydropower plants
63	wind energy converters
62	biogas plants
11,240	solar collectors
1,294	biomass combustion plants
1,769	heat pumps
15,306	**total installations**

The figures clearly show that the programme concentrates primarily on photovoltaics and thermal poker, the reason being that these sectors can only derive little if any benefit from the Electricity Feed Law, and that, for one reason or another, the target groups frequently make little use of available low-interest loan programmes.

According to present plans, this balance of emphasis is in principle to be retained beyond 1998, as the Economics Ministry is in favour of continuing the programme.

DM 23 million annually has been appropriated in the draft budget for the next four years.

Within the framework of the DM-100-million programme of market incentives, a "Sun-in-our-schools" project has been introduced. The Economics Ministry extends grants to schools and the Federal Ministry of Research supplies measuring installations. Schools must commit themselves to carrying out a measuring programme over a period of three years to be monitored by the Research Ministry. Young people are to be given the opportunity to get to know this forward–looking technology at an early stage. While only a single application was filed in 1995, there were as many as 69 in 1996 and 217 in 1997. So far, some 150 photovoltaics projects in schools have been supported.

With grants worth roughly DM 70 million, the Economics Ministry's programme has helped to launch investments in the amount of about DM 320 million. There is very strong demand. More than 60,000 applications have been received, in other words many more than could be approved.

This support instrument is particularly attractive as it produces direct liquidity effects. This is a decisive factor for investors in many cases, especially when less competitive technologies are involved (e.g. photovoltaics) and when investors are not accustomed to loan financing or hesitate for other reasons to finance by credits (e.g. private investors and small investment amounts). In any case: the funding must be extensive enough to prevent a "flash-in-the-pan" effect.

b) Capital subsidies are also given, e.g. in the form of operation cast grants for the energy produced.

This type of grant has been applied with great success in Germany to promote energy from wind energy converters (the 250 MW—wind programme of the Research Ministry). It has encouraged the development of reliable technologies, and of larger and more economic wind turbines.

These grants over the advantage that support is granted only to the extent to which facilities actually produce. The interest of the operator to keep the plant permanently functioning and to maintain and repair it appropriately is therefore particularly great. But this type of support requires more administrative effort than the one-off investment cost grant, owing to the accounting needed at least at regular periods and the associated ongoing recording of earnings.

c) Until the end of 1998, tax allowances for solar installations and heat pumps in newly built homes may be extended under the law governing allowances for owner-occupied homes, totalling up to DM 500 in the year the home is built or acquired and the same again for each of the next seven years. In contrast with gran, there is a legal entitlement to the tax allowances. Unfortunately, no figures are available on total support provided.

d) Farmers may obtain grants towards investment costs and low-interest loans under the agricultural investment support programme (AFP), a joint scheme of federal and under governments. This applies, for example, to solar units, biomass plants and heat pumps. Applications are to be filed with the responsible Office for Agriculture or the Chamber of Agriculture of the Land concerned.

C) Tax Measures : Special depreciation and allowances for special expenses under the "Law concerning special depreciation allowances and tax deductions in the assisted area" may be claimed in the eastern part of Germany since 1991. I do not have any figures on the shortfall in tax revenues.

But I can show you what it can amount to, by taking special depreciation allowances pursuant to Section 82(a) of the Income Tax Implementing Ordinance as an example. The section applied to investments that were completed by the end of 1991. It is no longer in effect for new projects. From 1978 (the year in which the Ordinance was amended to include renewable energies) until the end of 1996, tax revenue shortfalls amounted to more than DM 9.5 billion. The Economics Ministry is having a survey conducted to find out how much of this was accounted for by RE or by rational use of energy.

Deductions of income tax and of profit tax for corporations and higher write-off possibilities in the initial years for investors or users (accelerated depreciation allowance) are therefore likely to have proved powerful means to stimulate the use of renewables.

But tax benefits have no impact where taxes are not paid in any noteworthy amounts. Direct grants or low-interest loans are the more suitable means in these cases.

D) Capital subsidies, preferential loans and tax privileges may be granted singly or in combination. Germany's experience of this is shown in Table 2.

All three instruments have, for instance been used for wind turbines together with the higher buy-back rates pursuant to the

Electricity Fee Law. The result has bene the German "success story for wind energy". The installation of wind energy converters has boomed to such an extent that Germany now ranks first in installed capacity worldwide, with a rise from 62 MW at the end of 1990 to an installed capacity by the end of 1997 of over 2000 MW (Allnoch, 1997). At present, Germany is mass-producing the world's most powerful wind energy converters. 133 wind energy converters in the 1 to 5 MW capacity phase were operating at the end of 1997, more than in any other country. The main methods of promoting wind energy converters at present are the Electricity Feed Law and low-interest loans. Without this assistance, particularly in areas without high wind velocities, wind converters could not yet operate economically in Germany.

E) Guarantees could be the best option for particularly risky ventures, such as drilling operations in connection with the use of geothermal energy.

F) Finally, it should be pointed out that Germany also supports promotion programmes at EU level, such as Joule/Thermie and Altener.

iii) Regulatory Measures

On the one hand, regulatory measures may help RE to penetrate the market. On the other, existing statutory and regulatory provisions may impede the expansion of RE. The reason often is that legal and regulatory provisions frequently take insufficient account of RE's ecological and resource-saving features. But differences in the interpretation of legal provisions, lengthy approval procedures and some very restrictive conditions and compensation measures in the protection and conservation of nature must also be mentioned in this context.

Let me give you three significant examples of regulatory measures favouring RE:

- the introduction of the Electricity Feed Law
- adjustments to the Federal Building Code
- the amended version of the Energy Industry Act.

A) Act on feeding electricity from renewable energies into the utility grid (Electricity Feed Law)

The Electricity Feed Law (EFL) came into effect in 1991. It set minimum payment levels for electricity generated from wind, hydropower, biomass, biogas, landfill gas, sewage gas and solar energy (up to 5 MW installed capacity), where the plants are not operated by the electricity companies themselves. It currently obliges electricity companies to buy electricity from the plants situated in their area of distribution at a price between 65–90 per cent of electricity prices to all end users depending on the energy source (90% means in 1998: 16.79 Pf/kWh).

The strengths of the Law are the following:

- It has caused perceptible optimism towards some of the renewable energy sources in the electricity sector. It has made life much more economic for the operator of wind power, and it has improved the economic base for electricity from hydropower and biomass. Naturally, the law does not apply to the heating market (water/space), since other measures (investment grants, tax benefits etc.) are in force there.
- It enhances the readiness to operate such plants (regular maintenance, speedy repair work), as plants will only generate earnings when they are operating.
- Guaranteed minimum payments make it easier for the investor to estimate costs and for the banks to finance projects.
- The resultant steering effect is in line with the "polluter-pays" principle, but only in the electricity sector.
- The system is comparatively easy to apply and largely avoids the problems of implementation that are typical of subsidies.

However, as experience has shown, the Law also has its weaknesses:

- Given different geological and climatic conditions, the additional financial burdens imposed on system operators vary from region to region and may lead to distortions of competition. Hydropower, for instance, is used mostly in the south, while the best conditions for the use of wind power are in the north of Germany. In addition to this larger regional distribution, there are also local burdens, in particular for smaller system operators.
- As there are no limits to the total amount paid for electricity fed into the utility grid and there is no time limit, additional burdens for the system operator may increase considerably as the number of plants grows.

According to the VDEW (Association of the German Electricity Supply Companies), the added costs paid by electricity companies for electricity from renewables amounted to DM 280 million in 1995 and DM 780 million in 1996.

- As the Law also applies to installations that existed at the time it came into force, there is the risk of free-rider effects.
- The relatively inflexible method used to calculate payments may result in over-compensation of the costs of plants operated at particularly suitable locations.
- The obligation to purchase electricity at minimum prices reduces competition between the suppliers of electricity from regenerative sources.
- The hardship clause could curb developments in favourable sites in particular.
- Utilities have contested the constitutionality of the EFL, arguing that subsidies for renewables must be financed through public fixing, as with coal. Until now, the Federal Constitutional Court has not decided in substance on this matter.

Against the background of this experience, the Law was amended only a few weeks ago. The amendments are intended to eliminate the problems that have arisen, to clarify open questions and to take account of the liberalisation of the electricity market within the EU. The main elements of the amendments are:

- In future, it is no longer the electricity utility company that is obliged to purchase and pay for electricity but the system operator. This ensures that, under the new rules on competition, the independent producer wanting to feed electricity into the grid will still find a company that is obliged to purchase it.
- The Law will also apply to the use of all biomass.
- Support for off-shore units is specified.
- Electricity utility companies are requested to step up their efforts to use renewable energies (plus co-generation of beat and power) by way of voluntary agreements. After consulting with the parties involved, the Federal Government could lay down certain objectives if the need arises.
- The hardship clause in the Electricity Feed Law presently in effect is defined more precisely and extended as follows:

Where an electricity utility has to accept an amount of electricity under the Electricity Feed Law that exceeds 5 per cent of its sales, the economic burden has to be borne by the up-stream system operator. The utility remains responsible for purchasing and paying for these quantities as well, but it is entitled to reimbursement from the upstream utility (in other words, compensation among the electricity utilities will not affect independent producers).

Independent producers are only affected once the 5 per cent-limit is reached at the highest level of the grid, which normally is the interconnected grid. However, the payment obligation becomes ineffective only for new installations, namely for those where important parts have not been completed by the beginning of the next calendar year. Statutory payment remains fully in effect for an unlimited period for all plants that are operating or being built.

With respect to this second 5 per cent limit, the law provides that the Federal Ministry of Economics has to inform the Bundestag about the effects of the hardship clause in good time before the 5 per cent limit takes effect, at the latest in 1999.

Finally, I would like to mention that the Law only sets minimum prices. About 40 German municipalities are paying prices for electricity produced from photovoltaics that exceed EFL rates substantially and several municipalities are nearly matching the costs of photovoltaic operations (i.e. up to 1.15 $ US/kWh). Such voluntarily fixed cost-covering prices for a limited period (for example 2 DM/kWh for ten years offered by the Stadtwerke Munchen) are stimulating the German photovoltaics market perceptibly.

B) Two Provisions of the Federal Building Code Amended

a) Under German legislation, rather restrictive provisions apply to building projects in peripheral areas. Section 35(1), subpara.6 of the Federal Building Code stipulates that wind power and hydropower installations are considered privileged projects which may be approved in such areas. However, it grants local authorities a planning right allowing them to restrict the privilege or more or less cancel it.

b) Moreover the Code provides that the use of RE has to be taken into account in urban development planning, Section 1(5) subpara.7.

This is an important indication for authorities (usually the municipalities) preparing urban development plans. Building codes for the development of local real estate and the construction of buildings must therefore take into account the possibilities of the use of solar energy, both passively and actively, e.g. by ensuring that the orientation of buildings, the gradient of roofs, and the formation of shadows facilitate the optimal use of solar radiation. This involves no additional costs and is nevertheless a field of great potential for saving heating energy, for conserving fossil energy sources, and for taking pressure off the environment.

Urban development planning is all the more vital for RE as it may lay down building requirements for decades, if not centuries, with respect to an optimum use of the potential offered by RE in buildings.

C) Amendment of the Energy Industry Act

Structural change in energy markets has become considerably Inter in Europe. The EU Council Directive concerning Common Rules for the Internal Market in Electricity of February 1997 represents an important step in the energy policy field. It has to be translated into national law by February 1999. In Germany, this will be implemented by the Energy Industry Act that has just become effective.

The more competition-oriented rules are however not to be applied to the detriment of the environment. Therefore, environmental compatibility has been included in the objectives of the Energy Industry Act. Thus the Act places equal importance on objectives of energy policy and environmental requirements.

The Act deals with RE in various places:

- Section 2(5) of the Energy Industry Act correlates this law and the Electricity Feed Law, it is thus an integral part of the overall concept.
- Environmental compatibility has to be taken into account in state supervision (electricity prices, cartel).
- Feeding electricity into the grid of an electricity utility company no longer is subject to approval (up to the present, plants with an installed capacity of more than 1 MW had to be approved).
- Also, supplying customers outside the general supply system is no longer subject to approval if the electricity

supplied is for the most part generated from renewable energies.

- In cases where there is the risk of electricity generation from renewables being prevented, it is justified to refuse third parties access to the grid.
- Up to the year 2005, municipalities may refuse to grant the right of way for direct lines if there is a risk of power generation from renewable energies being prevented (refusal to approve line construction).
- Tariff customers may produce electricity from renewable energies for their own consumption without losing their entitlement to being connected and supplied at the terms and conditions for tariff customers.
- Cost incurred in least-cost-planning projects may be taken into account in approving electricity rates.

D) In speaking of regulatory measures one should not forget to mention an aspect which is often overlooked since it is hardly spectacular: I am thinking of standardisation.

Standardisation (national, regional, global) for energy equipment and fuel quality contributes considerably to the reduction of costs. Germany has, for example, proposed a standard for biodiesel. Producers of cars should get the possibility to add biodiesel to their products. This enlarges selling opportunities for this type of fuel.

E) Apart from the electricity sector, statistics on renewable energies are far from satisfactory. This is due, in the first place, to the fact that RE use is decentralised and it takes a lot of effort to compile the statistics. Nonetheless, it should not be underestimated that informative data may be useful in particular where the sensible employment of public support is concerned. For a number of years, the Federal Government has advocated an improvement of official statistics, thus far without any decisive effects, mostly for financial reasons.

iv) Information and Advisory Services

Technologies for the use of RE are for the most part employed in new and greatly varying sectors. Hence, more technical and financial information, consulting, and know-how of users, architects, technicians, engineers, and craftsmen is required. Goods-producing companies and service companies also frequently do not have sufficient information, and market transparency, that also hampers their initiative. The German government therefore engages in activities such as :

- Information and advice for individual consumers and small to midsize enterprises.

Since 1990, the Economics Ministry has spent more than DM 56 million on RE and rational use of energy, and approximately DM 84 billion on advisory services. In addition, the Research Ministry grants support for information and advisory services (BINE).

- Proposal of voluntary commitment by producers to better label energy data on appliances, vehicles, energy transformers etc. with the aim of improving overall energy efficiency.

All in all, there is a growing interest in renewable energies to be observed in Germany. But, on the other hand, there is also increasing criticism of installations. People involved in environmental protection and those in favour of the protection and conservation of nature take controversial stands. This results in considerably longer approval procedures for hydropower plants, it increases the cost of such installations and sometimes leads to refusal and legal proceedings. In the case of wind power, opponents argue that:

- the view of the landscape is spoiled,
- birds are disturbed in their breeding habits, in extreme cases birds may even be killed,
- converters are noisy.
- revolving rotor blades cause light reflections,
- people living in the vicinity of wind energy converters find their need for rest and recreation seriously impaired.

True, some of these arguments may seem exaggerated, but still the criticism should be taken seriously, after all, acceptance on the part of the population is indispensable. Hence, substantive information and advisory services are all the more needed, particularly where plants are in the planning stage.

Finally the federal government tries to set good examples:

It is seeking not only to reduce energy consumption in its property but to use renewable energies. Especially buildings being constructed in Berlin are to serve as models. The Economics Ministry, for instance, plans to have a 100-kWp photovoltaics facility built on top on one of its office buildings to set a visible examples.

v) Education and Training

RE will hardly be successful in the market without appropriate education and training measures for the occupations involved. In Germany education is the responsibility of the states (under). The Federal Government has, therefore, appealed to the states to include the treatment of renewable energies and other energy-environment issues at all education levels, Vocational training and education in the professions involved (training of architects, technicians, craftsmen) are in the curriculum of more than 40 German universities and colleges. Grants are given by the Federal Government for courses and presentations by specialised associations, craft chambers, and organisations.

4. ENHANCED INTERLINKING OF PROMOTION MEASURES

In conclusion, let me deal with an issue that, although it is specifically German, may become important in any federative system, and hence in the European Union, for example, In Germany, not only the federal government are involved in the promotion of RE, but the Länder as well. As far as financial assistance in the market is concerned, expenditure by the Länder even exceeds that of the federal government by far.

According to a survey by the Economics Ministry, market incentives for renewable energies in Germany amounted to about DM 2.22 billion in 1991–1996 (excluding tax revenue shortfalls due to tax benefits and excluding home-owners allowances). The Federal Government's share was approximately DM 165 million and that of the Lander about DM 1.06 billion.

It has turned out that producers, traders and investors find it very important that the transparency of measures is ensured that are intended to promote market penetration of RE. Therefore, they call for promotion measures at the various levels to be better harmonised and more easily identifiable.

This is very much in the interest of the Federal Government in view of enhancing the effectiveness of the support programmes.

Last year, the Economics Ministry discussed this issue with the experts in the ministries of the under who are responsible for RE: As a first step, an effort is to be made to exploit possibilities of harmonising pre-conditions for the promotion of photo-voltaics, solar collectors and heat pump projects. The Economics Ministry will prepare a proposal on this issue and discussions will continue.

Support for Renewable Energies by DtA 1990–1997 (Loan Commitment in Million DM)

	1990	*1991*	*1992*	*1993*	*1994*	*1995*	*1996*	*1997*	*Total*
ERP Environmental Protection and Energy Saving Programme									
Wind Energy	35.6	56.7	66.2	213.9	439.0	417.5	538.3	720.0	2,487.2
Hydropower	7.5	9.9	17.1	24.3	40.0	21.3	46.7	55.7	222.5
Biomass	1.4	5.5	3.9	73.2	39.5	34.4	32.8	14.5	205.2
Solar Energy	0.4	0.2	1.2	0.3	3.7	0	1>5	3.2	10.5
Other	2.9	4.2	3.7	0.1	10.2	0>3	1>3	1.8	24.5
Total	47.8	76.5	92.1	311.8	532.4	473.5	620.6	795.2	2.949.9
DtA Environmental Programme									
Wind Energy	5.3	20.1	35.9	89.5	172.3	168.7	209.8	290.8	991.6
Hydropower	0.9	2.1	6.8	10.1	19.8	6.7	10.7	25.7	82.8
Biomass	0.1	2.7	1.2	7.3	20.3	13.2	11.3	4.4	60.5
Solar Energy	0.1	0	0	0.0	0.8	0	32.5	51.7	76.1
Other	1.1	0.8	0.2	0.0	0.1	0.1	2.8	9.5	14.6
Total	7.5	25.7	44.1	106.9	21.3	188.7	257.3	382.1	1,225.6
Grant Total	**55.3**	**102.2**	**136.2**	**418.7**	**748.7**	**662.2**	**877.9**	**1,177.3**	**4,175.6**
Investments	**114.5**	**196.2**	**254.4**	**682.6**	**1,133.8**	**1,005.2**	**1,353.0**	**1,798.0**	**6,536.7**

20

Is There Convergence on Energy Efficiency Trends in Europe ?

D. BOSSEBOEUF
France

INTRODUCTION

- Post Kyoto: Towards a European solidarity and convergence for CO_2 emissions;
- The lack of technical knowledge for long term perspective of energy efficiency;
- Energy efficiency policies on demand side are a necessary option of CO_2 abatment strategies.

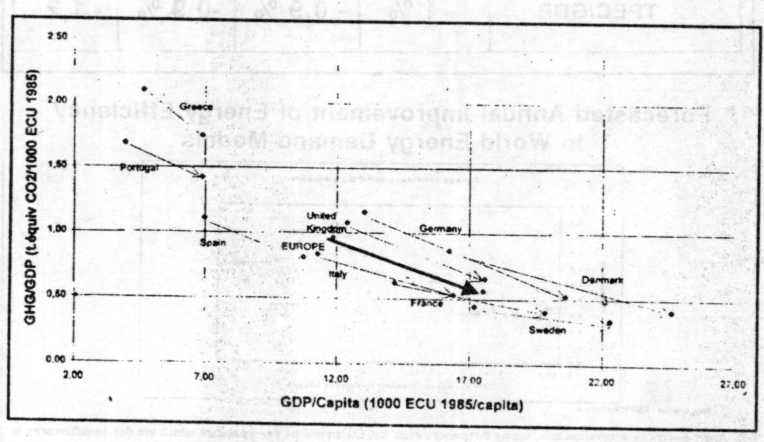

Greenhouse Gases Emissions/GDP (1990–2010)

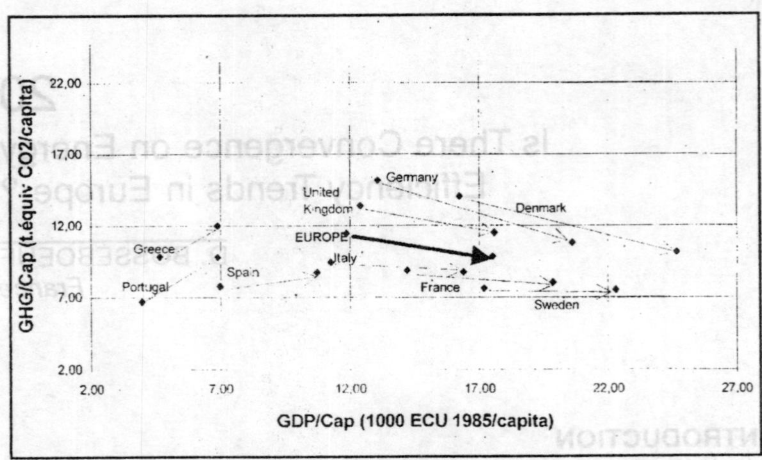

Greenhouse Gases Emissions/Capita (1990–2010)

	IEA 96	IEA 97	WEC	POLES
	1993 - 2010	1995-2015	1990-2020	1990-2030
Energy Intensities TPEC/GDP	- 1 %	- 0,9 %	-0,9 %	- 1,2

Forecasted Annual Improvement of Energy Efficiency in World Energy Demand Models

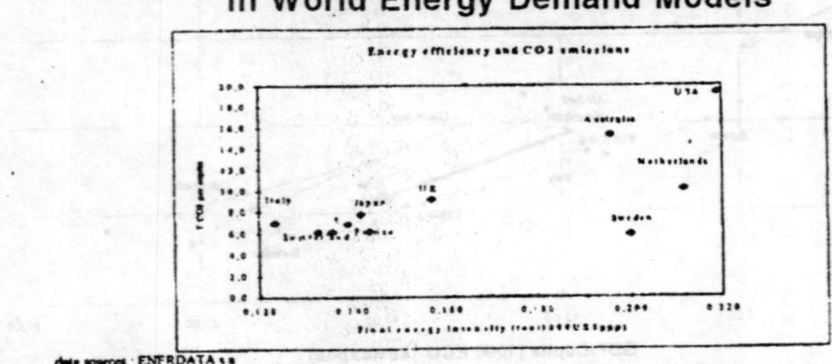

⇒ Apart electricity generation mix, strong differences exist in CO2 emissions per inhabitant which are due to differences in overall energy efficiency among countries

⇒The potential for future decrease in CO2 emissions is higher in countries with less energy efficiency and/or less progress in energy efficiency in the last 15 years

Apart Electricity Generation Mix, Energy Efficiency in the Main Factor for Improvement in CO_2 Emission Pattern

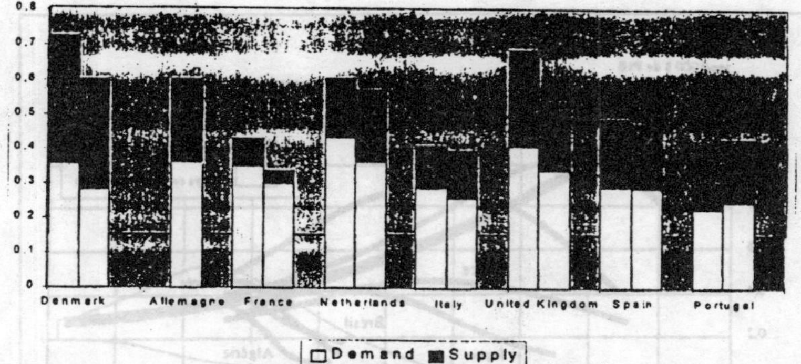

CO_2 Emission Abatement: Demand Versus Supply 1995

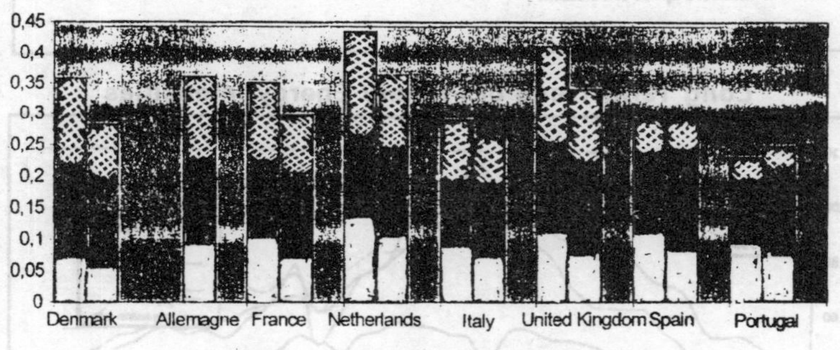

CO_2 Emission Abatement in Demand Side (1995)

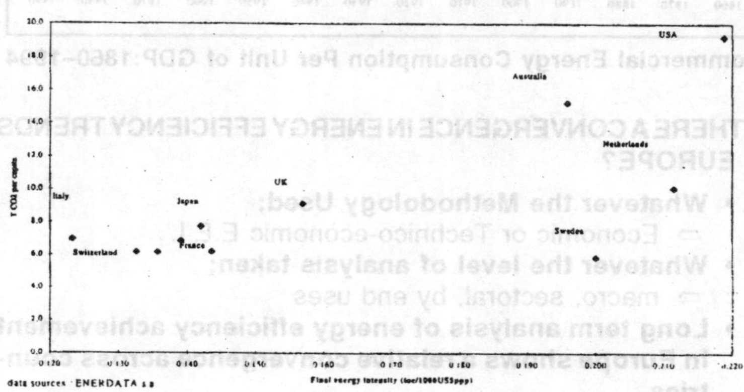

Energy Efficiency and CO_2 Emissions

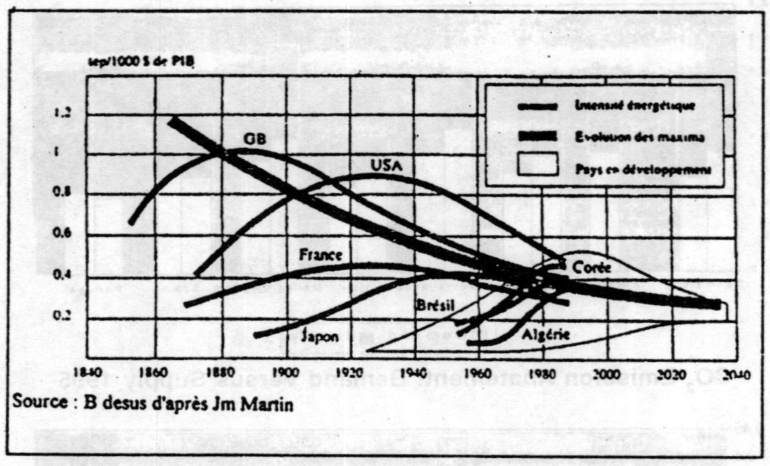

Long Term Trend of Primary Energy Intensities

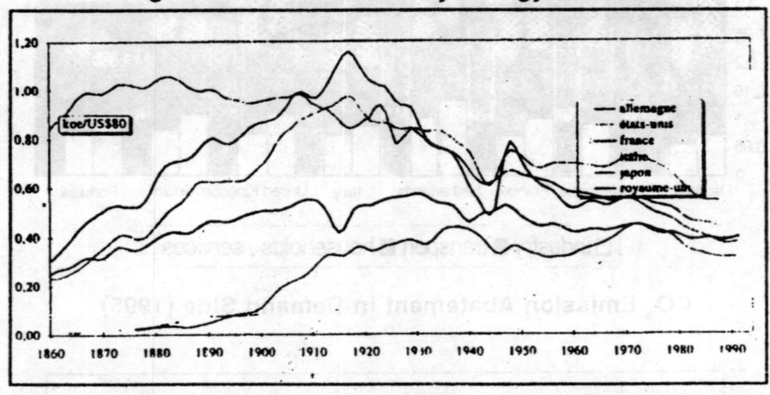

Commercial Energy Consumption Per Unit of GDP:1860–1994

IS THERE A CONVERGENCE IN ENERGY EFFICIENCY TRENDS IN EUROPE?

- **Whatever the Methodology Used;**
  - ⇨ Economic or Technico-economic E.E.I.
- **Whatever the level of analysis taken;**
  - ⇨ macro, sectoral, by end uses
- **Long term analysis of energy efficiency achievement in Europe shows a relative convergence across countries.**

Illustration

- Macro
- Industry
- Transport
- Residential
- Services

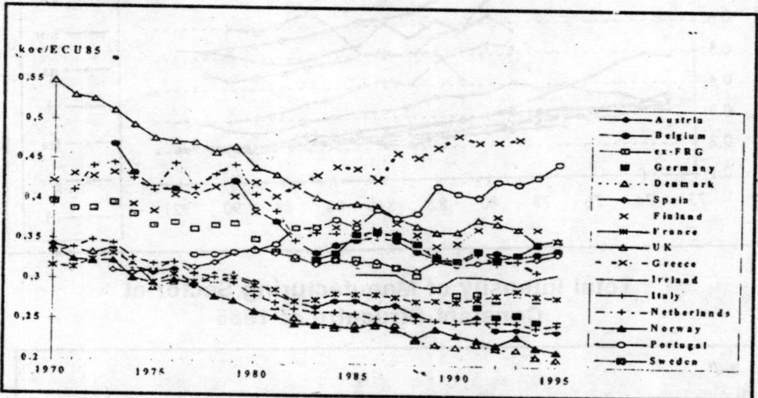

Primary Energy Intensity

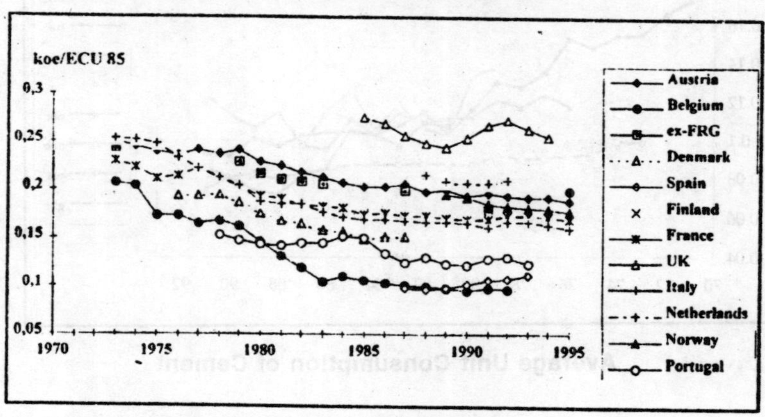

Final Energy Intensity at Constant Structure of GDP

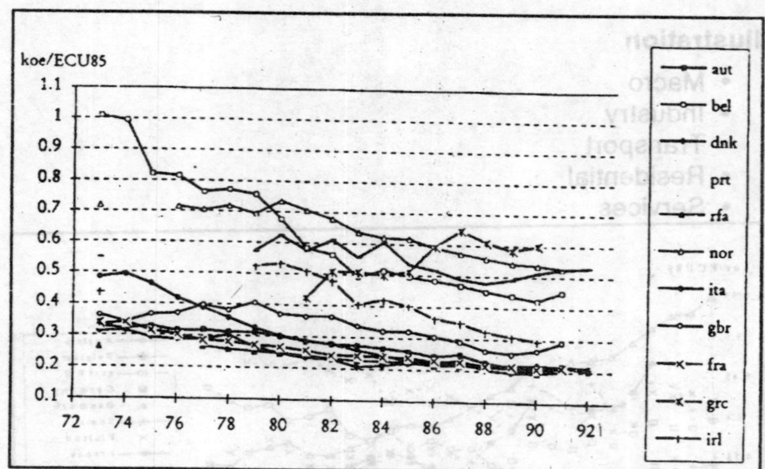

Total Intensity of Manufacturing Sector at Constant Strucutre of 1985

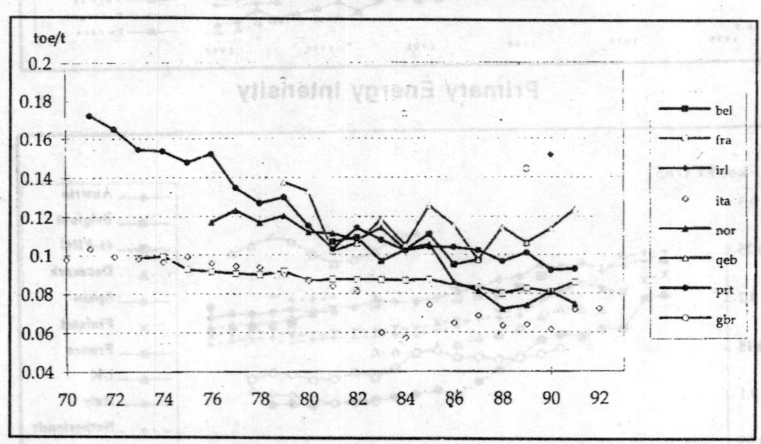

Average Unit Consumption of Cement

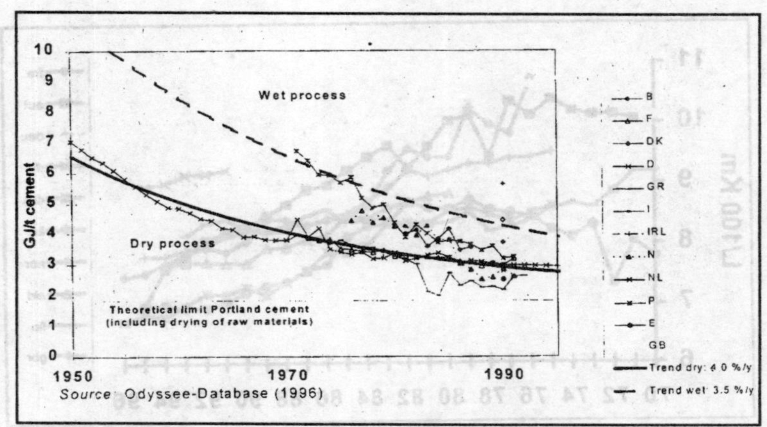

Specific Consumption of Fuels for Dry and Wet Cement Production (GJ/ton)

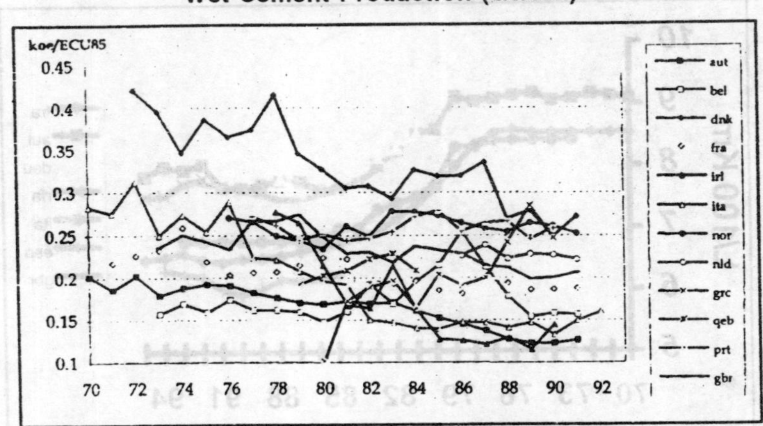

Energy Intensity of Food Industry

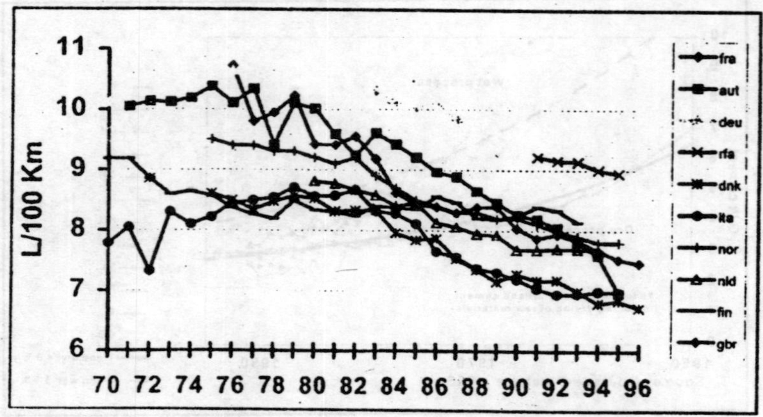

Average Specific Consumption of Cars (csvpc) (L/100 km)

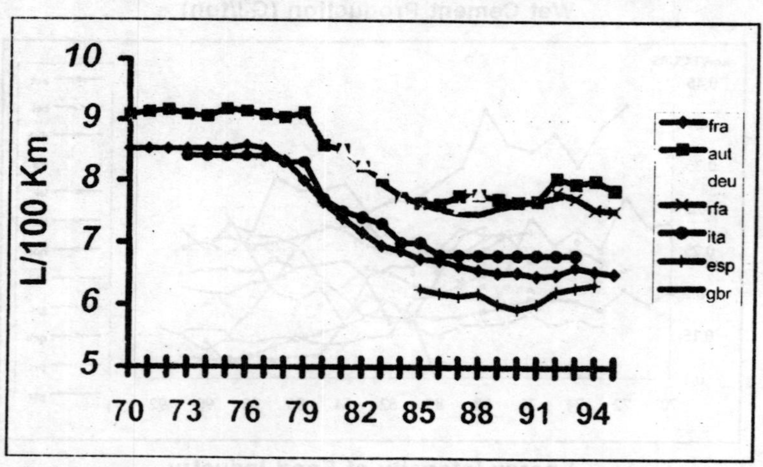

Tested Values of Specific Consumption of New Cars (cscbavpnth) (L/100 km)

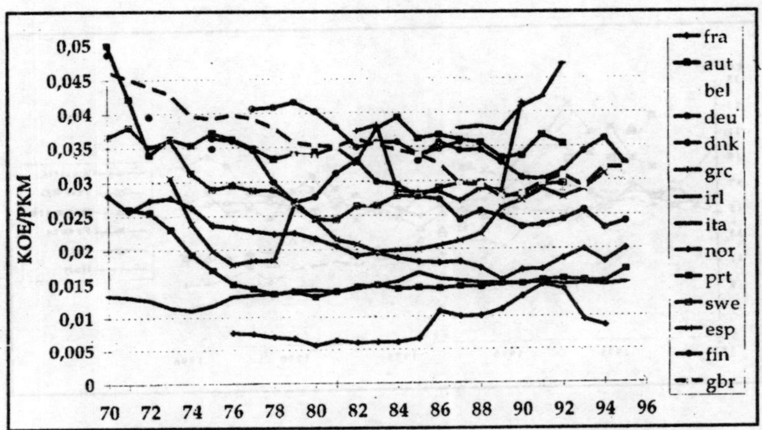

Unit Consumption of Passenger Rail Transport (cutocferpkm) (Koe/pkm)

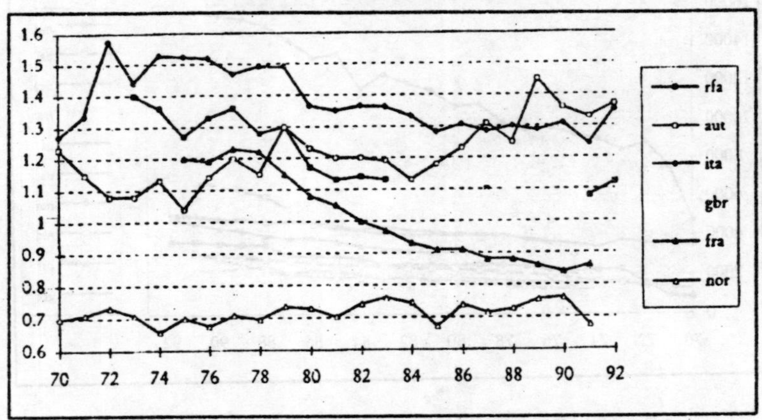

Average Unit Consumption per Dwelling for Space Heating, Scaled to European Average Climate

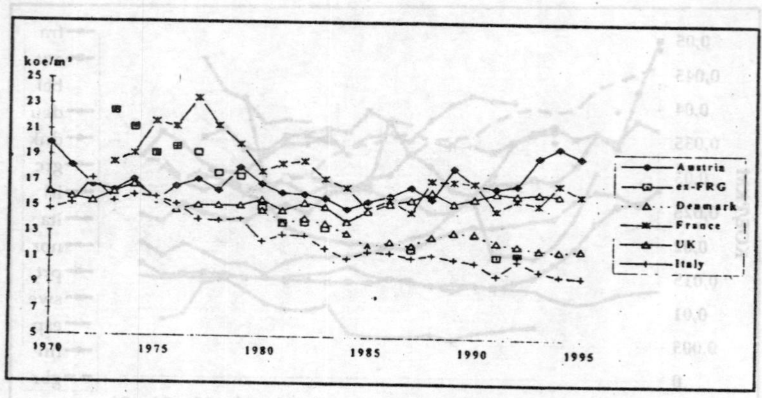

Average Specific Consumption per m² for Space Heating

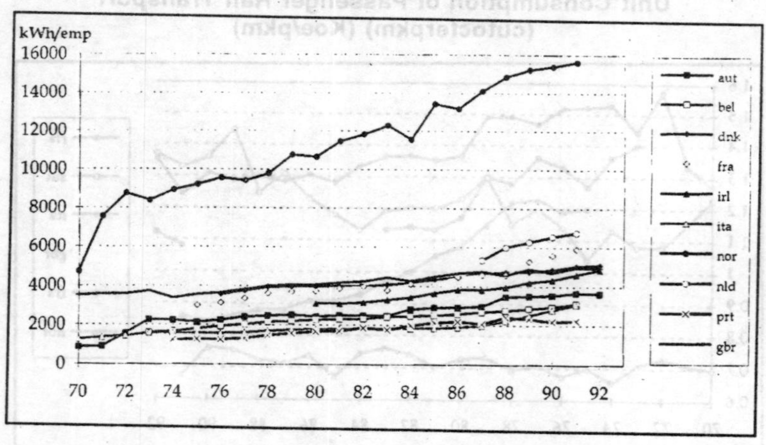

Average Electricity Consumption per Employee in the Tertiary Sector

A CONVERGENCE IN EE TRENDS IN EUROPE

Conclusion

- **We expect a convergence of energy efficiency levels among European, OECD and developing countries. This trend should occur because:**
  - ⇨ The Convergence in economic and life style context

⇨ The Convergence in the political statement of energy efficiency option (CO_2 targets) and technical progress.

⇨ The rhythm of convergence is uncertain particularly for non OECD countries

RECOMMENDATION

- In order to consolidate hypothesis on energy efficiency trend in the long term modelling:
    - **To improve the knowledge on the "characterisation" of the demand side technologies,**
    - **To design demand side scenario with technological and life style rupture.**

21

Joint Implementation of Energy Efficiency Improvements in the CIS: A Challenge for Greenhouse Emissions Abatement

G. ASLANIAN
Russian Federation

1. POPULATION GROWTH AND ENERGY DEMAND

Over the next half century global population is predicted to increase by almost 75 per cent, so that by 2050 nearly 10 billion people will inhabit the Earth. Population in developing countries are projected to increase by 80 million people each year, doubling by 2025 and tripling by 2050 in Africa. These increasing population, in turn, will increase global demand for energy. Despite the apparent abundance of fossil fuels, 40 per cent of the world population—over 2 billion people mostly in the developing countries, have no access today to commercial energy.

The International Energy Agency (IEA) has recently released its World Energy Outlook—1996 edition projecting by 2010 an increase in world global energy demand by 34 to 46 per cent over the current level. The World Energy Council Message for 1997 indicates at least a doubling in world demand by 2050. Most of this increase will be met by fossil fuels, with an estimated annual average increase in solid fuels of between 1,6 per cent, reaching over 3000 Mt o.e. in 2010, compared with less 2300 Mt o.e. in 1993.

Two alternative projections are presented based upon differing assumptions on energy use and prices of energy—the Capacity Constraints and Energy Saving cases. In the Capacity Constraint cases the pressure of rising energy demand is moderated by an assumed increase in primary energy prices. In the Energy Saving cases rising energy demand is assumed to be moderated by improvements in energy use that are more rapid than those history alone would imply.

Because much of our global energy provided through the combustion of fossil fuel (coal, oil and natural gas), this increased energy demand is projected to increase related emissions of pollutant and greenhouse gases. For example, by 2010 anthropogenic emission of carbon dioxide are projected to increase by 36 (Energy Savings case) to 50 per cent (the Capacity Constraints case) from their 1990 level.

Most of the increase in emissions is expected to occur—in the Rest of the World countries (ROW)—countries other than OECD: developing countries and those in transition in Central and Eastern Europe (CEE) and the Former Soviet Union (FSU). In the OECD, CO_2 emissions in 2010 are likely to exceed their 1990 level by almost 28 per cent in CC case and by 13 per cent in the ES case. In both the CC and the ES cases, the only region where emissions in 2010 may be below their 1990 level in FSU/CEE. (*Figure 1*).

Within the OECD, North America is projected to account for the largest share of the increase in CO_2 emission, as this region consumes the bulk of the energy in the OECD.

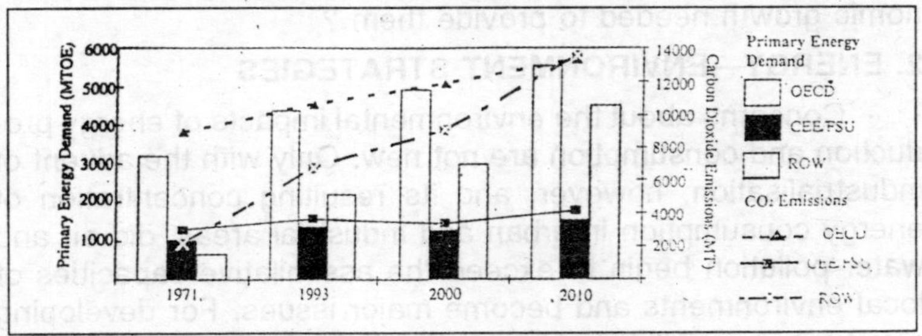

Fig. 1 : Projected Primary Energy Demand and Carbon Dioxide Emissions

Source: IEA World Energy Outlook, 1996

By 2010, in both cases, ROW is likely to have overtaken by OECD in terms of total CO_2 emissions. Within the ROW countries, China will remain the largest emitter of CO_2 and is projected to more than double its emissions by 2010. China's projected increase in emissions under the CC case of 2.7 billion tonnes is only slightly lower than the projected increase for the whole of the OECD of 2.9 billion tonnes.

Five dominant features emerge from the IEA's work:

- world primary energy demand is expected to grow steadily as it has over last two decades;
- fossil fuels will account for almost 90 per cent of total primary energy demand in 2010;
- the bulk of the world's energy will increasingly be consumed outside the OECD;
- to reconcile the increasing use of energy in support of economic and social development with protection of the environment, greater efforts are required to improve the environmental performance of fossil fuels, to raise more rapidly the efficiency with which all forms of energy are provided and used, to increase the use of non-fossil forms of energy;
- the reconciliation of economic and social development based on increased energy use and protection of the environment calls for truly global effort led by the industrialised countries.

As a global community, the challenge we face is: how can we provide the sustainable and shelter demanded by these populations, yet reduce the total energy required for the economic growth needed to provide them ?

2. ENERGY—ENVIRONMENT STRATEGIES

Concerns about the environmental impacts of energy production and consumption are not new. Only with the advent of industrialisation, however, and its resulting concentration of energy consumption in urban and industrial areas, did air and water pollution begin to exceed the assimilative capacities of local environments and become major issues. For developing countries, air pollution from industrialisation has now joined traditional indoor air pollution as a major challenge. At the same time, environmental concerns tin the industrialised countries have shifted more toward very long-term and global issues, most prominently climate change.

The type and extend of pollution are closely related to the degree of economic development and industrialisation. This is illustrated in *Figure 2*. To some extent, economic development enables societies successfully to address environmental problems of both poverty, like inadequate sanitation or indoor air pollution from traditional biomass use, and industrialisation, like

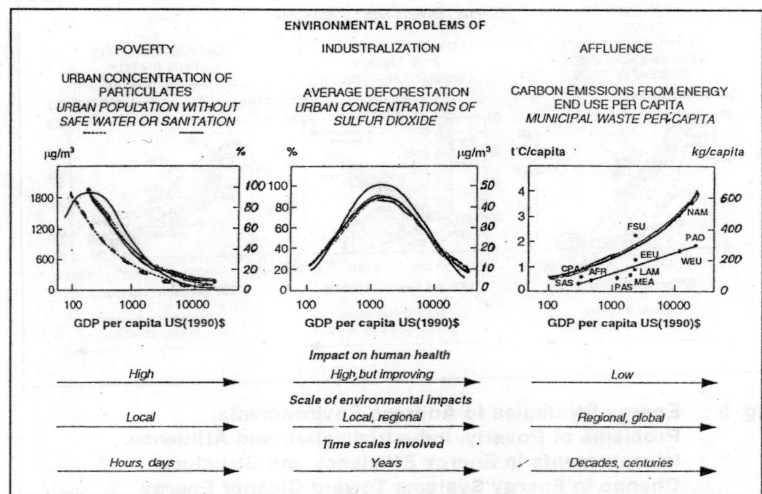

Fig. 2 : A Typology of Environmental Problems as They Evolve with Economic Development.
Source : Adapted from WEC—IIASA, 1995

sulfur emissions. But economic development and affluence can also generate new environmental problems, ranging from waste disposal to possible global warming.

Local impacts, including indoor air pollution, have been and continue to be the major environmental concern related to energy use. Indoor and outdoor air quality are equally important for many megacities of the developing world today. The cause is basically the same, emission-intensive fuels, such as coal and wood, burned in inefficient open fireplaces and cook-stoves. In industrialised countries, these problems have been successfully resolved in three ways. First, by the use of more efficient end use devices (e.g., central-heating systems and fuel-efficient stoves). Second, by the more complete combustion within these devices. And third, by the shift to clean (often grid-dependent) fuels such as electricity and gas (*see Figure 3*).

Compared to the historical reductions in environmental impacts through these sorts of changes in efficiencies and fuels, “end-of-pipe” environmental solutions applied to large point sources, are recent phenomena. They also have a more limited effect. They tend to reduce one pollutant in one particular

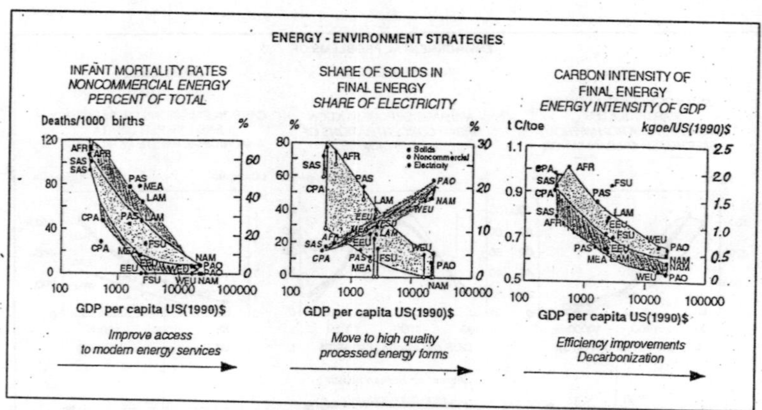

Fig. 3 : Energy Strategies to Address Environmental Problems of Poverty, Industrialisation and Affluence. Improvements in Energy Efficiency and Structural Change in Energy Systems Toward Cleaner Energy Carriers ("Decarbonisation") are Important Generic Long-term Strategies to Tackle Environmental Problems.

Source : Adapted from WEC–IIASA, 1995

location, rather than have the broader impact of a switch to cleaner fuels, or the adoption of more efficient end-use devices.

Despite major reductions in local air pollution problems in the industrialised world, these problems remain acute in the developing world. More than one billion people in developing world cities are exposed to unacceptably high ambient concentrations of suspended particulate matter and sulfur dioxide, significantly exceeding WHO guidelines. The situation with transport-related air pollution is also dramatic in many cities of the developing world. It is estimated that people in the poorest countries are exposed to air pollution levels orders of magnitude higher than in the industrialised countries, despite the fact that the latter are using orders of magnitude more energy .Due to cooking on open wood fireplaces, indoor air pollution in poor rural areas is more than 20 times higher than in industrialised countries. Particulate concentrations in urban areas of the developing world can be more than five times those in OECD cities.

That more time is spent outdoors because of the climate and inadequate housing conditions, only serves to compound the problem. Consequently, pollution exposure can be up to a factor of 20 higher in developing countries than in industrialised ones.

Local environmental problems should thus have first priority. As in the past, the most effective solutions will be those that are most comprehensive, relying on a mix of efficiency improvements, cleaner fuels, and "end-of-pipe" control technologies. Both efficiency improvements and the shift to cleaner fuels will yield triple dividends: lower resource use, lower overall energy system costs and lower emissions. Past structural changes in the energy system have moved in exactly these directions, although not yet quickly enough to offset the vast expansion of human activities. This is illustrated in *Figure 4* where the carbon intensity of primary energy over time is taken as a proxy for other pollutants. (Sulfur and, in many cases, nitrogen oxide emissions, are generally higher for high-carbon fuels, such as coal). Although the decarbonisation of the world's energy system shown in *Figure 4* is comparatively slow at 0.3 per cent per year, the trend is consistent and in the right direction. Measured in terms of final energy (*see Figure 3*), progress has been quicker, due to the increasing use of clean, grid-dependent energy carriers, such as electricity, district heat and gas.

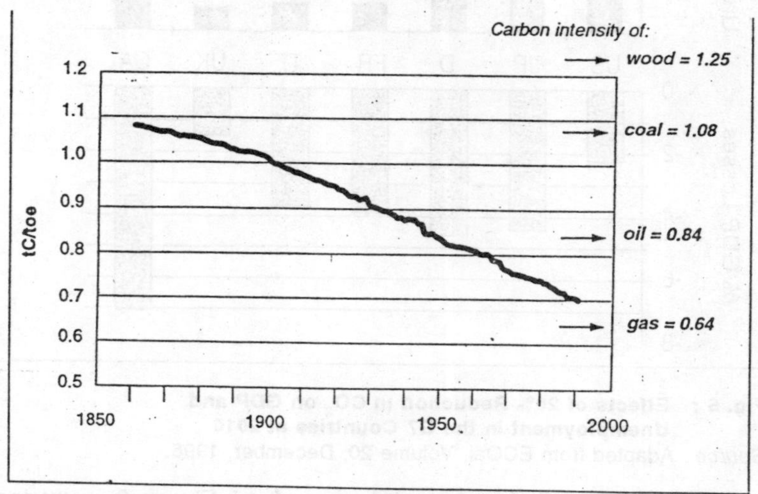

Fig. 4 : Carbon Intensity of World Primary Energy, 1850–1990, in tC per toe, including Emissions from Unsustainable Uses of Fuelwood

Source : Adapted from WEC–IIASA, 1995

3. ECONOMICS OF CLIMATE CHANGE MEASURES

Current proposals for near term and middle terms emission reductions in developed countries, which imply curbs on fossil fuel-based energy use, would result in large costs that inhibit economic growth now and would negatively affect trade, investment, competitiveness, employment and life-styles for both developed and developing countries, in all individual nations and regions. *Figure 5* presents estimates of GDP and unemployment losses if 20% reduction in CO_2 from 1990 greenhouse gas levels by 2005 will be committed. Comparison of effects on GDP of the both CO_2 emission 20 per cent reduction and stabilisation commitment aimed at reduction of national emission to 1990

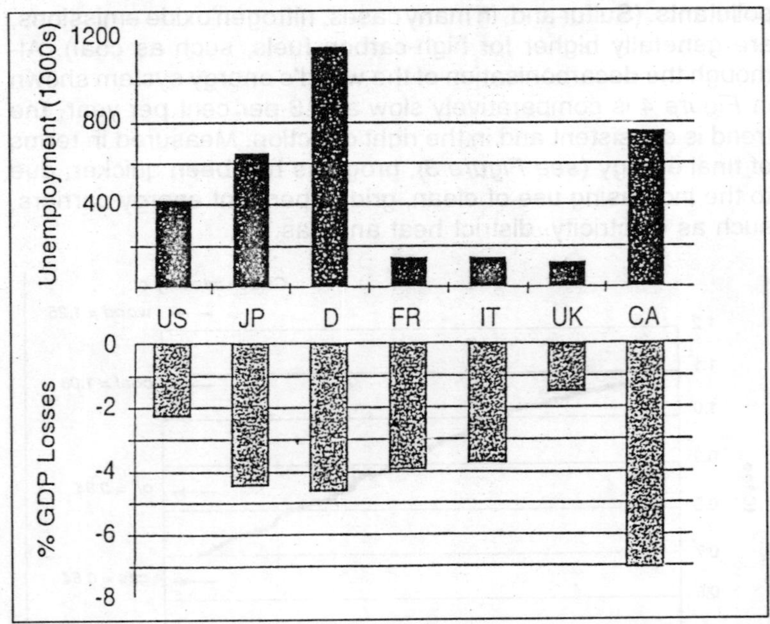

Fig. 5 : Effects of 20% Reduction in CO_2 on GDP and Unemployment in the G7 Countries at 2010

Source : Adapted from ECOal, Volume 20, December, 1996.

level, is given in *Figure 6*. As it is seen from Figure 6 countries more heavily reliant on fossil fuels, such as New Zealand, Australia and Canada, are estimated to experience the largest

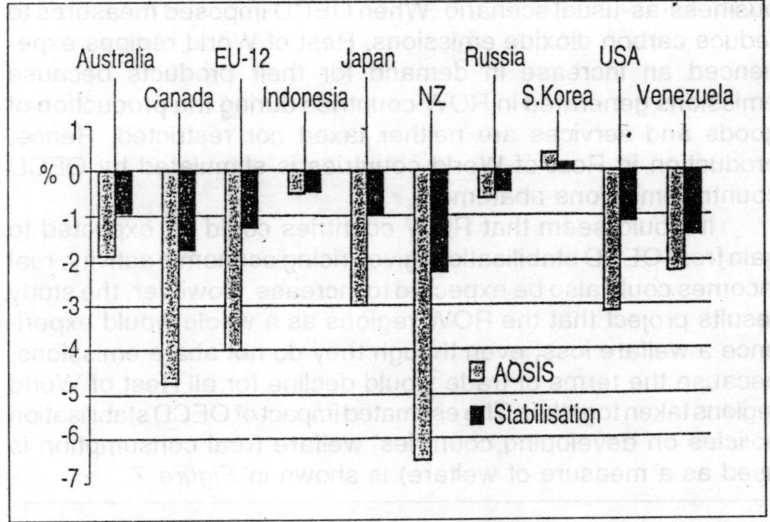

Fig. 6 : Effects of 20% Reduction in CO_2 on GDP in Selected Countries at 2010

Source : Adapted from ECOal, Volume 20, December 1996

losses in GDP. Taking into account the estimated welfare losses the benefit of these proposals are highly uncertain and would not be realised for years. It is also unlikely that over the next few decades fossil fuel use will be phased out in the developed world—as this would be necessary to achieve some of the proposals presented to the Climate Change Secretariat.

It is clear that many developing nations and countries in transition are not likely to constrain their economic growth to mutigate a potential but unproven problem largely attributed to the developed world. It is obvious that economic and social development and poverty eradication are the first and overriding priorities of the development countries.

Although developing countries are unlikely to be required to undertake any new emissions abatement commitments, emissions abatement in developed countries would still impact significantly on the welfare of developing countries because OECD emissions abatement will have an impact on world trade. Stabilisation policies in OECD countries are estimated to lead to an increase in total production for all Rest of World regions relative to business-as-usual scenario. When OECD imposed measures to reduce carbon dioxide emissions, Rest of World regions experienced an increase in demand for their products because emissions generated in ROW countries during the production of goods and services are neither taxed nor restricted. Hence, production in Rest of World countries is stimulated by OECD country emissions abatement.

It would seem that ROW countries could be expected to gain from OECD stabilisation: given rising economic activity, real incomes could also be expected to increase. However, the study results project that the ROW regions as a whole would experience a welfare loss, even though they do not abate emissions, because the terms of trade would decline for all Rest of World regions taken together. The estimated impact of OECD stabilisation policies on developing countries' welfare (real consumption is used as a measure of welfare) is shown in *Figure 7*.

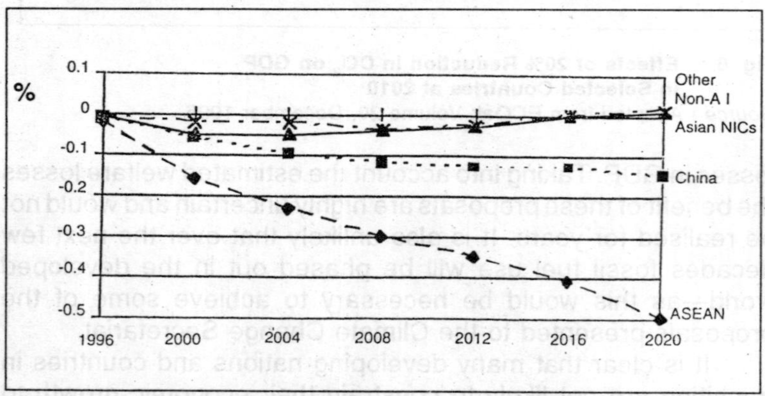

Fig. 7 : Change in Welfare in Non-Annex I Regions from Stabilisation
Source : Adapted from ECOal, Volume 20, December 1996

To conclude, emission stabilisation in OECD economies is projected to reduce welfare in all abating countries. The effectiveness of such an abatement policy on global emissions is minimal largely because projected rapid emissions growth in developing countries remains unconstrained. Stabilisation policies in OECD countries may lead to a reduction in welfare in some developing countries. Global cooperation is required to ensure an effective and cost minimising strategy to combat global warming. This includes participation by developing countries.

With all uncertainties related with global warming and huge economic cost required for OECD countries to undertake the necessary stabilisation action the question remains—What should be done?

A range of economically attractive responses could be recommended:

- continued research to improve information both about the impacts of climate change and the assessment of response option.
- promotion of voluntary agreements to implement no or low regret options, such as economically justified energy efficiency improvements and carbon sequestration.
- R&D to lower the costs of technological options to limit the future greenhouse gas emissions or climate change impacts.
- replacement of existing equipment and infrastructure in each sector with more efficient replacement at the time of normal turnover of capital stock.

4. JOINT IMPLEMENTATION

Analyses show that significant economic "savings" can be made through flexible collaborative arrangements between countries, by which abatement measures can be made less expensively in other locations—referred to as the "*where flexibility*" alternative. For example, many OECD countries could achieve far greater emissions reduction in many developing countries that they could for the same cost in their home countries—this essentially is the concept behind joint implementation (JI). Looking purely at the economics, it is clearly the way to proceed, but the political and commercial dimension behind this strategy is an extremely complicated issue which should be discussed in the Como's forums. Some experts indicated that the "*where flexibility*" alternative could decrease cost to developed countries by 70 per cent from the estimated 2–8 trillion USD cost of stabilising greenhouse gas emissions by 2000.

Considerable savings can also be achieved through "*when flexibility*"—alternative—allowing flexibility in the timing of policy implementation. The studies shows that policy response costs would be far lower and potentially as effective if emissions restrictions where phased in more slowly over time. Allowing reductions where phased in more slowly over time. Allowing reductions to be taken in step with planned turnover of capital stock (rather than to bear the costs of replacing equipment prematurely) and allowing for efficient technology to become available could decrease costs to OECD countries by about 40 per cent.

It is interesting to note that by combining "*where flexibility* and "*when flexibility*", the economic cost of stabilising greenhouse gas emissions at 1990 levels by 2000 could be reduced by 90 per cent—that is a saving to the industrialised world of between 2–7 trillion USD.

A possible way to combine "*where flexibility*" and "*when flexibility*" is through Activities Implemented Jointly (AIJ). The terms "*Joint Implementation*" is used to describe a wide range of possible arrangements between interest in two or more countries, leading to the implementation of cooperative development projects that seek to reduce or sequester greenhouse gas emissions. JI is based upon classical economic principles—measures to abate greenhouse gases should be taken where they are the cheapest. This means that, for any given reduction target, the most cost-effective alternative to achieve target should be implemented. In this case, countries with high reduction targets may invest in abatement projects in other countries where the returns per unit of investment will be higher than at home. Good prospects are for the JIs among countries of European Union and the Commonwealth of Independent States (CIS) considered as the world's largest energy related CO_2/GDP emitters (*see Table 1*) are of high interests. As it was stressed the subject of JI should be economically justified energy efficiency. We will consider the ways, means and incentives to initiate Activities Implemented Jointly by the EU and the CIS on energy efficiency, mainly at end use.

5. ENERGY EFFICIENCY IN THE CIS COUNTRIES

The countries of the Commonwealth of Independent States are characterised, by comparison with the Western industrialised countries, by a large energy related emission per unit of gross domestic product (GDP) and by very high energy intensity,

Table 1 : World's Largest Energy-Related CO_2/GDP Emitters (1993)

Country *(tons/1990 US$1,000)*	*CO_2/GDP*
Bahrain	3.89
Kazakhstan	**3.42**
Iraq	3.24
Estonia	2.70
Ukraine	**2.58**
Russian Federation	**2.51**
Qatar	2.36
United Arab Emirates	2.35
Turkmenistan	**2.26**
Trinidad and Tobago	2.09
Bulgaria	2.05
Uzbekistan	**1.99**
Romania	1.94
South Africa	1.94
Lithuania	1.87
Poland	1.85
Czech Republic	1.78
Brunei	1.75
Tajikistan	**1.67**
Slovak Republic	1.46
Belarus	**1.41**
Libya	1.39
Moldova	**1.32**
Georgia	**1.32**
Saudi Arabia	1.31

i.e. the ratio of energy consumption (primary or final) to GDP. The situation varies from country to country but, as a general rule, the energy intensities of the CIS countries are two to three times higher than those of the European Union countries (*see Table 2*).

This high level of energy intensity in the countries of the CIS in mainly due to three factors:

- the structure of economic activities in these countries, which gave high priority to energy-intensive industrial development;
- the inefficiency of the energy sector: low performance and high losses in energy extraction, production, transportation

Table 2 : Energy and Electricity Intensities in 1990 and 1993 of the Largest Energy Consumers in the Industrialised World

	Year	*CIS*	*Russia*	*E.U.*	*Germany*	*U.S.A.*	*Japan*
Primary Energy intensity (toe/1000 $)	1990	0.71	0.69	0.23	0.26	0.34	0.20
	1993	0.90	0.91	0.23	0.25	0.35	0.20
Final Energy intensity (toe/1000 $)	1990	0.47	0.51	0.16	0.19	0.24	0.13
	1993	0.62	0.60	0.16	0.17	0.24	0.13
Final Electricity intensity (kWh/$)	1990	0.65	0.67	0.29	0.34	0.48	0.33
	1993	0.88	0.88	0.30	0.33	0.50	0.33

Source : Energy indicators for the countries of Europe and the CIS; Synergy Programme, Ademe, Enerdata, ICE.

and distribution, often due to obsolete equipment and poor maintenance;

- the inefficient use of energy in all activity sectors.

This untapped energy efficiency potential is extremely high and lies as a heavy burden on the already depressed economies of the CIS countries. In the former USSR, the energy system was considered as a whole and the difference between the "energy rich' and the "energy poor" Republics was less apparent than now. With the new rules prevailing in the CIS and the need (at least theoretically) to buy energy products at international prices and in hard currencies, the energy gap between "energy producer and exporter" and "energy importer" is widening. The difficulties of the importers are increasing, without great improvement for the exporters since their global energy system is in deep crisis.

Faced with this situation, the governments of almost all the CIS countries are conscious of the vital role and/or absolute necessity of elaborating and implementing energy efficiency policies, to allow the countries to reconstruct their economy and pave the way towards sustainable development These governments have in general proclaimed that energy efficiency is a priority of their energy policy (where the latter has been formulated).

For reasons which are either commonly found throughout the world or which are specific to the former–USSR countries, the actual decisions and achievements are relatively poor in comparison with the needs, the potential and the benefits to be expected from a consistent energy efficiency policy. Fortunately, the situation is not homogeneous and several countries—often helped by the international cooperation (notably by the technical assistance programmes of the European Union)—have taken decisive steps to set up legislative, institutional and sometimes financing instruments to develop energy efficiency programmes.

Energy, environmental and economic factors of energy efficiency

Energy factor—the possibility to realise the energy saving potential estimated within Russia at 460–540 mln.t.c.e.

Environmental factor—the experience of developed countries shows that about 85 per cent of decreased environmental degradation may be reached due to energy saving implementation

Economic factor—energy intensity reduction leads to increasing competitiveness of commodities and services as well as to consumers cost cutting. The 1 per cent energy economy causes the increase of gross domestic product within Russia by 0.03 per cent.

An insight to level of existing total energy efficiency potential could give estimates for the Russian Federation which having the largest national oil, gas and coal resources, Russia at the same time has the largest energy efficiency potential. Energy intensity of Russia's economy is 3 times higher than in developed countries. In terms of economic recession energy intensity per unit of GDP has even increased by 30–40 per cent.

As of the beginning of 1994 the total potential of energy saving is estimated at 460–540 mln.t.c.e. (*Table 3*). This potential may be defined as the possible reduction in energy consumption in due course of the implementation of the most effective conservation measures. Fossil fuels make up more than a half of the potential, the rest part belongs to electricity and heat energy. About one third of the existing potential is concentrated in the Fuel and Energy Complex, 20 per cent—in residential and commercial sector, 15 per cent—in industry. Given the importance of energy resources, or the lack of them, on their domestic and foreign trade balances, it seems reasonable to believe that

Table 3

Potential saving of energy resources in Russia billion m^3	*Natural gas million tons*	*Petroleum products, million tce*	*Coal, coke*	*Electricity billion kwh million GJ*	*Thermal energy, tons*	*Total million*
The fuel-energy complex	50–60	15–17	33–39	38–46	670–760	150–180
Including:						
oil extraction	8–10	1–15	—	3–4	10–15	10–15
coal extraction				8–10		3–4
transport of energy carriers	8–9	—	7–8	30–36	630–710	52–59
electrical and thermal engineering	32–42	10–12	26–31	—	—	80–97
oil refining	1–15	4.5–5	—	1–1.5	40	9–11
Communal and household sector	10	0.6–0.8	21–23	65–70	500–600	75–83
Agriculture	1.4–1.5	14–15	1.5–1.7	8–10	15	27–29
Transport		29–34				42–50
Industry	34–42	6–7	12–14	220–265	700–870	158–190
Including:						
industry-wide measures	10–13	0.5	—	150–185	310–420	73–92
metallurgy	12–15	2	10–11	20–24	20–25	34–39
machine-building	–(3–4)	0.5		55–60		15–16
construction materials	10–11.5	1.7–2	2–2.5	–(8.5–10)	170–190	20–23
chemistry and petro-chemistry	5–6	—	—	4–5	50–60	9–10
forestry, and the paper and pulp industry	0.3–0.7	1–2	—	—	150–170	8–10
Total figures	**100–100**	**65–75**	**70–80**	**330–390**	**1880–2250**	**460–548**

there is little hope the CIS will overcome the present economic crisis unless this energy wastage is ceased. Yet, the general economic picture is in such a state of morosity, that experts from both the CIS and from Western industralised countries often seem to despair of any significant progress in the near future.

Unfortunately, there is no legislative and economic background for implementation of the energy saving potential in Russia. The economic non-stability and lack of investments do not enable to foresee the perspective for energy efficiency.

However it is possible to speak about the optimistic energy efficiency scenario in case there is no investment limitations on the one hand and on the other hand—the lack of technical feasibility at the definite period of time (*Figure 8*). The total volume of energy saving in Russia will be determined by the fixed prices on energy carriers. If the internal prices correspond to the world prices the possible energy saving will have amounted to 180 by 2000 and 480 mln.t c.e. by 2010 (fair right curves).

It is obvious that the given scenario is a hypothetical one. The position of a possible scenario is to the left from the right curves on the Figure 8. Its implementation depends on economic incentives and availability of investments in energy efficiency. One of such scenarios was elaborated by the Institute for Energy Research of the Russian Academy of science. It is a possible scheme of the economic development which implies the moderate growth rate of prices on energy carriers but it is ahead the growth of other prices. According to the scenario the price pattern on energy carriers will have agreed by 2002–2005 with the price pattern fixed in developed countries. The scenario includes the measures on economic inciting of energy efficiency.

The intersection of the left fair curves corresponding to the possible scenario of energy efficiency in the zone A (Which corresponds to the achievement of the world price pattern) determines the probable annual energy economy in the amount of about 90 by 2000 and about 300 mln.t c.e. by 2010. The energy efficiency policy objectives are presented by the stepped curves. These measures are generally aimed at reaching the energy conservation but not at the technical retrofitting. The space between the stepped and right fair curves corresponds to probable energy saving.

The vital question today is as follows: how to provide the

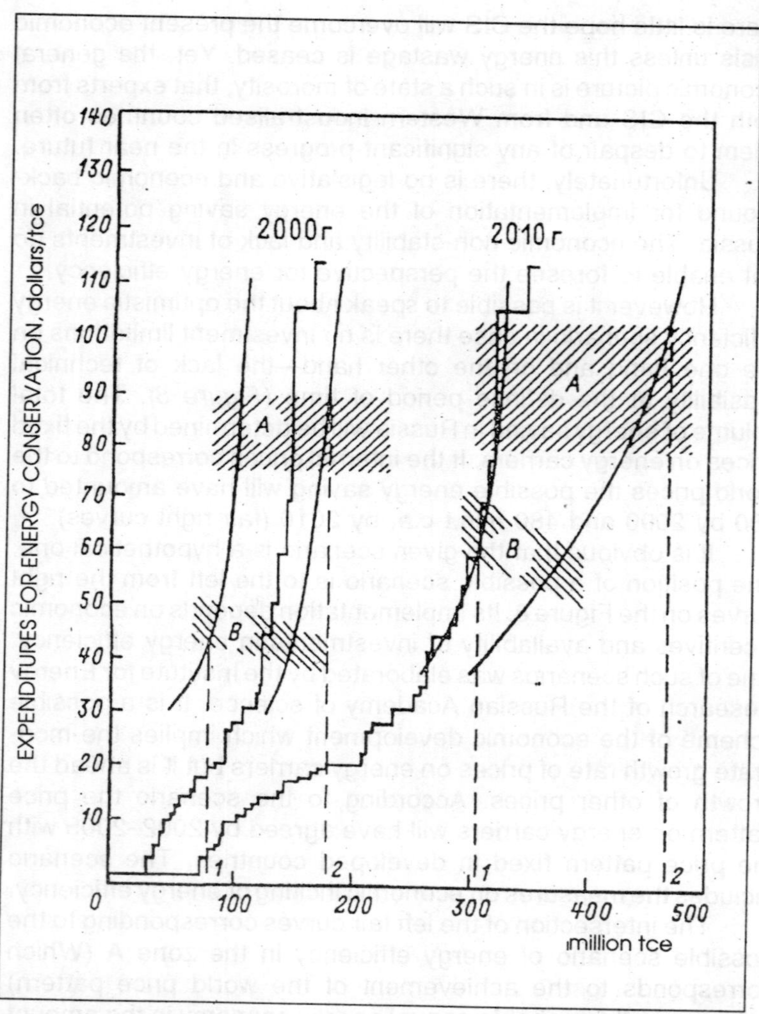

Fig. 8 : Scenarios for Energy Conservation. A—world energy prices; B—domestic prices of production: (1) probable savings; (2) optimistic savings

background for reaching the highest energy efficiency. In figure 2 it is observed that energy efficiency can reach 150–160 by 2000 and 360–390 mln t.c.e. by 2010 in terms of domestic prices (zone B), 170–190 and relatively 430–470 mln.t.c.e. in terms of the world prices (zone A).

6. FINANCING ENERGY EFFICIENCY

Thinking about financing energy efficiency the first question to be clarified is: Who are Players and what are their Motivation?

A great number of public and private structures are interested in energy efficiency in view of potential profits they can gain from energy efficiency. The participation of players in energy efficiency investment can be achieved both at macro and micro level (*Figure 9*).

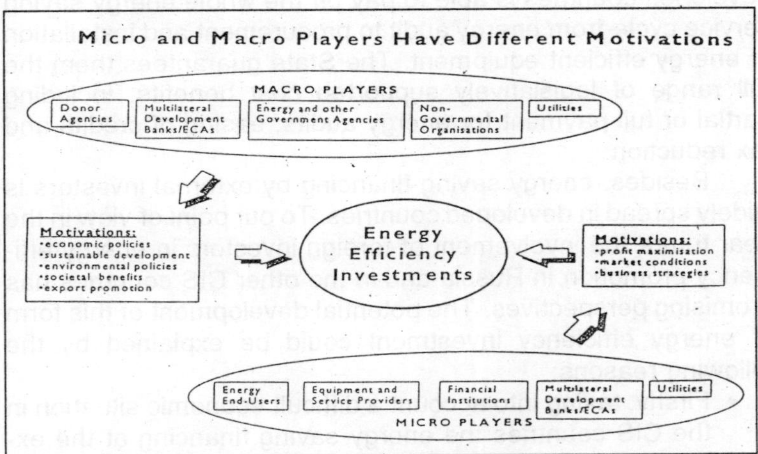

Fig. 9: **Many-sided (social, political, economic and environment) nature of energy efficiency implies the potential invovlement of both macro and micro players in financing energy efficiency**

Macro players are mostly concerned with the perspective to influence on the energy efficiency policy as well as economic policy as a whole

Micro players are generally motivated by the potential to maximise profit

Mainly the motivation defines which level this or that player belongs to. Macro-level players are generally concerned with the overall energy efficiency policy, energy and environment security of the society and its living standards. Micro-level players are generally motivated by the potential to benefit from energy efficiency, for instance to reimburse the invested money (for external investors) or reduce spending on energy (for energy consumers).

Despite outlined motivation on energy efficiency implementation, in terms of financial and economic non-stability and crisis of non-payments internal consumers have difficulties in finding opportunities for investments in implementation of energy efficiency programmes in the CIS. Not every consumer even in developed countries is able to pay off the whole energy saving service cycle-from energy audit to procurement and installation of energy efficient equipment. The State guarantees them the full range of legislatively supported cost benefits, including partial or full payment for energy audits, easing of credits and tax reduction.

Besides, energy saving financing by external investors is widely spread in developed countries. To our point of view in the near future the involvement of foreign investors in energy efficiency promotion in Russia and in the other CIS countries has promising perspectives. The potential development of this form of energy efficiency investment could be explained by the following reasons:

- Firstly, taking into account a difficult economic situation in the CIS countries the energy saving financing at the expense of local budgets may be considered as problematic despite the fact that energy saving was declared the top priority in the scheme of the State energy policy.
- Secondly, the amount of investments in energy efficiency projects is considerably less compared to that in the field of fuel production. The pay-back period of energy efficiency projects is rather quick. It is caused by constant growth in domestic prices on energy carriers.
- Thirdly, investor-consumer financial relations are properly worked out in foreign countries. The try-out of the scheme of energy efficiency projects implementation with participation of investment energy service companies and the scheme of return on investments due to saved energy proved to be reliable. Taking into account the difficult

investment climate in the CIS countries it is noteworthy that investments in energy efficiency are quite perspective in terms of a wide equipment and service market, small costs on projects and other positive factors.

Attractiveness of investments in energy efficiency projects in the CIS

Relatively small costs, therefore–low financial risk.

Potentially quick pay-back (taking into account a constant growth in prices on energy carriers)

Extended market of energy saving equipment and services

Proved schemes of reimbursement of energy efficiency investments due to saved energy

This "two-step" financing approach has an essential advantage: while it is difficult at national level or at international level, for example, to pinpoint "energy efficiency convergence points", local teams and companies can develop schemes at grassroots levels, taking into account the specific local contexts, and negotiating adapted arrangements with potential participants—the local energy producer and distributor, the public authorities, the energy consumers.

For larger investments, two mechanisms must be explored: bank loans and the setting-up of joint-stock energy saving investment companies (or 'energy service company': ESCO). In both cases, we are faced with the problems of shareholder and the guarantee of loan or initial investment reimbursement by the consumer. These difficulties, which are found in all countries, are aggravated in the CIS countries by frequent situations of non-payment and remaining state subsidies in a number of sectors (for understandable social reasons). We think that emphasis should be put at all levels—and in particular through international cooperation—to create simple and reliable mechanisms to overcome these barriers. The initiatives undertaken by the EBRD seem the most promising at present.

Definition and Forms of ESCO, Principles of its Activities

Energy Service Company (ESCO)—an energy saving company which delivers energy efficiency services both in private and public sectors.

ESCO provides equipment installation and operation at its expense to achieve the guaranteed savings.

ESCO reimburses the investments and gains profit due to energy savings.

After that the equipment is transferred into the clients property and the client gains all the profits from the equipment operation.

As ESCO guarantees energy savings (excluding technological risk) the client is ensured in reduction of his electricity bills.

One such theme presently under development with the Energy Efficiency Unit of the EBRD is the creation of energy service companies in Russia and Ukraine. Approach is threefold: building the capacity of a local team of experts in designing and/or identify and evaluating "good" energy efficiency projects, organising negotiations between the potential investors in the future energy service companies, and finding energy efficiency convergence settings where such a company would have the best chances of success.

7. AN INITIATIVE OF CENTRO VOLTA

Recently Centro Volta in cooperation with Research Centre for Energy Economics, Transport and Environment (CEETA), Lisbon, Centre for Energy Policy (CEP), Moscow made a proposal to the Foundation of a Training Centre for the CIS Countries, with the Cooperation of European and the CIS's Institutions and Firms.

The projects aims to train decision makers at the municipality, firm and public levels to carry out projects on energy efficiency using adequate technologies and applying to the financial sources, as to establish fruitful ties with European firms and agencies dealing with energy instruments and technologies.

The idea is to build a permanent training centre in Como, Italy, to assist the CIS countries in training and in joint ventures

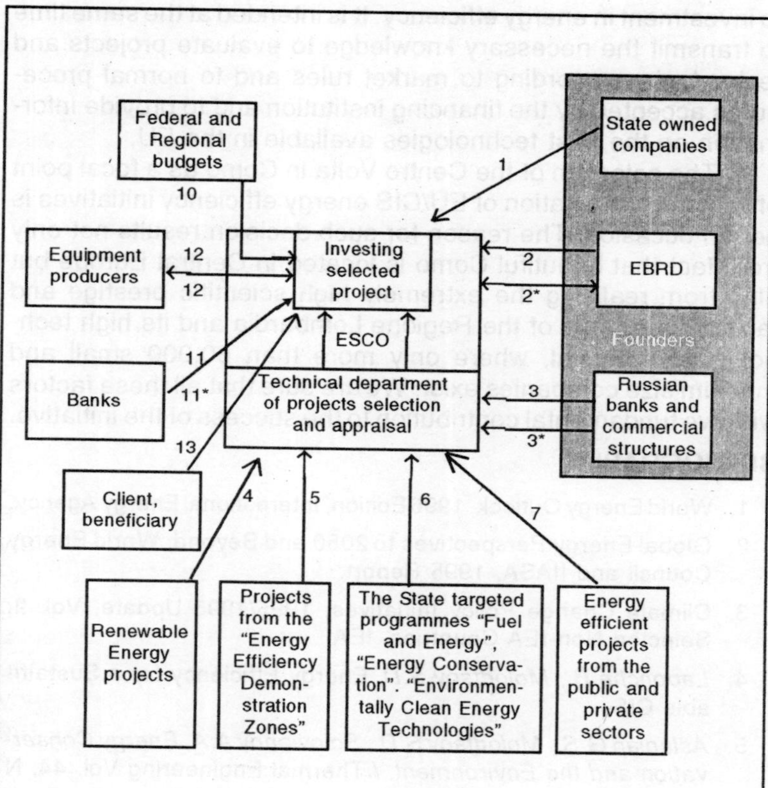

1.	Equity from the State owned companies
2.	Equity from ERD (not more 40%)
2*	Credit line for ESCO from EBRD
3.	Equities from Russian banks
4, 5, 6, 7	Project flows
8	Project selected by ESCO for inverting
9	Investing the projects from ESCO resources
10	Centralized State investments given as a grant or as credit free of change
11	Credits from foreign and local investors
11*	Return of credits
12	Investment in form of leasing
12*	Pay-back for leasing
13	Partial financing of the project by client

to investment in energy efficiency. It is intended at the same time to transmit the necessary knowledge to evaluate projects and technologies according to market rules and to normal procedures accepted by the financing institution and to provide information on the best technologies available in the EU.

The selection of the Centro Volta in Como as a focal point of joint implementation of EU/CIS energy efficiency initiatives is not an occasion. The reason for such decision results not only from fact that beautiful Como is located in Central Europe but also from realising the extremely high scientific prestige and facilities available of the Regione Lombardia and its high technological potential, where only more than 60,000 small and medium size companies exist. We are sure that all these factors will give fundamental contribution to the success of the initiative.

BIBLIOGRAPHY

1. World Energy Outlook, 1996 Edition, International Energy Agency.
2. Global Energy Perspectives to 2050 and Beyond, World Energy Council and IIASA, 1995 Report.
3. Climate Change Policy Initiatives, 1995/1996 Update, Vol. 2, Selected Non-IEA Countries, IEA.
4. *Laponche B., Molodtsov S.D.* Energy Efficiency for a Sustainable CIS.
5. *Aslanian G.S., Molodtsov S.D., Solovianov A.A. Energy Conservation and the Environment.* //Thermal Engineering Vol. 44, N 11, 1997.
6. *Aslanian, G.S., Molodtsov, S.D.* Energy Efficiency Investment Financing. //International Energy Efficiency Workshop, Cheliabinsk, 24–26 September, 1996.
7. *Aslanian G.S., Laponche B.* Energy efficiency Measures. Papers at the Training Course "Analysis, Evaluation, Feasibility Studies and Promotion of Energy Efficiency Projects, Moscow, 16–21 December, 1996.

22

Globe Report on Energy in Central and Eastern Europe and the CIS: An Urgent Need for International Cooperation on Energy Efficiency Improvement for Sustainable Development

B. LAPONCHE AND H. BAGUENIER
Portugal

1. THE REGIONAL ENERGY CONTEXT

i) High Energy Intensities, Low Energy Efficiency

The countries of Central and Eastern Europe (CEEC) and of the Confederation of Independent States (CIS) are characterised, by comparison with the European Union countries, by a very high energy intensity i.e. the ratio of energy consumption (primary or final) to gross domestic product (GDP).

The situation varies from country to country but, as a general rule, the energy intensities of the CEEC and CIS are two to three times higher than those of the European Union, as is shown in the Appendix, which gives the main energy consumption and global indicators of these two groups of countries for the year 1990 (the milestone year between the period of centrally planned economy and the period of transition toward market economy).

This high level of energy intensity is due to three factors:

a) The first factor is the structure of economic activities of these countries, which gave high priority to industrial development, in particular of basic industries (steel, chemicals, cement, etc.) which are high consumers of energy. The consequence of this structural difference is reflected by the difference in the shares of final energy consumption between the various consuming sectors: apart from a few countries, which are in general more advanced in the transition scheme, industry is by far the largest energy consumer.

In recent years, this situation has been rapidly changing in most countries, due to a decrease or even collapse of large parts of industrial production. As a consequence, energy consumption has strongly decreased.

But the impact on the energy intensity of this change imposed by the economic crisis remains low since a rational restructuring of the industrial sector has not yet been implemented in most countries and many factories are operating at very low capacity, increasing their energy consumption per output (volume or value).

b) The second factor is the inefficiency of the energy sector itself: low performance and high losses in the extraction, production, transport and distribution of energy products. The poor performance of power plants and refineries and the heavy losses in the district heating systems are often due to obsolete equipment and poor maintenance.

Some countries are faced with an over-capacity of an obsolete energy production and transformation system, coupled with a high global energy intensity and shortages in the energy supply to the consumer due to the lack of fuel or of peak capacity.

c) The third factor, which presents the major problems is the energy end-use inefficiency in all sectors of activities, confirmed by a large number of sectorial analyses.

The main sectors which have at present the highest share in this global inefficiency are industrial activities and space heating. With the changing economy, the transport sector and the domestic use of electricity will be very probably, if nothing is done, an increasing burden for the economy and the environment.

ii) On the Supply Side: A Variety of Situations

On the supply side, the extent of which energy requirements are satisfied in the different countries depends on the level of domestic energy reserves and is closely linked to the legacy of the former Soviet system.

a) In some countries, where coal and lignite resources exist, the use of these fuels is still dominant, notably for the production of electricity and heat (often by co-generation plants).

In most cases, these reserves have been substantially depleted and the capacity for economically viable energy production has fallen as a result.

b) Concerning oil and gas, Russia is by far the highest producer and exporter. Most of CEEC and CIS countries are heavily dependent upon supply of oil and gas from Russia, Kazakhsthan and Turkmenistan. Export deliveries from these republics, particularly Russia, have fallen drastically over the last few years. This has largely been the result of the importing countries inability to pay in hard currencies the much higher prices demanded as export subsidies have been progressively withdrawn.

For a few countries, the drastic shortage in imported fuel supplies are also linked to specific hazardous political situations.

c) Both for the total energy requirements and for electricity production, the use of fossil fuels is in general predominant. Concerning electricity production, the share of hydropower was 13.5 per cent and the share of nuclear power was 12.5 per cent in the CIS in 1990 and respectively 9 per cent and 15 per cent in CEEC.

d) Although the share of nuclear energy is relatively small on average in the total electricity production of the region, and, consequently, much smaller in the total energy requirements (around 4%), the safety of the nuclear reactors in operation is of great concern, both for the countries themselves which utilize this energy and for all European countries.

Due to the importance of the nuclear safety problems a specific policy has to be defined on the future of the nuclear power plants.

Further more, in particular because of the former USSR policy, a few countries are relying heavily on nuclear power for their electricity supply.

e) Although the potential for some renewable energy sources, other than large hydropower plants, is substantial in almost all countries, resources like small hydro, wood, rural and urban waste, solar, wind and geothermy have not been consistently developed.

2. AN ENERGY EFFICIENCY STRATEGY

i) An Urgent Need for Energy Efficiency Policies and Cooperation

Energy production and use are particularly inefficient in the CEEC and CIS. If this situation continues, it is unrealistic to hope to achieve the conditions required for sustainable development.

a) The needs in capital for the investments in energy production, transformation and transport, and the need for hard currencies to import fuels for the importing countries (or the losses of hard currencies for the exporting countries) to fulfill the energy requirements are too high for the economies of these countries if this situation of inefficiency is perpetuated.

   Energy inefficiency is one of the major factors hampering the economic recovery of these countries.

b) Energy production, transport and consumption causes considerable problems in terms of risks (accidents, in particular in nuclear reactors), air and water pollution (use of low-grade coal, poorly maintained installations, dissemination of hazardous waste, etc.).

c) One has also to be aware that, at world level, the prices of oil, gas and coal are particularly low on the international market. Any increase of these prices, which could happen for a number of reasons, would drastically deteriorate the already extremely fragile economic situation of the importing countries, which are a majority in the zone, in particular for oil and gas. It is now, by taking advantage of this period of low international prices, that the measures for improving energy efficiency have to be launched in order to be well prepared when the foreseeable future increases occur.

The waste of energy imposes an enormous, financial, environmental and health burden on the CEEC and CIS: energy efficiency must be a priority of their economic and energy policies. It offers an enormous potential for domestic actions, with very positive effects on economic growth, quality of environment, increase in jobs, and for the international cooperation.

ii) Energy Efficiency and Sustainable Development

Faced with this situation, an economic policy must:

— restructure industrial production by reducing heavy industry and developing light, value-added industry and services;

— promote energy efficiency and the rational use of energy in all sectors so that identical services and levels of production can be achieved with much lower energy consumption.

To develop economically and socially, CEEC and the CIS need to achieve an overall standard of energy efficiency in its economy comparable with Western European countries.

Energy efficiency is achieved by renewing existing facilities, introducing more energy-efficient procedures and equipment and redirecting some activities.

Renewing existing facilities covers factories (repairs, heat recovery, maintenance, etc.) and buildings (insulation, regulation of heating systems, etc.).

The greatest potential for energy efficiency lies in new facilities: housing, and buildings in general domestic appliances, lighting, new factories, new transport equipment. In all these sectors, experience shows that energy savings of 30 per cent–50 per cent can be made for an equivalent service, *with no major increase cost*.

What is needed is a coordinated effort of information, promotion and regulation to encourage the manufacture and sale of more energy-efficient, less polluting equipment and appliances.

Some activities cannot merely improve the performance of their equipment; they need to be reorganised for lower energy consumption and protection of the environment. The priority sector here is transport, where the Western development model has led to disastrous situations (pollution, congestion accidents, energy consumption). Energy efficiency, environmental quality and quality of life all point towards the development of public transport and railways.

Energy efficiency is a factor in economic development, since experience shows that often it is cheaper to save energy, or avoid consuming it, than to produce it. Financial resources that would have been spent on energy production, say, for building power stations, or energy imports, using foreign currency, can be devoted to other activities, improving the standard of living, domestic comfort and the quality of life. Industries can be modernisation, housing built, public transport developed, health, education and sports services improved.

Over and above the overall effect of this transfer of resources to more relevant and profitable activities meeting people's needs better, the direct effects of energy efficiency on production are considerable:

— industry's productivity and competitiveness are improved;
— and industry develops the produce efficient equipment, which can be exported.

The positive consequences for the environment are even easier to grasp.

The least polluting energy is that which is not consumed, or even produced.

Every time energy consumption is reduced for a given activity (insulation of housing, greater engine efficiency), the quantity of pollutants is automatically reduced in proportion. This is the argument used in the advertisements, and perfectly true, that sulphur dioxide, carbon dioxide or radioactive waste are reduced when, for the same level of lighting, fluo-compact bulbs are used instead of incandescent ones.

Action to improve energy efficiency is also the cheapest way to improve the environment, since it pays for itself by energy savings. Thus the improvement to the environment costs nothing, compared with action to clean up pollution.

In CEEC and the CIS transition period, the social benefits of an energy efficiency policy are particularly important:

a) *Effects on Employment*: Indirect effects via the overall stimulation of the economy by the reduction of energy costs; direct effects through the emergence of new activities advantageously dispersed throughout the country: energy audits, building insulation, adjustment and maintenance in factories and blocks of flats, heating networks.

   Energy efficiency at distribution (notably district heating) and consumption levels offer opportunities for work and benefits in almost all activities and sectors of consumption (industry, buildings, transports, rural activities) and all locations: it is possible to develop in every town, or factory, or rural area a programme of new public lighting, of insulation of buildings, of improvement of public transportation, of recovery and use of wasted energy.

b) *Effects on Household Income*: At a time of rising energy prices, energy efficiency measures will reduce the energy bill for households (heating, electricity). If efficient appliances are on sale, households will be able to increase their spending on consumer goods without excessively increasing their energy bills.

iii) The Necessary Tools for Energy Efficiency Implementation

Despite all the arguments for energy efficiency and the priority given to it in public statements by governments, few programmes have yet been developed. Neither the countries concerned, nor their international partners, have done enough in terms of legislation, institutions or finance.

But changes are now occurring in all countries at three levels: nationally, proposals, for legislation and organisation are being made; a number of decentralised initiatives are happening in towns and regions, with the active cooperation of regional teams from Western Europe, especially a number of regional delegations of Ademe; international aid is supporting energy efficiency through several programmes, in particular PHARE, TACIS, THERMIE, and the SYNERGY programmes of the European Commission; loans from the World Bank and EBRD (already granted for renovation of heating networks) are available for energy efficiency programmes if appropriate funding mechanisms are applied.

To give a greater impulse to energy efficiency policies, a number of reforms and initiatives are necessary, which are outlined below:

Energy Pricing: A Necessary but not Sufficient Condition

The countries in transition are at present moving, not without difficulty, towards the market economy. For this economy to work properly, prices have gradually to be based on costs. This is particularly true for the consumer prices of energy products, which need to reflect international primary energy prices or production costs, and, if possible, include externalities such as impact on the environment.

The subsidies for energy, as are still practised in some countries, cannot continue: they are a burden on the State budget, interfere with the normal running of energy companies and inhibit the rational use of energy by the consumer, who has no incentive to avoid waste.

It is true that higher energy prices exercise pressure on the consumer to save. But in a number of countries, energy prices, particularly of electricity, are already very high compared with incomes: consumers are forced to reduce their consumption, not by using energy more efficiently but by restricting its use. This is because the effort has not been made to promote energy efficiency programmes.

Furthermore, since price adjustment can only be gradual for social and political reasons, especially for heating, the State continues de facto to subsidise energy. So it is in the direct interest of public finances to implement energy efficiency programmes, without waiting for "market prices", which will not alone solve the problem.

The example of Western industrialised countries, well versed in the market economy, shows that the successful energy efficiency policies, which stabilised consumption while GDP grew 30 per cent from 1975 to 1987, only began with the rise in the oil price. For them to be effective, it took a whole range of state intervention: regulations, information campaigns, institutions and other bodies, research and innovation programmes, financial incentives to investors (tax concession and subsidies).

Energy Efficiency Programmes

Programmes are defined on two main principles:

a) Reduce existing waste: renovation of heating networks, monitoring energy consumption, regulation systems, housing insulation, heat recovery in industry, etc.
b) Produce new energy-efficient plant: low-consumption electrical appliances, public transport, well-designed, insulated housing, efficient industrial processes, etc.

The main features of a national energy efficiency programme are the following;

- new regulations, especially for new buildings and household appliances;
- information campaigns for consumers, middle and senior managers (particularly politicians who veer between an old style obsession with production and full-blooded market economics, which is just as devastating);

- training programmes for engineers, technicians, architects, managers, company heads, town technical services;
- energy audits and demonstrations in industry and housing estates to define action programmes;
- technical support (consultants) and financial support for R&D institutes and industries to develop, produce and market energy-efficient equipment;
- support for the creation of design and engineering offices able to provide energy efficiency services, define programmes, and carry out projects;
- support for industrial agreements (purchase of licences, shareholdings) and technology exchanges,
- national and international financial support (grants, loans for investment in energy efficiency, nationally, and locally, especially for towns, which are at present the most reliable players in such a policy;

Apart from their effects in terms of energy, economy and the environment, these programmes could have a major impact on employment in all activities and regions, and thus provide essential help in this transition period.

Bodies to Promote, Motivate and Coordinate

The State needs to intervene; programmes have to be defined, coordinated and promoted; a network of partners needs to be set up. Public bodies are necessary to elaborate, promote and coordinate the implementation of energy efficiency programmes at national and local levels.

The implementation of energy efficiency programmes is by its nature a diversified and decentralised activity: it concerns companies—either in the management of energy consumption, or the production and sale of efficiency equipment—local authorities, service companies or administrations, and households.

A body responsible for the national energy efficiency programme is not intended to carry out projects itself, but to create the necessary conditions for those projects and ensure that they are as effective as possible, technically, economically, socially and environmentally. This is a public service role, new for the administration, especially in Eastern countries, and requires the ability to motivate, discuss respond quickly and understand the problems and constraints of various partners.

In the light of experience elsewhere, the solution that is slowly being adopted in a number of countries is as follows:

a) Set up decentralised teams, particularly in large countries, in regions, cities and industrial centres. Local teams have the advantage of being close to consumers, working on the ground, and developing programmes that fit both with the national programme and local conditions. Depending on the organisation of local government, these teams can be linked to regional or local authorities or depend on the national body. The former solution is better, since it reduces the size of the national body, and respects the principle of decentralising initiatives.

b) Set up a national body with responsibility for: integrating energy efficiency objectives into the economic industrial and energy policy of the country; defining a national energy efficiency programme; proposing national standards and regulations; encouraging and coordinating programmes of research, innovation and demonstration; organising and coordinating national and international financial support; coordinating international aid.

Finance for Energy Efficiency Action

Finance is needed for running costs (salaries, equipment, offices) and direct actions (information, training, economic studies, international activities) for all the bodies responsible for the programmes and also to support the programmes themselves: innovation, demonstration, energy audits, investment support, industrial development support.

For CEEC and the CIS, international aid is obviously necessary for energy efficiency and needs to be increased from its present level. In particular, the international development banks (World Bank, EBRD, EIB) need to adapt their procedures to investments that are new to them; they need to learn how to finance programmes of varied action rather than large-scale projects, by using national contacts, with banks (such as development banks) and technical bodies (energy efficiency agencies are essential to set up and monitor programmes).

The countries concerned, however, must also provide the financial means for their political ends. In a number of countries (including Russia and Romania), it has been suggested that a national fund for energy efficiency be set up, financed by a 1 per cent levy on all final energy "bills".

The economic and social development, the improvement of the quality of environment and the quality of life in CEEC and the CIS requires the implementation of energy efficiency programmes in all sectors of activity. These countries should show their political commitment by decisions concerning institutions, regulations and finance. International aid should support these programmes more effectively than is now the case: this is one of the major conditions for a successful transition and sustainable development of these countries.

iv) On the Supply Side

The driving mission of international cooperation must be to help the CEEC and the CIS to improve the efficiency of the energy sector by cooperating for policy definition, institutional and capacity building, regulatory establishment and by basing all financial interventions in the sector on a least-cost and integrated resource planning approach that supports the transition to efficient, competitive and as environmentally sound as possible energy production and use.

b) The international cooperation must support the introduction of market prices for energy, in line with the quality of the services provided to the consumer (for instance, if space heating has to be paid at its cost, the production, transport and distribution system must be efficient so that the consumer pays a fair price and not the inefficiency of the producer or the seller);
c) And the adoption of least-cost planning by governments and utilities that explicitly includes both supply and demand options.
d) And the support of programmes and projects for the development of renewable energy sources (biomass, small hydro, solar, wind, geothermy).
e) Concerning the safety of operating nuclear reactors (and subsequent issues like radioactive waste management), the following measures must be taken:
    i) Independent (from the nuclear lobby) assessment of the role and importance of the nuclear power plants for the electricity supply at country and region level, on the basis of least-cost and least environmental risk or damage by comparison with alternative solutions based

on a different energy carrier (natural gas for example) and, or energy end-use efficiency measures.

ii) Independent assessment of the safety situation of the reactors in operation and of the reactors which might be planned, by comparison with the Western countries nuclear safety standards.

iii) Support of least-cost non-nuclear investments that assist in limiting or cancelling the need to operate the reactors which are not fulfilling the safety requirements (including the management of radioactive waste), with a view to closing them down.

iv) Support for nuclear retrofitting, like for any other generating source, should be provided only if this is the least cost option, with explicit, full cost comparison made to energy efficiency measures, and if this retrofitting really ensures that the safety of the reactor will be satisfactory.

APPENDIX

Main Energy Data and Global Indicators

1. PRESENTATION

The following tables give the main energy consumption data and global indicators for Central and Eastern European countries (CEEC) and the countries of the Community of Independent States (CIS), and their comparison with the European Union corresponding data, for the year 1990.

i) The Year 1990 has been Chosen for Several Reasons:

The availability of data on primary and final energy consumption are available for this year for all countries concerned, which is not always the case for some countries for the following years. Obviously, if an when a complete set of data is available, the same tables will have to be updated, which will allow to study the evolution of the energy consumptions, their level and their structure.

b) The global economic data which are used for the international comparisons on energy intensities, i.e. the gross domestic products (GDP) calculated with purchase power parities, are available for 1990 for all countries, and not for the following years.

c) For all CEEC and the CIS countries, 1990 is the milestone year between the centrally planned economy period and the transition period toward market economy. To understand the energy system patterns inherited from the Soviet system, the data of this year are very useful. They are often more significant for most countries on the type of energy consumptions than more recent data, which are those of deeply depressed economies or even, for some countries, of situations of war or high political tensions (in former Yugoslavia and the Caucasus region in particular).

Of course, the situation is different from country to country vis-a-vis the steps already taken towards restructuration, privatisation and market economy, but one finds a lot of common situations and problems in all these countries, in particular concerning the common necessity to increase energy efficiency.

d) Obviously, for concrete action, in particular on the supply side and for emergency measures, the present energy situation of each country has to be known as precisely as possible. For this aim, the data given in this Appendix are not very useful in the countries where the situation has deeply changed and for short term measures.

But they are an indispensable information to elaborate energy strategies for the medium and long term and define the priorities of the international cooperation with these countries.

ii) The Three Groups of Countries that We Consider and Compare are

— The Central and Eastern European countries (CEEC): Poland, former Czechoslovakia (CSFR), Hungary, Bulgaria, Romania, Estonia, Latvia, Lithuania, former Yugoslavia, Albania;

— The CIS countries (CIS): Armenia, Azerbaidjan, Belarus, Georgia, Kazakstan, Kirghistan, Moldova, Uzbekistan, Russia, Tadjikistan, Turkmenistan, Ukraine;

— The European Union (U.E.)

Data are also given in a separate column for Russia, which is by far the largest energy consumer of the whole area.

For each of these groups, the following data are provided, for year 1990:

— Area and population;

— Gross domestic product per capita, calculated with purchase power parides;

— Primary energy, final energy and final electricity consumptions per capita,

— Primary, final and electricity, intensities (ratio of the energy consumption to GDP);

— Share of each energy source in the primary energy consumption;

— Electricity production by sources;

— As an example of the recent evolutions, the electricity production of Russia from 1990 to 1993.

iii) The source of data for this Appendix is the ENERDATA Data Base (Grenoble).

a) The GDP calculated with purchase power parities come from CEPII (Centre d'études prospectives et d'informations internationales, Paris) and are given in US dollars of 1980.

b) For the primary energy consumptions, the fossil fuels contribution is expressed in toe (tone of oil equivalent) with the classical equivalence rules; the primary electricity production from renewables (almost exclusively large hydropower) and from nuclear power, as well as "imported electricity" which appear in the primary energy balance, are expressed with the same equivalence: 1000 kWh are counted as 0.26 toe.

c) For the final consumption of energy products, the fossil fuels consumptions are expressed in toe with the classical rules and the electricity consumption with the internationally used equivalence: 1000 kWh=0.086 toe.

d) Concerning renewable energy sources, only large hydropower is taken into account for the "commercial" energy data and the calculation of the global indicators.

The other contributions (mostly wood) are not indicated in the tables concerning the primary energy consumption. As a general rule, the data for the renewable energy sources are very questionable and, in particular, the use of wood and rural waste is always underestimated (even in the data concerning the European Union).

2. TABLES

Table 1 : Global Consumption Indicators for 1990

	E.U.	*CEEC*	*C.I.S.*	*Russia*
Area (1000 km2)	2369	1343	22102	17075
Population (million)	339	134.2	281	148
GDP per capita (1000 $)	9.8	4.0	5.4	6.6
Primary energy consumption per c.(toe)	3.6	2.7	5.2	6.5
Final energy consumption per c.(toe)	2.5	1.8	3.6	4.4
Final electricity consumption per c. (1000 kWh)	4.7	2.8	4.9	5.8
Primary energy intensity (toe per 1000 $)	0.37	0.69	0.95	0.99
Final energy intensity (toe per 1000 $)	0.26	0.45	0.66	0.67
Final electricity intensity (kWh per §)	0.48	0.69	0.90	0.88

a) The energy consumptions are those of commercial energy sources and energy products.

b) For lack of data on the GDP of Albania, the GDP per capita and the energy and electricity intensities for CEEC are calculated withoug Albania.

Comments

a) The comparison between the European Union and the "Eastern" groups of countries on the consumptions per capita shows that they are of the same order of magnitude; lower for CEEC and higher for CIS and Russia.

b) The GDP per capita are much lower for the "Eastern" groups, around a factor of two (and probably more in reality due to the difficulty of appreciate the GDPs of countries having so different economic structures and standards of living).

c) The result is that the energy intensities and the final electricity intensity are much higher for the "Eastern" groups: around a factor of two between CEEC and European Union (a factor 1.4 for electricity) and a factor of three between the CIS and Russia and the European Union (a factor 2 for electricity).

Table 2 : Primary Energy Consumption by Sources

		U.E.	*CEEC*	*CIS*	*Russia*
Total commercial energy sources (Mtoe)		1236	369	1460	968
of which (%)	Coal	24	44	24	22
	Oil	43	27	29	28
	Natural gas	17	20	40	43
	Hydro	3	3	4	4
	Nuclear	13	5	3	3
	Imported electricity*	0	1	–0.2	–0.2

** the sign—indicates net exports*

Comments:

a) In the three groups of countries, the fossil fuels have by far the largest share: 91 per cent in CEEC, 93 per cent in the CIS, 84 per cent in the European Union.

In the CIS, the second position is for the renewable energy: hydropower (4%).

Nuclear energy ranks second in the European Union (13%) and third in the CEEC (5%).

b) The dominant primary energy sources are coal (and lignite) in CEEC, natural gas in the CIS and oil in the European Union. For the two "Eastern" groups, the explanation lies in the economic and political will to rely on the predominant domestic resources and also, in the low level of automobiles. For the European Union, which is poor in domestic oil, it is the importance of the transport sector in the final consumption and its heavy dependence on petroleum products which explain the importance of these products in the energy balance.

Table 3 : Total Final Energy Consumption

Total final consumption (Mtoe)	*U.E.*	*CEEC*	*CIS*	*Russia*
	8062	242	1012	651

Comments

a) This table is interesting because it shows that the CIS and the European Union have comparable final energy consumptions, and that the final energy consumption of all CEEC countries is relatively small: It is equal to the final energy consumption of Germany.

b) The energy intensities of CEEC and CIS being respectively two and three times higher than that of the European Union, that means that, if these countries had the levels of energy intensities of the European Union, their final energy consumption would be much lower: of the order of 120–130 Mtoe for CEEC and 500–600 for the CIS.

c) Which shows that the potential for energy efficiency improvement is enormous, even if a large part of the high level of energy intensities is due to the structure of economy (with emphasis on heavy industries).

Table 4 : Final Energy Consumption by Sources and by Sectors in 1990 for the European Union and Two Representative Countries: Poland and Russia

	European Union	*Poland*	*Russia*
Total final consumption (Mtoe)	862	64	651
of which (%)			
Coal	10	26	15
Petroleum products	51	18	31
Gas	22	16	12
Electricity	16	13	11
District heat	1	27	31

Contd...

Table 4 : (Contd...)

	European Union	Poland	Russia
of which, by sectors (%)			
Industry (1)	36	48	50
Transport	28	12	13
Others (2)	36	40	37

1) including the non-energy uses.

2) domestic, commercial and tertiary, agriculture.

Comments

a) The data on the share of each sector (industry, transports, others, that is domestic, commercial and tertiary and agriculture) are not available for each country of the "Eastern" groups. Table 4 gives these shares for the two largest countries of each group, Poland and Russia, which are representative of the situation in most countries, and compare them to the corresponding values for the European Union.

b) The first striking difference between the two "Eastern" countries and the European Union is, in 1990, the shares of the three large sectors in the final energy consumption: industry is predominant with about 50 per cent in the East, with a very low share for the transport sector; the 35–40 per cent share of "others" is mostly due to the high level of energy consumption for space heating, in particular by district heating, due partly to the severe climate conditions but above all to the low efficiency of the heating systems, from production to consumption.

The tendency in the West is to reach a sharing of about 1/3, 1/3, 1/3, between the three sectors with an increase of the transport sector and a decrease of the industrial sector.

c) This situation is rapidly changing in most Eastern countries due to a sharp decline of industrial production and, in some countries, the increase of electricity consumption in household and commercial sectors and an increase of the consumption of the transport sector due to the increase of the use of private cars.

d) This situation is largely reflected in the sharing of the energy consumption by energy products. The share of petroleum products is lower in the Eastern countries than in the European Union, and the direct use of coal and the use of direcct heat are much higher.

The share of electricity is lower for the Eastern groups: CEEC are in an intermediate position (13%) between CIS (11%) and the European Union (16%).

One must be cautious when comparing the percentages: we have seen that the electricity intensities were higher in the Eastern countries than in the European Union, but that the difference was less important than for energy intensities. Then, the lower share of electricity comes primarily from the considerable waste and non-rational use of the other energy products and not from and non-sufficient electricity supply: the potential for electricity savings is also very important.

Table 5 : Production and Consumption of Electricity in 1990

		U.E.	*CEEC*	*CIS*	*Russia*
Total production (TWh):		1903	494	1728	1082
From (in %):	Fossil fuels	58	76	74	74
	Hydro	9	9	13.5	15
	Nuclears	33	15	12.5	11
Net imports (TWh)*		20	13	–33	–9
Total final consumption (TWh)		1600	373	1372	859

* *The sign—means net exports.*

Comments

a) On the production side, the fossil fuel power plants are dominant in the three groups of countries, but with a lesser share in the European Union.

Hydropower has the same share in CEEC and the European Union where it ranks third after nuclear power, but is above it in CIS and Russia.

Nuclear power represents 1/3 of the total electricity production in the European Union (with great discrepancies between the member states); the proportion is much lower in CEEC and the CIS (where only Russia and Ukraine have nuclear power plants).

b) On the consumption side, the total final electricity consumption is of the same order of magnitude in the European Union and the CIS, and the total final electricity consumption of CEEC is between that of France (302 Twh) and Germany (455 Twh).

Table 6 : The Evolution of Electricity Production in Russia in the Recent Years

Source	*1990*	*1991*	*1992*	*1993*
Fossil fuels	797	780	717	661
Hydro	167	168	172	175
Nuclear	118	120	120	119
Total	**1082**	**1068**	**1009**	**955**

Comments

a) The decrease of the electricity production from 1990 to 1993 (13% decrease over four years) is mainly due to the decrease of the demand of the industrial sector, due to its declining production.

b) The fossil fuels are by far dominant but slightly declining (74% in 1990 to 69% in 1993); the second electricity production source is hydropower with a slight increase in absolute value and an increase in percentage (15.4% in 1990 to 18.3% in 1993). The nuclear contribution is constant in absolute value and its share increases from 10.9 per cent in 1990 to 12.4 per cent in 1993.

c) It is interesting to note (this is not in the table) that for former USSR, the energy sector alone consumed, in 1990, around 20 per cent of the total consumption. The consumption of the energy sector is not included in the final energy consumption gives in Table 5.

PART — 5

POWER GENERATION TECHNOLOGIES FOR FUTURE:

Fuel Cells, Non-conventional Motor Fuels and Safe Nuclear Power and Perspective of Fusion for Energy

23

The Restructuring of the Italian Electricity Sector: Proposals for the Inclusion of Objectives for Increased Energy Efficiency and Renewables Exploitation

L. PAGLIANO
Italy

SYNOPSIS

Ongoing restructuring of the electricity industry and tariff re-regulation offer a unique opportunity for embedding objectives of energy efficiency and renewables exploitation in the new structure, Important results have already been achieved.

ABSTRACT

A profound restructuring of the electricity industry is underway in Italy. Privatisation of ENEL, the national, vertically integrated, state owned electricity utility, is planned to start in 1997, and a final decision about the degree of vertical and horizontal integration is expected shortly. A Regulatory Authority has been created for the first time, a new price regulation mechanism (price-cap) will be adopted, and new concessions and contracts will be issued. Municipal utilities are also undergoing changes in ownership and a strong debate is underway as to the price to be paid to IPPs.

No one yet knows in detail what the future structure will look like or how long the evolution from the present structure will take. In this paper, we rely on emerging trends to speculate on some of the possible transition and end-states for this evolution. We first consider whether additional promotion of energy efficiency and renewable energy would be warranted in any of these states We then describe the incentives and disincentives for energy efficiency and renewable energy programmes that would be created in each of these states. Finally on the basis of these considerations we propose, where appropriate, additional public policies to promote energy efficiency and renewable energy development in the evolving Italian power sector.

INTRODUCTION

A profound restructuring of the electricity industry is underway in Italy. Privatisation of ENEL, the national, vertically integrated, state owned electric utility, is planned to start in 1997, and a final decision about the degree of vertical and horizontal integration is expected shortly. A Regulatory Authority ("Autoritá per l'energia elettrica ed il gas") has been created for the first time, a new price regulation mechanism (price-cap) will be adopted, and new concessions and contracts will be issued. Municipal utilities are also undergoing changes in ownership and a strong debate is underway on the price to be paid to IPPs. No one yet knows what the future structure will look like in detail or how long the evolution from the present structure will take. A key concern for those interested in curbing emissions of greenhouse gasses is the compatibility of the new structure with efforts to increase end use energy efficiency and the exploitation of renewable energy resources.

In the first part of the paper we examine the current system of ownership, organisation, and control of the Italian electricity sector. We follow with a description of the changes to this system that have already occurred or are expected to occur in the near future. On the basis of these considerations we discuss three elements of an integrated strategy to encourage end use energy efficiency and the exploitation of renewable energy resources: (1) the accommodation of energy efficiency within the proposed price-cap system of rate regulation; (2) the inclusion of least-cost planning principles in the incorporation of electricity distribution franchises; and (3) the use of existing electricity taxes to support the development of a more competitive energy-efficiency and renewable energy industry. We develop the three arguments independently because they rely on different (but not mutually exclusive) assumptions regarding the future Italian electricity system. As such it would be possible to pursue a singular implementation of any one element. However, we believe they can and should be designed to reinforce one another as part of an integrated strategy.

1. THE RESTRUCTURING PROCESS IN ITALY: MAIN FEATURES AND DIFFERENCES WITH RESPECT TO OTHER EU COUNTRIES

i) Current status and Organisation of the Italian Power Sector

Since the merging/nationalisation of a oligopolistic cartel in 1962, ENEL has had the almost exclusive right to operate in the power sector. Other companies allowed in the market were municipal companies, generators consuming at least 70 per cent of produced electricity (autogenerators) and minor private companies (<15 GWh/y).

The first important reform occurred in 1991, when Law 9 imposed on ENEL the duty to buy the power produced by IPPs, subject to qualification. Bill 6/92 established the price to be paid by ENEL to IPPs, setting a minimum level equal to long run avoided costs for ENEL, based on a hypothetical combined cycle plant, plus an incentive differentiated according to the type of generation facility: cogeneration (over a certain overall efficiency threshold), waste incineration, renewables (*Fonti Rinnovabili e Assimilate* or *FRA*).

The largest share of electricity generation is due to ENEL (80%), followed by autoproducers and IPPs (15%), Municipal utilities (4%), and minor private companies (less than 1%). Something less than 93 per cent of energy is distributed by ENEL, 7 per cent by Municipal utilities, and 0.2 per cent by minor private companies. Consider also that imports account for around 15 per cent of total consumption, sourced mainly from France and Switzerland. In 1994 the fuel mix of ENEL generation was roughly: 19 per cent hydro, 8 per cent coal, 12 per cent gas, 58 per cent oil (Italy shows the largest production of electricity from oil in Europe: 113 TWh in 1993, compared with 24 in UK and 10 in Germany), 4 per cent geothermal and other Renewables.

ii) Current Tariff Regulation

Presently for each customer group, the tariff structure is uniform throughout the country irrespective of geographic area and distribution company (ENEL or municipal/local companies). The Italian rate setting system has till now been overseen by the Ministry for Industry, in the form of a loose cost-plus regulation, but the rationale for rate setting has never been completely made explicit, and beyond recovery of costs also other objectives of social policy and inflation control have been pursued.

The total bill to customers is composed of a number of elements:

a) a tariff (*"tariffa energia"*) made up of two components: a monthly fixed component (*"quota fissa mensile"*) correlated with the maximum power demand set for the specific contract type (L/kW), and a variable component correlated with the energy consumption (L/kWh), called energy price (*"prezzo dell' energia"*);
b) a distinct charge also correlated with the energy consumption (*"sovrapprezzo termico"*, L/kWh)
c) tax revenues to national, municipal and country (provincie) level (L/kWh)
d) value added tax on top of everything.

The tariff (a) is meant to remunerate fixed costs, while element (b) is meant to cover variable costs, that is fuel costs, and electric energy acquired from other companies/generators. In fact for some customers, including households, besides the fuel component proper (*sovrapprezzo ordinario*) the *sovrapprezzo termico* (b) includes a collection of otherwise unrelated surcharges: *sovrapprezzo nuovi impianti, maggiorazione straordinaria sovraprezzo, maggiori imposte di fabbricazione sugli oli combustibili.* Of particular interest here is the *sovrapprezzo nuovi impainti*, a small surcharge (L/kWh) used to pay for the incentives given to IPPs for energy produced from renewables, cogeneration and waste incineration (*Fonti Rinnovabili e Assimilate* or *FRA*) for the first 8 years of production, as established in the bill CIP 6/92.

The surcharge for funding incentive payments to energy produced through congeneration and renewables is one of the key elements of the present debate. Since 1992, the greater number of plants realised after approval in the specific graduatoria established by ENEL with state oversight have been large industrial cogeneration plants, with only a few generation plants based on renewables (mainly biomass and wind: 20 MW of wind turbines installed in 1995–96). Large cogeneration plants started first and grew faster, thus absorbing much of the funds available. More recently many projects for renewables exploitation have been presented (around 700 MW wind and 1200 biomass have been included in the graduatoria at June 95). But at the end of

1995 the fund showed a deficit of 280 billion lire (147 million ECU), and ENEL declared its unwillingness to anticipate money to IPPs. In July and November 1996 the government established by law that only those plants already included in the *graduatoria*[1] in June 1995 would be granted the incentives, if built within a certain time horizon. At the same time the surcharge has been increased (see Table 1), but at present it's unclear how many MW from renewables will be really installed, if the fund will reach a balance and if this level of funding will be sufficient for any further development.

Table 1 : Surcharge on kWh Sold for Funding Congeneration and Renewables

	April 92 (Lire '92/kWh)	*from July 96 (Lire '97/kWh)*	*from July 96 (ECU '97/100 kWh)*
low voltage customer	0.7	3.2	0.168
medium voltage customer	0.5	2.7	0.142
high voltage customer	0.4	2.3	0.121

Three features of the present tariff system are worthy of note since they differentiate the Italian situation from that of other European countries and/or have an effect on conservation programmes and renewable exploitation:

1) the accepted principle of tariff uniformity over the whole territory.
2) the price of kWh is strongly progressive with consumption for domestic customers
3) the attempt to set tariffs according to marginal costs at least for large customers through time of use tariffs.

The principle of tariff uniformity for electric energy (but not for gas or other fuels) was stated, though with opposition, in law 481 of 1995. If this decision to maintain uniform tariffs proves conclusive this might increase the complexity of the mechanisms required to reward utilities for DSM programmes.

The price of kWh to household customers with a contract with power limit at 3 kW is highly progressive with consumption, as shown in Figure 1. The general characteristics of this contract established in 1974, have led to its uptake by 90 per cent of

1. The official list of projects seeking funding.

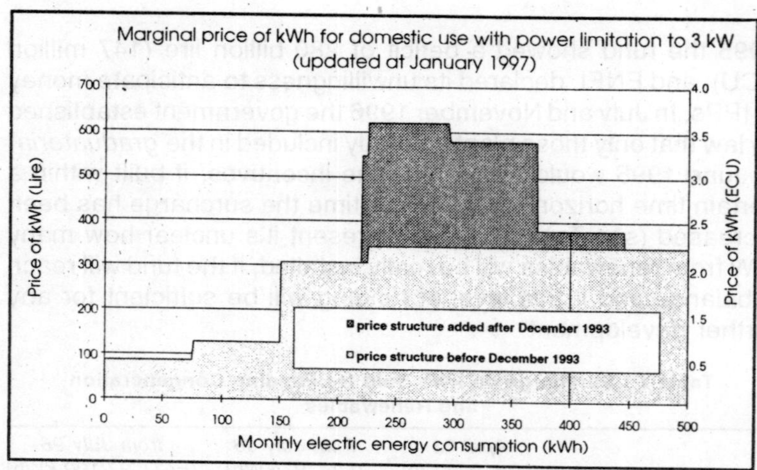

Figure 1 : Structure of Prices for Domestic Customers, before and after December, 1993.

household customers. This has probably had the effect of limiting the total power demand of the domestic sector (virtually no central electric heating is in place). The tariff based on night/day prices for energy is available only for a power of 6 kW or more and due to the high level of the quota fissa (32000 to 51000 L/month) and is economically attractive only for every high consumption levels. In December 1993, as a part of a price increase requested by ENEL in order to "preserve financial and economic balance", and backed also by the Government in order to facilitate ENEL's privatisation, the structure became even more progressive. However since customers have not been informed of the considerable financial savings to be made by lowering consumption levels, the large potential for energy savings has been little exploited.

Industrial customers (>50 kV and 500 kW) have access to a 'time of use' tariff (based on four time blocks), which roughly reflects marginal costs. In 1993, 68 TWh, or more than 70 per cent of energy sold to the industrial sector was delivered through 'time of use' contracts. Tariffs for other customer classes are, in principle, based on average costs.

Taxation is a relatively important component of the overall price of energy. For example for household customers with 3 kW contracts taxation accounts for 15–30 per cent of total expenditure, depending on consumption levels. The average price paid by different customer classes in 1995 is shown in Table 2.

Table 2 : Average Prices for Different Customer Classes

	L/kWh	*ECU/100 kWh*
household customers	216.1	11.37
external public lighting	174.0	9.16
non household <30 kW	273.1	14.37
non household 30 kWΠ500 kW	180. 3	9.49
non household > 500 kW	111.6	5.87

iii) Energy Efficiency and Environmental Protection

ENEL has implemented some DSM programmes in the past (solar hot water panels in 83–86, power factor correction in 79–92, heat pumps in 89–92, CFLs in 1990. In most cases, from the few data made available from ENEL, programmes appear to be cost effective, when cost of conserved energy is compared with avoided costs. No explicit mention to new DSM programmes was made in the investment plan approved in 1994. In the latest "Contratto di Programma", which regulates the State concession to ENEL, the utility is required to invest 10 BL/year in information campaigns on efficient use of energy, that is 0.03 per cent of total revenues (in the order to 35000 BL/year).

In 1994 ENEL presented a scenario for the evolution of the system up to 2010 (G. Carta et al. 1994) where the capacity needed in 2010 was forecasted to be 73 GW, a roughly 50 per cent increase on present capacity. According to this scenario, if no nuclear plants are built, CO_2 emissions produced by the electricity sector will grow by 48 per cent (reaching 215 Mt/year) in the years 2000 to 2010. ENEL proposes that this increase in emissions could be limited to 40 per cent if 4 GW of nuclear could be gradually commissioned starting in 2005. In 1990 the contribution of the electricity sector to total CO_2 emissions was 30% (120 of 413 Mt). Not surprisingly in conclusion the report states that "Such projections are strongly diverging from the theoretical 'CO_2 stabilisation" target at the 1990 level, and show that for Italy this is *unrealistic*."

2. Restructuring Developments. Steps Already Achieved in the Direction of Energy Efficiency and Renewables Exploitation

The following laws and documents are shaping the new "operational environment" for the electricity industry in Italy:

- in July '92 ENEL became a Ltd. company (Law 359/92) although currently the sole shareholder is the Treasury Ministry. Privatisation, even if long announced will not take place until the lines of the reform are established. Some municipal utilities are also to be privatised, in particular the process is advanced for AEM Milano, and ACEA Roma.
- Law 481 issued in November 1995 has created for the first time in Italy a Regulatory Authority for energy (*Autorità per l'Energia Elettrica ed il Gas*), and chosen a price-cap formula as the basis for the new tariff regulation
- in December 1995 a concession was granted to ENEL for the next 40 years
- a new *Còntratto di Programma* expected soon, is to define more precisely the objectives and duties of the companies granted the concessions. The Ministry for the Environment hopes that the concessions will include DSM obligations.
- a commission (*Commissione Carpi*) set up by the Ministry for Industry published in January '97 draft guidelines for the restructuring of the electricity industry

Following, we briefly summarise these developing trends with an emphasis on current status and areas of uncertainty, and on the effects for future utility sector energy efficiency and renewable energy policies.

Law 481/95 has created for the first time in Italy a regulatory authority for energy, independent from the Government (members are appointed for seven years), with adequate funding (20 billion Lire/y, 102 million ECU/y) and staff (80 people). The president and the two members were appointed at the end of 1996, and the first public hearing with Consumer Associations was held on 11th February '97. Their powers range from advise to the government on the structure of the market; renewal, modification or withdrawal of concessions, control over quality of service, customer protection, and decisions about tariff structure and level. The reform of the tariffs according to a price-cap mechanism will be the first task for the Regulatory Authority, and the Government is expecting this task to be performed by June

1997, pending a new tariff increase of 1.5 per cent requested by ENEL.

The draft of the law approved by the Senate in April 1995 declared as its main objectives: quality of the service and low costs to individual customers; reasonable profits to new shareholders; revenue maximisation to the State from the sale of the utility. It also chose a new tariff making mechanism (price-cap) to foster increased economic efficiency. Other societal goals, such as environmental protection or efficient resource use were not even mentioned in the draft.

We (Eto and Pagliano) after evaluation concluded that the draft provided a number of disincentives for energy efficiency and environmental protection. Therefore we prepared a series of amendments to be presented during a seminar held in the Parliament House in May 1995, in the presence of the Ministry of Industry, A. Clò, and representatives of several Political Parties. During the following months the second branch of the Parliament (Camera) approved 4 of the proposed amendments. Especially important is the inclusion, detailed in Art. 1, of environmental objectives among the goals/duties of the Authority and of the new tariff regulation: *"The tariff making mechanism must harmonise the economic goals of Utilities with the general societal objectives of environmental protection and efficient use of resources".* As a first implementation of this principle the new version of the law contains a price-cap formula modified according to our suggestions. Under the new formula, utility investments in DSM activities can be recovered through the tariffs.

The resulting mechanism, which will be applied both to *electric energy and gas* and to *telecommunications*, can be summarised with the following price-cap formula where:

$$P_t = P_{t-1} \times (1 + \%change\ in\ \text{CPI}-X) + Z$$

P_t = an index of maximum prices or tariff to captive customers in year t (the law defines tariffs as "the maximum unitary price of services, at the net of taxes" and does not specify if this definition has to be referred to a basket of goods, at every customer class unitary price, or else...)

P_{t-1} = an index of utility prices or tariff at year t–1

CPI = a specified inflation index such as the Consumer Price Index (*parameter*)

X=an assumed rate of productivity improvement (*parameter*)

Z=Z factor not subject to indicization

The Z factor has to take into account the following *elements*:

a) change in the *quality of service* with respect to standards established over a period of at least 3 years
b) adjustment for *unforeseen events* beyond the management's control
c) costs incurred for the implementation of programmes for the control and management of demand, through efficient use of resources (what we will call in the following *DSM component of price-cap*).

The law distinguishes two procedures by which tariffs are updated to cover respectively two sets of cost elements. With respect to changes in the cost of fossil fuels and electric energy bought from abroad or from national IPPs tariffs will be updated automatically, based on criteria set by the authority and correlated to market trends. For the remaining costs the utilities will prepare a proposal for updating the tariff every year before September 30 on the basis of inflation (CPI) and productivity increases(X), and in consideration of changes to quality of service, unforeseeable events and energy efficiency programmes. The value of the parameters CPI and X are determined by the Authority. The Authority will reply to the proposal within 60 days; after which time the proposal is to be considered automatically approved. The new tariffs will come into force on 1st January of every year.

Even if not explicitly stated in the law, we can therefore conclude that the price-cap indicization is limited to the fixed cost components: capital recovery, labour operation and maintenance, transmission and distribution, billing and metering, while a number of voices are kept out of the cap:

a) change in the quality of service
b) adjustment for unforeseen events beyond the management's control
c) costs incurred for the implementation of DSM programmes
d) fossil fuel costs
e) electric energy bought from abroad or IPPs.

As with every price-cap, this scheme will prompt utilities to increase the productivity of the factors under the cap, eventually shifting costs toward the elements outside the cap, which are directly passed on to customers through tariffs. Careful scrutiny from the regulatory authority on the prudence of use of these latter elements, mainly fuel and electricity brought from other generators, will be needed.

The recently published report produced by the Commissione Carpi proposes a possible structure for the electricity market, designed to accommodate for partial liberalisation of the market, and customer and environmental protection. The report suggests the establishment of two parallel markets.

- An open market, where largest customers, representing about 30 per cent of energy sales will progressively be allowed to choose directly from whom to buy.
- A franchised market, accounting for about 70 per cent of sales, made up of small and medium customers. They will continue to be served by distribution companies which will exercise a monopoly over a geographical area.

The co-ordination between the three markets (open, franchised and renewables) will be ensured by the wholesale electricity market (*Mercato Elettrico all'Ingrosso* or *MEI*). The commission proposes that the MEI should be operated by the company owing the transmission grid and exerting the dispatching function. This company should be independent from the generation and distribution companies, owned and controlled by the state, and eventually will own the hydro and pumping plants, in consideration of their function of equilibrium between demand and supply. ENEL should be transformed initially into a holding company, owing a number of generation companies and one or more distribution and service companies, (see figure 2).

Among the elements of the picture still missing is a decision as to whether to maintain the incentive payments to energy generated through renewables/cogen/waste-to-energy (Fonti Rinnovabili e Assimilate, FRA) and if so at what level. The report of Commissione Carpi states most importantly that incentives to these sources are still justified for their contribution to environmental protection (locally and globally), to fuel diversity, and to employment; and that extra-payments to these sources should be collected from customers. The previously existing rules will continue to apply to plants approved until June 1995 (6º graduatoria). New rules have to be set for the future and the report makes a number of suggestions. Incentives should be, as in the past

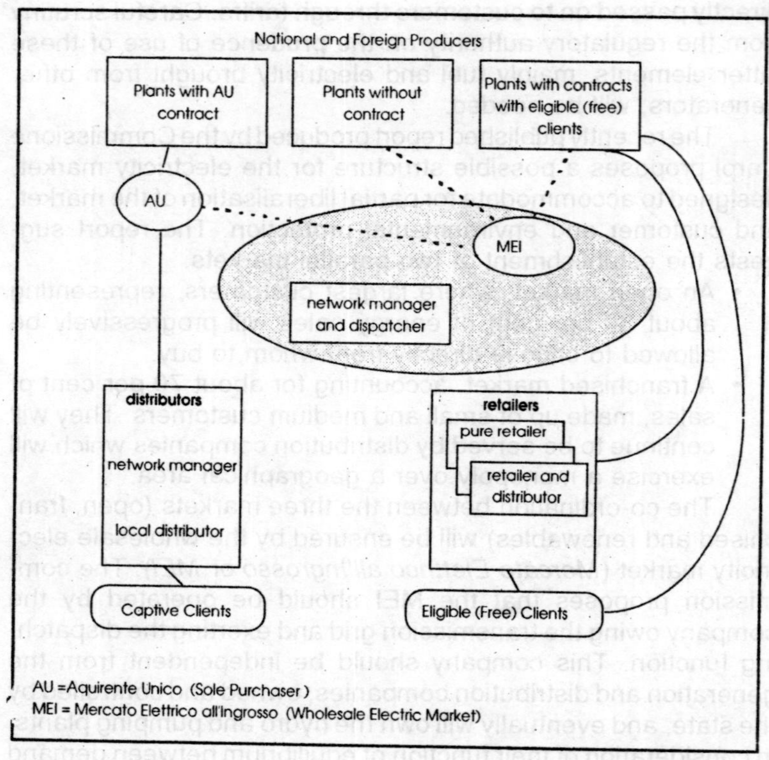

Fig. 2 : Structure of the Electricity Industry as Proposed in the Conclusive Report of the Commissione Carpi, January 1997.

proportional to the amount of energy produced, but the total amount of incentives per year and the share for each sources/ technology should be predetermined; the selection of new plants to be built and unitary the incentive (L/kWh) will be based on bidding procedures for each source/technology. The payments should be made through a fund, administered by the MEI, and overseen by the Authority, which will establish the amount of the surcharge to be paid by customers, with the obligation to keep the fund in balance. Plants generating electricity from FRA should have priority in the dispatching procedures.

3. SHOULD THERE BE PUBLIC POLICIES FOR ENERGY EFFICIENCY AND RENEWABLES IN A RESTRUCTURED ITALIAN POWER SECTOR ?

Power sector restructuring throughout the world is predicated as a means to capture economic efficiencies resulting from price setting by market forces. Market-based pricing, if it is not unduly affected by abuses of market power or other influences, could lead to prices closer to the marginal cost of production. Thus, market-based pricing would begin to address an early rationale for utility energy efficiency and renewable energy programmes, which was that regulated prices did not accurately reflect the true marginal cost of production, leading to inefficient production and consumption decisions.

However, it seems unlikely that power sector restructuring will, by itself, address the myriad additional failures that plague energy service markets. These failures include imperfect information, which manifests in the high transaction costs consumers face when making energy use décisions, as well as externalities (notably, those associated with the environmental consequences of electricity generation) that are unlikely to be reflected in market-based prices for electricity. Hence, we believe that, despite the prospects (but not necessarily the guarantee of) improved allocative efficiencies promised by electricity restructuring, the continuing presence of these market failures is compelling justification for continued government intervention.

As for Italy, evidence for the value of continued policies to promote energy efficiency comes from technical potential studies that suggest substantial cost-effective energy savings at current electricity prices. The Energy Plan for The City of Rome, compiled in 1994/95 by Ambiente Italia and the Fisica Tecnica Ambientale group of Politecnico Milano, showed that 30 per cent of electric energy could be saved through measures whose cost of conserved energy is lower than the energy price to customers of the domestic, service and industry sectors.

Traditional rationales for utility funding of energy-efficiency DSM programmes have included: (1) ratepayer funding is fair because the "problems" addressed by the programmes are unique to electricity use; (2) it is more practical than alternative public-policy responses; and finally (3) it is more consistent with other social objectives. We now briefly expand on these rationales, which we maintain are unaffected by electricity restructuring.

It's fair. The environmental consequences of electricity generation are significant and electricity consumers have a unique responsibility for the consequences of their purchase decisions. Ratepayer funding for energy-efficiency programmes, which are a partial solution to these environmental problems, is consistent with this responsibility. Whether such programmes or ratepayer funding of them are the most appropriate ways to fulfil this responsibility is separate from accepting the basic principle that the polluter should pay.

It's practical. Because the existence of environmental externalities in many activities is widely accepted, there is substantial debate about the appropriateness of policies that specially target the utility sector. For example, economic theory has been used to argue that a tax levied uniformly on all forms of greenhouse gas emissions according to their relative contributions offers a more efficient approach to address one significant environmental consequence of activities that include electricity production. However, to the extent that such a tax or even agreement that this type of approach is appropriate is unlikely or may only partially internalise these costs in the short term, additional efforts may be warranted because electricity generation is a major contributor to the problem.

It's consistent with other social objectives. A final justification for ratepayer-funded energy-efficiency programmes is pragmatic: these programmes promote public support and acceptance for policies that rely on voluntary participation. From the consumer's point of view, DSM programmes, unlike government product standards and building codes, represent a non-coercive approach to promoting energy efficiency. Moreover, these programmes can be designed to provide a stimulus to the private sector that, in the long run, may decrease the need for them.

4. CHALLENGES FOR FUTURE ENERGY EFFICIENCY AND RENEWABLES

In the first and second sections we described how the Italian power sector is changing. In the third section, we considered the continuing need for energy efficiency and renewable policies. We now examine three issues which we believe are the key to the future development of energy efficiency and renewables exploitation.

1) tax policy and the recycling of taxes to support energy efficiency and renewable programmes
2) modifications to the price cap to reduce disincentives to utility DSM
3) introduction of explicit LCUP/IRP principles into the regulation of distribution companies for franchise monopoly customers.

We develop the three arguments independently because they rely on different (but not mutually exclusive) assumptions regarding the future Italian electricity system. As such it would be possible to pursue a singular implementation of any one element. However, we believe they can and should be designed to reinforce one another as part of an integrated strategy.

i) Tax Policies

Current Italian electricity prices are relatively high when compared to other European countries. A primary reason is the high level of taxation (see earlier discussion). To some extent, these taxes currently can be seen as reflecting in prices the effect of internalising (to some, as yet unknown, degree) environmental externalities. Yet, as noted above, substantial opportunities remain at current market prices which is indicative of additional market failures that high prices alone have not led the market to overcome (and hence a role for intervention).

There are two possibilities. The first is that taxes go away, and prices move closer to their true market value; in this case whatever effect prices (made higher due to taxes) have had on consumption in the past will be lost creating more pressure for other means of internalising externalities and improving the efficiency of the market. The second is that taxes remain; in this case we would recommend that a proportion of them be redirected to fund public purpose programmes. Examples include the Energy Savings Trust scheme in UK, wires charges in the US (California), and the levy collected in Holland by distribution companies to fund energy efficiency and renewables, under a voluntary agreement with the government. With this structure utilities will no longer be requested to collect taxes in the name of the state for generic purposes, but simply to contribute to solve some of the environmental problems connected with their activities.

ii) Price Caps

By themselves, price caps provide strong disincentives to utilities to promote energy efficiency (and renewables). There are two reasons: first it is advantageous to the utility to minimise all costs associated with production (including elimination of activities not necessarily associated with production, such as energy efficiency programmes or higher cost renewables). Second it is additionally advantageous to expand sales whenever marginal cost of production is less than the price cap; lowering sales (improved end use efficiency) generates less revenues. In Italy the first disincentive has been addressed by allowing recovery of DSM programme expenses outside of the price cap. The second disincentive can be resolved through some sort of net lost revenue adjustment.[1]

In the following we provide an example based on data taken from the ENEL end of year financial balance for 1995. Considering total revenues at 36,000 BL, and total sales of 212 TWh, we arrive at an average unitary revenue of about 171 L/kWh. Under the new price-cap regulation the fixed margin for the utility will be given by average revenue from elements within the cap less their short run marginal costs. Average revenues from elements within the cap can be calculated as total revenues from sales less total fuel and electric energy expenses, divided by total sales. See Table 3.

Table 3 : Average Revenue from Elements within the Cap Calculation

	Lire		ECU	
A) Total Revenues from Sales A)	36111.8	BL	19.006	BECU
B) Fuel Expenses	7270.4	BL	3.827	BECU
C) Expenses for buying electricity from other generators	5062.5	BL	2.664	BECU
A – (B+C)	23778.9	BL	12.515	BECU
Total Sales	211.6	TWh	—	—
Average revenue from elements within the cap	112.4	L/kWh	0.059	ECU/kWh

Short run marginal costs for elements within the cap will be nearly zero for capital and labour, and small for O&M and T&D

(mainly losses), coming to a total of about 10 L/kWh. The fixed margin on each kWh sold is thus in the order of 112–10 ª 100 L/kWh. Due to the high value of this margin, there is a strong incentive to increase sales, regardless of whether the sales are economic or could be displaced by a less costly energy efficiency measure. The goal of harmonisation between the interests of the utility and the objectives of efficient use of resources is far from being achieved as required by art. 1 of law 481.

A way to offset this incentive to increase sales is a net lost revenue adjustment (NLRA). The net lost revenues, calculated by multiplying utility's fixed margin with the net loss of sales attributable to its efficiency programmes is recovered by distributing the sum across those kWh which are still sold. The combined result is a slight increase in the price of the kWh for the user, no profit losses for the utility, and energy and hence economic savings for society.

In order to assess the effect of NLRA in Italy we developed some short term (1996–2000) scenarios about the evolution of energy demand on the ENEL grid. The business as usual scenario is based on the growth rates assumed in the ENEL report for year 1995, that is a growth in energy demand of 2 per cent in 1996 and 2.5 per cent per annum till year 2000. The economic analysis made in real terms, that is without the effect of inflation, using 1995 prices.

Assuming that DSM programmes could offset at least part of this forecasted growth, we calculate the impact on tariffs and the reduction to the total energy bill of the nation, when applying the NLRA. A phased introduction of energy efficiency measures commencing in 1996 arriving at a 2.6 per cent reduction in consumption in the year 2000 with respect to the BAU scenario would result in an increase of 1.6 per cent in tariffs. In the same year the energy bill would be reduced by 420 BL, or if we consider the incremental gains over the phasing in period (1996–2000) a cumulative reduction of 1050 BL. See Figure 3.

For the purpose of comparison, we also considered a very aggressive DSM programme, again implemented by phased introduction, which would completely offset growth in the year 2000 to provide a 7 per cent reduction in consumption with respect to the BAU projections. Even in this case tariffs would only increase by 4.3 per cent. The saving on the total bill would

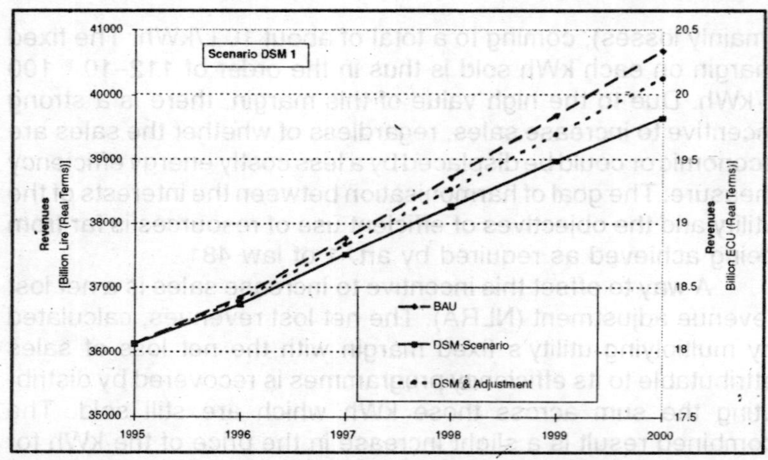

Fig. 3 : Evolution of Revenues for BAU, DSM, and for DSM with Net Lost Revenue Adjustment. Limited Intervention Case, Resulting in 2.6% Reduction in Consumption by the Year 2000.

be 1100 BL in year 2000 or 2700 BL cumulative for the years 1996–2000, (see Figure 4). However such an aggressive programme would be both politically and technically difficult to achieve.

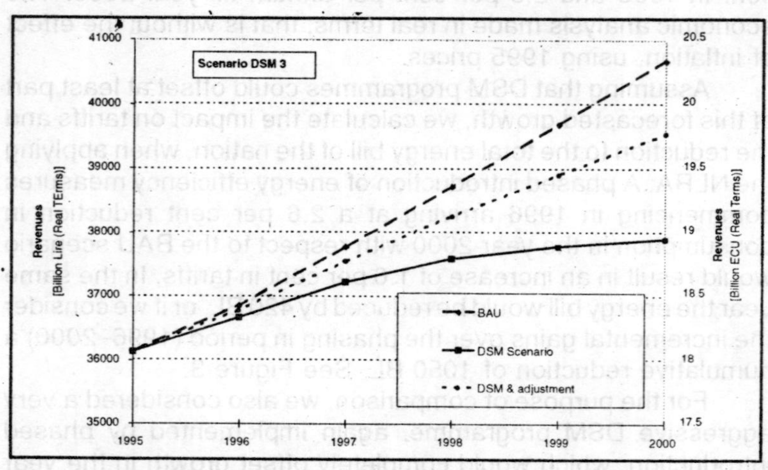

Fig. 4 : Evolution of Revenues for BAU, DSM, and for DSM with Net Lost Revenue Adjustment. DSM Intervention are Such as to Offset Growth by the Year 2000

In both cases, the resulting impact on tariffs, even in the case of an extremely aggressive programme is very limited when compared to other recent changes; a 20 per cent increase for some customers in 1993, a 3 per cent increase in November 1996 for fuel cost escalation, and an ENEL proposed rate increase of 1.5 per cent (1997) to cover supply side investments.

While DSM cost recovery, and NLRA will remove present disincentives if properly designed and implemented, it might be necessary to ensure that DSM investments will result as profitable as others available to the utility, through for example shared savings or other similar mechanisms. The detailed assessment of how big this additional incentive should be is outside the scope of this paper, and will also depend on the definitive structure of industry following the reform. If distribution is to be separated from production and transmission, then distribution companies will be the candidate for DSM actions and incentives will have to be calibrated against the return these companies can earn on alternative investments.

iii) Regulation of Distribution Companies

It is worth reiterating that the price-cap method will apply to captive customers only who will be serviced by a number of distribution companies which will exercise a monopoly over a geographical area. The Commissione Carpi suggests that these companies could be derived from the 14 management subdivisions (*Direzioni*) currently operated by ENEL and would inherit 70 per cent of the present distribution activities.

The Commission also suggests that in urban areas where presently distribution franchise are attributed to different companies (for example in Milano 50% of customers are served by ENEL and 50% by AEM), it will be necessary, in order to obtain higher economic efficiencies, to redefine the concessions and to create single distribution companies for the whole area.

Distribution companies will be able to offer additional services connected with the distribution of electricity (e.g. end-use energy efficiency services), though it will be necessary to maintain separate accounts and management structures.

Organised in this way the electricity industry will provide an ideal structure for the promotion of energy efficiency and renewables exploitation. Distribution companies with captive customers are in an ideal situation to conduct LCUP/IRP. Further, confined to distribution activities, with for example, no capital recovery of generating plant to cloud choice, management of demand will prove at least as attractive as increased supply.

We would recommend that the regulatory authority incorporates LCUP/IRP principles into the regulatory framework of the distribution companies and ensure that other legislation affecting the operation of distribution companies (such as the modifications to the price cap, suggested above) be adapted to ensure harmony with the use of the same principles.

References

1) L. Baxter. 1995. "Assessment of Net Lost Revenue Adjustment Mechanisms for Utility DSM Programmes", Oak Ridge National Laboratory, ORNL/CON–408.
2) C. Marney and G.A. Comnes, 1990, "Rate marking for Conservation: the California ERAM Experience." Berkeley, CA: Lawrence Berkeley Laboratory. LBL–28019. March.
3) J. Eto, S. Stoft and T. Beldon, 1994, "The Theory and Practice of Decoupling," CA: Lawrence Berkeley Laboratory. LBL–34555. January.

24

Biofuels: Renewable and Environmentally Friendly Liquid Fuels for the Transport Sector

W. KÖRBITZ
Austria

ABSTRACT

As used already by Rudolf Diesel in 1912, plant oils represent not only a new alternative fuel compared to fossil sources, but by the force of the oil supply shocks in the 70s a new development of Biodiesel was triggered. This paper gives a review of the political background, the historical development since the beginnings in Austria and the volumes produced today in the world, the main raw materials used, key fuel properties and standards. It highlights the fuel's environmental advantages and different marketing strategies applied as well as key factors of micro and macro economic considerations.

KEYWORDS

Beef tallow; biodegradability; biodiesel; bioenergy; biofuels; fatty-acid-methyl-ester; rapeseed-oil-methyl-ester; non-food crops; oilseeds; waste oil.

POLITICAL TRIGGERS FOR BIODIESEL

The strongest impulse was given by the crisis in supply of mineral oil as the major source for energy in the 70s and again by the Gulf War in 1991. Being highly dependent on huge imports of fossil oil as a finite energy source the European Union has to face today again an increasing risk in *security of energy supply* for the transport sector caused by the following issues as emphasised by the International Energy Agency (IEA): a) the production-demand gap of fossil oil is declining world-wide, b) North Sea oil will be finished by the year 2010 latest, and c) the energy demand of the non-OECD world is growing dramatically e.g. in China.

According to the IEA there will be a need for *all* alternative fuels for the transport sector, and Biodiesel will be one of them (European Commission = EC, October, 1996). The EC proposes in the White Paper on Renewable Energy a 2 per cent market share for biofuels by the year 2010 (EC, December 1997) and in the FORUM-scenario a 12 per cent market share for biofuels by the year 2020 (EC, Energy to the year 2020, Spring 1996).

Concerning *environmental damage* the transport sector has a clear responsibility. Within the last 10 years its part in global warming potential has increased from less than 20 per cent to more than 25 per cent, now bigger than those of the domestic and industrial sector, while its contribution to acid pollution constitutes 75 per cent of total emissions of this pollution type. As one reaction the European Commission has developed a Directive on the Quality of Fuels with new environmentally driven fuel specifications (European Commission, October 1997).

Substantial and costly *overproduction of agricultural crops* for food has led to a reformed Common Agricultural Policy introducing a set-aside percentage for food-crops but allowing to produce for non-food crops, e.g. rapeseed. The initial percentage for set-aside of 15 per cent declined step by step over the years to 5 per cent today, putting the young Biodiesel-industry at a substantial risk of raw material shortage.

The future agricultural policy however will have to consider, that with the enlargement of the European Union by the Countries of Central Europe (CCE) tremendous opportunities for Biodiesel production are opening up, as those countries have presently double the acreage per citizen compared to the EU-15 with an enormous upward potential in agro-productivity.

MILESTONES IN THE DEVELOPMENT OF BIODIESEL

First Biodiesel initiatives were reported in 1981 in South-Africa and then in 1982 in Austria, Germany and New Zealand. Already in 1985 a *small pilot plant* in Austria tested the production of RME with a new technology (ambient pressure and temperature) and in 1990 the first farmers' co-operative started commercial production of Biodiesel. In the same year the completion of *large fleet tests* led to engine warranties by most of the tractor producers as e.g. Case, Deutz, John Deere, Ford, Massey-Ferguson, Mercedes, Same, Steyr as a big step forward towards a successful market introduction of Biodiesel.

Another important step was the *first fuel standard* ON C 1190 for Biodiesel in 1991 by the Austrian Standardisation Institute assuring a high quality of the fuel. Detailed tests on product properties such as engine performance, emission reductions, biodegradability and toxicity were followed, while process economics improved as well continuously.

Biodiesel plants were started, mainly in the European Union but also in East Europe, Malaysia and in the USA; the actual capacity and production figures are given in Table 1.

Table 1 : Biodiesel Capacities and Production Volumes

estimated in 1.000 mt:	*1996 capacities*	*1996 production*	*1997 production*	*future projects*
Austria	38	17	22	0
Belgium	200	20	20	0
France	310	227	250	100
Germany	287	63	83	30
Italy	199	141	109	0
Others	14	6	11	70
EU - 15	1.048	474	495	200
Czechia	63	22	45	0
Rest of Europe	10	8	10	5
U.S.A.	38	5	8	300
Canada	1	1	1	100
Malysia	10	10	10	20

(Austrian Biofuels Institute, 1997)

The year 1996 was a big step forward marked by the start-up of *large industrial scale plants* in Rouen/France and in Leer/Germany;—and by the milestone of *overall warranty* for all models of Audi, Mercedes and Volkswagen as trailblazers in the personal car sector. In the same year the foundation of the European Biodiesel Board as a professional organisation of all major Biodiesel producers took place, indicating the further growth of a young industry.

MAIN RAW MATERIALS

In the beginning was rapeseed or canola. With the high

content of the monounsaturated oleic acid (C 18:1) of about 60 per cent, the rather low level of saturated fatty acids (palmitic and stearic acid <6%) and also acceptable levels of linolenic acid (C 18:3) *rapeseed-oil is a rather ideal raw material* for the European climate with sometimes cold winter conditions and of reasonable product stability expressed by an Iodine Value (IV) of <115.

Other raw materials used were palm-oil in Malaysia (Schäfer, 1991; Ahmad, 1997) and sunflower-oil in France and Italy, while soybean-oil became the raw material of choice in the USA. In Nicaragua the locally available oil of *Jatropha curcas* plant is processed.

Today low cost sources of triglyceride raw materials as e.g. used frying oils collected at restaurants or even low grade beef tallow are used for Biodiesel production with improved process technologies. At the end however it must be high quality standardised Biodiesel as demonstrated in related tests (Sams, 1996).

KEY FUEL PROPERTIES OF BIODIESEL

In testing plant oils as a fuel it was the first lesson to learn, that pure oils, even of fully refined quality, do not fit the modern fast running Diesel engine of high efficiency and with a low emission profile. The methyl-esters were the plant oil derivative of choice, simple in production and coming very close to the fuel properties of Diesel (Table 2). There are slight but acceptable differences in density and viscosity, the higher flash-point is a beneficial safety feature, and the sulphur-free plant oil is the reason for the excellent SO_x–emissions of Biodiesel. Generally the Cetane no. is higher for Biodiesel resulting in a smoother

Table 2 : Physical-Chemical Properties

standardised properties:	unit	Diesel EN 590:1993	Biodiesel (FAME) DIN E 51.606:1997
density at 15°	kg/m^3	820–860	875–900
viscosity at 40°	mm^2/s	2.00–4.50	3.5–5.0
flash-point	°C	>55	>110
sulphur	%(m/m)	<0.20	<0.01
Cetane No.		>49	>49
other properties :			Biodiesel (RME)
oxygen content	%(m/m)	0.0	10.0
caloric value	MJ/dm^3	35.6	32.9
efficiency degree	%	38.2	40.7

(Walter, 1992)

running of the engine with less noise.

Biodiesel (RME) is by nature an oxygenated fuel with an oxygen content of about 10 per cent. Oxygen is causing all the favourable emission results, but it is also the reason for a 7 per cent lower caloric value.

Concerning winter operability as expressed by the Cold Filter Plugging Point (CFPP) RME by nature can be used down to –80°C, with additives down to –22°C. (Wörgetter, 1995).

BIODIESEL STANDARDISATION

As a condition for a successful market introduction and supportive acceptance by engine producers and end-users as customers a fuel for Diesel engines must be specified with carefully selected criteria as a common tool for quality assurance.

The very first standard was the Austrian RME specification ON C 1190 for Biodiesel based on 00-rapeseed-oil. It was followed by the Austrian Fatty-Acid-Methyl-Ester (FAME) specification ON C 1191 allowing a wider range of triglycerides–virgin or waste oils and fats of plant or animal origin–provided the required high quality for Biodiesel is assured.

Other countries followed with similar standards, e.g.: France and Italy in 1993, Czechia in 1994; Sweden in 1996 (SS 15 54 36), and in the USA with an ASTM committee established. (Schindlbauer, 1996).

The latest Biodiesel standard today is the German FAME draft specification DIN E 51.606. This standard has obtained a strong European dimension, as it is the basis for warranties given by major Diesel engine producers such as Audi, BMW, Ford, Iseki, IVECO, John Deere, Kubota, MAN, Mercedes-Benz, Seat, Skoda, Valmet, Volkswagen, Volvo, a.o. all over Europe. The next step already initiated is the completion of a CEN-specification in 1998 within TC 19, which is supported through a EC mandate supported by the ALTENER-programme of DG XVII.

ENVIRONMENTAL BENEFITS

Tremendous efforts have been put into life cycle analysis exercises in many countries; Biodiesel appears to be one of the best researched products. In summary one can state, that there is a clear contribution to the *reduction of greenhouse gases* by at least 3.2 kg CO_2-equivalent per 1 kg Biodiesel (Scharmer,

1993); those results have been improved since then by lower inputs in raw material production and by more efficient process technology.

It is as well established, that there are significant locally impacting emissions e.g. a 99 per cent reduction of SO_x-emissions, and -20 per cent for CO, -32 per cent for HC, -50 per cent in soot and -39 per cent for particulate matter, while there is a slight increase of NO_x-emission,—with delay of injection timing however a decrease of 23 per cent can be obtained (Sams 1996). Biodiesel appears to be also an ideal synergistic partner for the oxidation catalytic converter (oxicat).

Not surprisingly Biodiesel as a plant oil derivative has a very low toxicity as a compound being the reason for the high biodegradability of more than 90% within 3 weeks and for substantial reduction of toxicity risks to lead water organisms like trout, daphnia, water cress and algae, advantageous in case of accidental spillage. (Rodinger, 1994).

MARKETING STRATEGIES

Bringing a new product to the market requires careful consideration of the market conditions and customer needs. Different strategies have been applied so far:

Italy has one of the highest levels of mineral oil tax in Europe, both on Diesel fuel and heating oil. Given full detaxation it was therefore a logic step to penetrate the easier accessible market for heating oil.

France has chosen another strategy by delivering Biodiesel to refineries, where it is blended with 5 per cent to fossil Diesel and distributed through the existing system, mainly by Elf, Shell and Total. However the customer is not in a position to identify the difference in the fuel. This strategy is avoiding to build a separate and costly infrastructure and big volumes can enter the market immediately on the one hand, the advantages of Biodiesel are applied only in a diluted way and without visibility on the other hand.

Another blend strategy is tried in the USA, where 20 per cent soy-oil-methyl-ester are mixed with fossil Diesel, mainly because of price reasons. The 80/20 fuel blend in combination with a catalytic converter has recently obtained EPA certification for the Urban Bus Retrofit programme.

Taking all the benefit all the benefits to Biodiesel a 100 per cent and undiluted to the market is the strategy of choice in Germany and Austria. Target applications identified can be *environmentally sensitive segments*, e.g. water protection on lakes and quality groundwater areas, irrigation pumps, skiing areas, forestry operations, national parks, as well as taxis, city buses in smog endangered locations and the "green" driver.

MICRO AND MACRO ECONOMIC CONSIDERATIONS

Without going into much detail the key sensitivities in the micro economics calculation should be highlighted: by far the most important factor in the production is *cost of raw material*. Oilseeds from set-aside fields have been traded in Europe at acceptable prices at around DEM 300,— /ton seed so far, cheap used frying oil and other waste oils and fats can improve the calculation.

The second most important factor—and often not recognised as such—is the *yield* in the process, i.e. to what degree trans-esterifiable triglycerides and free fatty acids are turned into high value methyl-esters; it should not be lower than 99.7 per cent (Koncar, 1996).

Usually the selling price is oriented at the fossil Diesel price and can compete under the condition of mineral oil tax relieves.

The exercise of internalising all external costs gives the following macro-economic picture:

Following a recently published study of the IFO-Munich, which completed an input-output study assuming 300.000 ha of rapeseed in Germany, a Biodiesel production from this acreage would create 5,000 jobs and this *job-creation-bonus* would justify already 70 per cent of the detaxation given (Schöpe, 1996).

Additionally there is a *less-global-warming-bonus*, which is calculated as CO_2 -avoiding cost of US$ 0.64 per 1 kg Biodiesel in industrialised countries (Hohmeyer, 1993).

For the application in environmentally sensitive areas the relevant less-risk-bonus depending on local risk levels has to be evaluated individually.

In addition there is an *energy-supply-security-bonus*: in a rather easy investigation a US-government study calculated the cost of US strategic presence in peace times in the Gulf with US $ 9.68 per barrel imported fossil oil into the USA (Ravenal, 1991).

Biodiesel has a positive energy balance of approx. 1: 3.2 (Schäfer, 1996) and has therefore a fossil energy saving effect of ca. 0.85 kg. mineral oil saved per 1 kg Biodiesel (Scharmer, 1993). Obviously there is a substantial *renewability-bonus* growing in value for the years to come.

REFERENCES

1) Ahmad, S. (1997). *Oleochemicals and other Non-Food Applications of Palm Oil and Palm Oil Products*. PORIM, Kuala Lumpur, Malaysia.
2) Austrian Biofuels Institute (1997). *Biodiesel Development Status World-wide*, Report for the International Energy Agency. Vienna, Austria.
3) European Commission—DG XVII, (Spring 1996). *European Energy to 2020*, A Scenario Approach. Brussels, Belgium.
4) European Commission,, (October 1997). *Council Directive to the Quality of Petrol and Diesel Fuel*. Brussels, Belgium.
5) European Commission—DG XVII, (October 1996). *Proceedings Improving Market Penetration for New Energy Technologies*: Prospect for Pre-Competitive Support. Brussels, Belgium.
6) Koncar, M. (1996). Criteria for the Development and Selection of Low Cost and High Quality Technologies for Biodiesel. *Proceedings 2nd European Motor Biofuels Forum*, Joanneum Research, Graz, Austria.
7) Körbitz, W. (1995). Utilisation of Brassica OIls as Biodiesel Fuel. *Brassica Oilseeds: Production and Utilisation*. CAB International, Wallingford, Cambridge, United Kingdom.
8) Mittelbach, M. and Körbitz, W. (1995). Biodiesel: A Regenerative Fuel for Unmodified Diesel Engines Produced from a Variety of Fresh and Used Plant Oils. *Earth Conference on Biomass for Energy, Development and the Environment*. Havana, Cuba.
9) Ravenal, E.C. (1991). *Designing Defence for a New World Order*. CATO Institute, Washington D.C., USA.
10) Rodinger, W. (1994). Neue Daten zur Umweltverträglichkeit von RME im Vergleich zu Diesel-kraftstoff. *Bundesanstalt für Wassergüte*, Vienna, Austria.

11) Sams, T. (1996) Use of Biofuels under Real World Engine Operation. *Proceedings 2nd European Motor Biofuels Forum,* Joanneum Research, Graz, Austria.
12) Schäfer, A. (1991). Pflanzenölfettsäure-Methyl-Ester als Dieselkraftstoffe/Mercedes-Benz. *Proceedings Symposium Kraftstoffe aus Pflanzenöl für Dieselmotoren.* Technische Akademie Esslingen, Germany.
13) Schäfer, A. (1996). Environmental and Health Concerns at Mercedes-Benz. *Proceedings Commercialisation of Biodiesel: Environmental and Health Benefits,* University of Idaho, Moscow, USA.
14) Scharmer, K. (1993). Umweltaspekte bei Herstellung und Verwendung von RME. In: *RME Hearing. Ministry for Agriculture,* Vienna, Austria.
15) Schindlbauer, H. (ed.) (1995). *Proceedings 1st International Conference on Standardisation and Analysis of Biodiesel.* FICHTE/Technical University, Vienna, Austria.
16) Schöpe, M. (1996). Economic Aspects of Biodiesel Production in Germany/IFO-Institute, Munich. *Proceedings 2nd European Motor Biofuels Forum,* Joanneum Research, Graz, Austria.
17) Walker, K. and Körbitz, W. (1996). Biodiesel—Production and Exploitation. *Energy from Crops.* Semundo Ltd., Cambridge, United Kingdom.
18) Walter, T. (1992). Untersuchungen des Emissionsverhaltens von Nutzfahrzeugmotoren am Prüfstand bei Betrieb mit RME/Swiss Research Institute EMPA. *Proceedings Symposium RME—Kraftstoff und Rohstoff.* FICHTE/TU Vienna, Austria.
19) Wörgetter, M. (1995). Improved Winter Operability with an CFPP—Improved Biodiesel. *Proceedings 1st International Conference on Standardisation and Analysis of Biodiesel.* FICHTE/Technical University Vienna, Austria.

25

Development and Perspectives of Fuel Cell Technology

R. VELLONE

Principles of Fuel Cell (Phosphoric Acid Type)

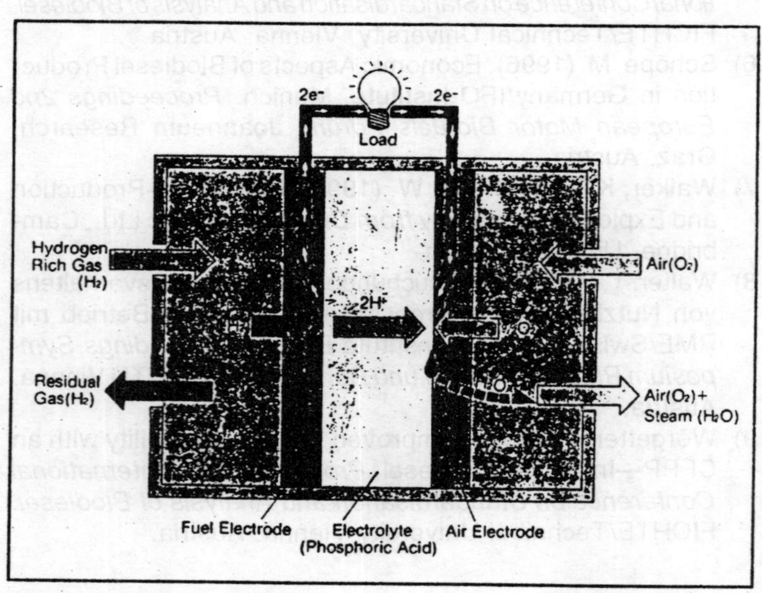

Basic Construction of Cell

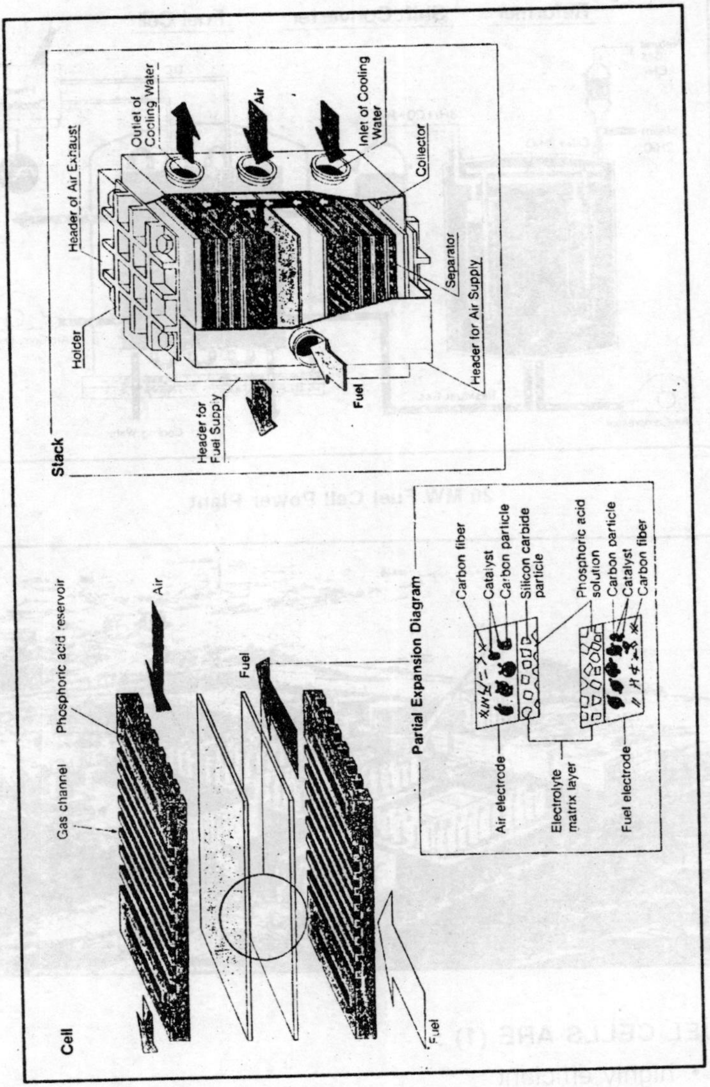

Fuel Cell Power System

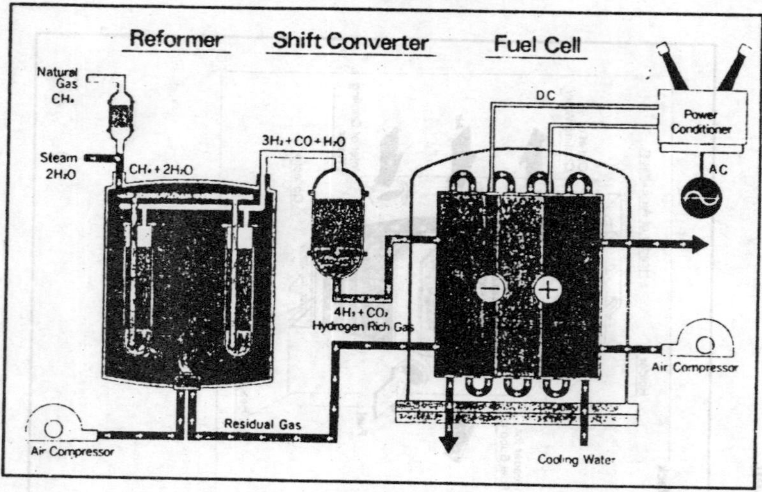

20 MW Fuel Cell Power Plant

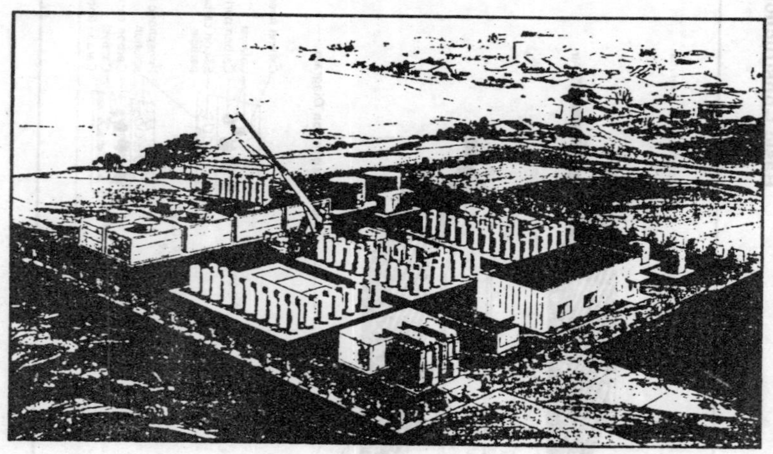

FUEL CELLS ARE (1) :

- highly efficient
- suitable for both stationary and transportation applications
- the high-temperature fuel cells are especially suited for cogener-·ion applications

- characterised by highly efficient operation that is almost independent of system size
- well suited for optimal distribution of generated power with short transmission lines, therefore they offer good oppor-, tunities to reduce transmission and distribution cost.

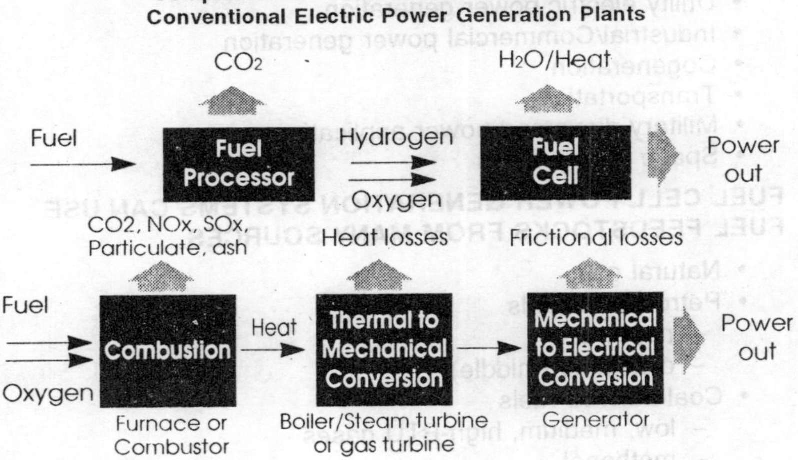

FUEL CELLS ARE (2):

- compatible with renewable energy because of the preferred use of hydrogen

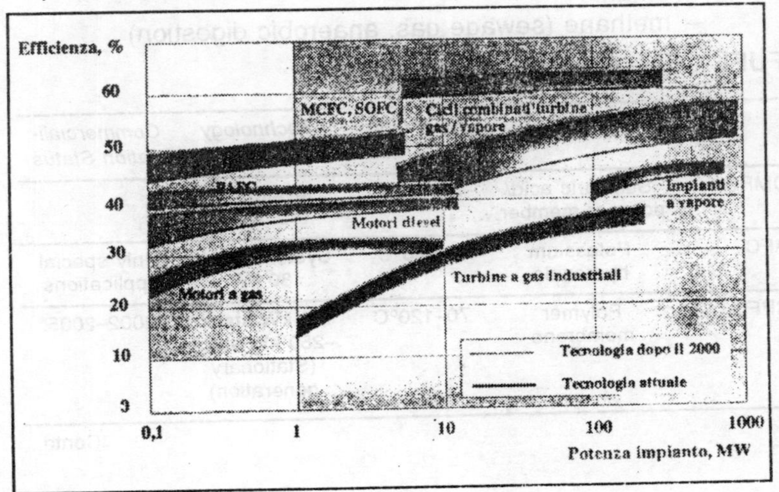

- adapted to provide flexibility since they may be designed to utilise any primary fossil-fuel resource
- suited for flexible planning because of their small modular sizes
- clean and non-polluting

FUEL CELL POWER GENERATION APPLICATIONS

- Utility electric power generation,
- Industrial/Commercial power generation
- Cogeneration
- Transportation
- Military-dispersed power applications
- Space programmes

FUEL CELL POWER GENERATION SYSTEMS CAN USE FUEL FEEDSTOCKS FROM MANY SOURCES

- Natural gas
- Petroleum liquids
  - naphtha
  - distillates (middle)
- Coal-derived fuels
  - low, medium, high-BTU gases
  - methanol
  - coal liquids
- Biomass/Waste-derived fuels
  - methanol
  - ethanol
  - methane (sewage gas, anaerobic digestion)

FUEL CELL CHARACTERISTICS

Cell Type	*Electrolyte*	*Operating Temperature*	*Technology Status*	*Commerciali-sation Status*
DMFC	Sulphoric acid/ polymer member.	45–100°C	Development phase	
AFC	Potassium hydroxide	60–90°C	Systems up to 80 kW	Only special applications
SPFC	Polymer membrane	70–120°C	30 kW stack –250 kW stack (Stationary generation)	2002–2005

Contd...

PAFC	Phosphoric acid	160–220°C	30 kW stack Power plants up to 11 MW	1995–2000
MCFC	Alkali carbonates	600–700°C	250 kW stack 2 MW demon. plants	2002–2010
SOFC	Stabilised zirconium oxide	850–1000°C	30 kW stack-(tubolar type) 10 kW stack (planar type)	2005–2010

SOLID POLYMER ELECTROLYTE FUEL CELLS

The technology is in the stage of full-scale demonstration. SPFC have characteristics (high power density, rapid start up; quick response to changing load) that make them of interest for:

- on-site cogeneration (20–250 kW, natural gas as fuel)
- transportation application (car, commercial vehicles, buses powered by hydrogen and methanol)

SOLID POLYMER ELECTROLYTE FUEL CELLS MAIN DEVELOPMENT PROGRAMMES (1)

North America

- **Cell and Stack Technology**
  - Ballard (Canada), H Power, Energy Partners and International Fuel Cells (US) have programs in various stages of development.
- **Stationary systems**

  Ballard formed a joint-venture company with GPU International in the US to manufacture and commercialise power plants.
  - first 250 kW stationary cogeneration prototype started operation in 1997
  - development of a 250 kW fuelled by natural gas is on going. Fully commercial units on the market around 2000.

SOLID POLYMER ELECTROLYTE FUEL CELLS MAIN DEVELOPMENT PROGRAMMES (2)

- **Systems for transportation applications**
  - Ballard Power systems in a R&D programme began in 193, with substantial funding from Daimler Benz, has

produced stacks of sufficient power density for use in automotive field.

– Ballard power units are operating in commercial-scale zero-emission transit buses fuelled by hydrogen. The cities of Chicago and Vancouver are demonstrating prototype buses in their public transport system by 1997; commercial vehicles in 2000.
– Development of experimental vehicles fuelled with hydrogen and methanol in the frame of the PNGV programme (Ford, General Motors and Chrysler; experimental fleet by 2000).

BALLARD POWER SYSTEMS: ZERO-EMISSION BUS

Vehicle type: Commercial transit but (12.2 m length, 75 passengers)
Power output: 205 kWe
Fuel: Compressed gaseous hydrogen
Operating range: 400–560 km (with break energy recovery)
Maximum speed: 95 km/h

Fleet Tests: 6 buses in Chicago and Vancouver from 1997

SOLID POLYMER ELECTROLYTE FUEL CELLS MAIN DEVELOPMENT PROGRAMMES (3)

Japan

- **Cell and stack technology**
  Fuji Electric Mitsubishi Electric, Sanyo, Toyota
- **Systems for Transportation Applications**
  Some car automarkers are active through NEDO/MITI contracts
  – Toyoto has developed RAV-4 fuel cell vehicles fuelled with hydrogen and methanol
  – Experimental systems with Ballard cells (Nissan, Honda)

SOLID POLYMER ELECTROLYTE FUEL CELLS MAIN DEVELOPMENT PROGRAMMES (4)

Europe

- **Cell and Stack Technology**
  - Siemens and De Nora
- **Systems for Transportation Applications**
  - Daimler-Benz is developing fuel cell powered passenger vehicles in a joint R&D programme with Ballard Power Systems (NECAR and NEBUS Projects); commercial units available by 2004.
  - Development of experimental vehicles fuelled to hydrogen in frame of EC Programmes (Ansaldo, De Nora, Renault, Peugeot, Neoplan).
- **Stationary System**
  - Development of demonstrative prototypes up to 15 kW (De Nora).

DAIMLER BENZ/BALLARD POWER SYSTEMS NECAR PROJECT

NECAR I (1994)

- fuel: hydrogen
- 12 stacks, 230 V
- power 50 kW
- 21 kg/kW

NECAR II (1996)

- fuel: hydrogen
- 2 stacks, 280 V
- power 50 kW
- 26 kg/kW

NECAR III (1997)

- fuel: methanol
- power 50 kW

Daimler Benz/Ballard
NEBUS

Vehicle type: Commercial city bus (12m length, 58 passengers)
power output: 250 kWe (10 stacks)
Fuel: Compressed gaseous hydrogen
Operating range: 250 km
Maximum speed: 85 km/h

SOLID POLYMER ELECTROLYTE FUEL CELL PROGRAMME

The National Solid Plymer Electrolyte Fuel Cell Program has the aim of developing a proprietary technology for stacks and system suitable for transportation applications.

SOLID POLYMER ELECTROLYTE FUEL CELL PROGRAMME1

- 1990–1994 Development of a PEM stack based on a proprietary Italian technology (*ENEA and CNR-TAE in collaboration with De Nora*)
  5kW stack @ 0.04 kW/kg in 1994
  founding: ~2 million USD
- 1995–1998 Development of a PEM stack with higher specific power suitable for transportation application (*ENEA and CNR-TAE in collaboration with De Nora*)
  10 kW stack @ 0.10 kW/kg in 1996
  30 kW stack @ 0.25 kW/kg in 1998
  founding: ~ 4.8 million USD

SOLID POLYMER ELECTROLYTE FUEL CELL PROGRAMME2

- Study of low-cost proton membrane
  Selection of new polymers/Investigations on products of new synthesis (*ENEA in collaboration with Polytechnic of Milan, Universities of Milan and Brescia*).
- Development of a POX reformer prototype for SPFC powered vehicles (power about 10 kW), in 1998
  (*ENEA and CNR-TAE in collaboration with De Nora*)
  founding: ~2 million USD

De Nora Stacks of Different Sizes

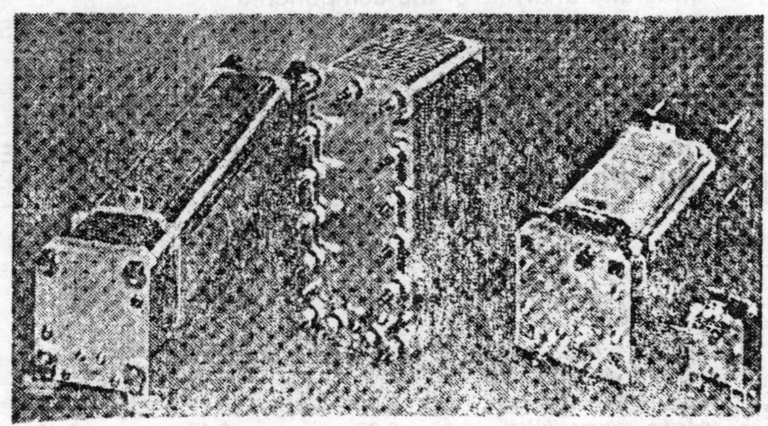

SOLID POLYMER ELECTROLYTE FUEL CELL PROGRAMME ACTIVITIES OF PARTICIPANTS

De Nora

- Development of stack technology
- Manufacturing and testing of single cells and stack (active area 900 cm^2)
- Design and construction of improved 10÷15 kW unit (0.1 kW/kg)
- Design and construction of high power density 3 kW unit (about 0.25 kW/kg)
- Mathematical modelling of stack design

CNR-TAE

- Preparation and evaluation of new electrocatalysts
- Development of electrode technology (*low Pt loading anodes (<0.2 mg/cm^2); CO-tollerant anodes; low pressure cathodes*)
- Study of manufacturing processes of electrodes and M&E
- Single cell modelling to predict performance over a full range of operating conditions

ENEA

- Physico-chemical and electrochemical characterisations of materials, electrodes and M&E assemblies
- Single cell tests on laboratory scale
- Simulation models of the fluid dynamic behaviour of the stack
- Optimisation of the stack specific power performance through structural analysis of the components
- Design and construction of test facilities for stacks up to 15 kW
- Stack qualification under a wide range of operating conditions

SOLID POLYMER ELECTROLYTE FUEL CELL PROGRAMME ITALIAN PROGRAMME

	Standard Unit (1993)	*Improved Unit (1996)*	*Advanced Unit (1998)*
• *power (kW)*	5	10	30
• *voltage(V)*	62	30	60
• *current (A)*	81	333	500
• *operating pressure (bar)*	3	4	1.5
• *operating efficiency (%)*	50	57	57
• *power density (kW/kg)*	0.04	0.10	0.25
• *specific volume (kW/l)*	0.06	0.17	0.43

De Nora 10 kW stacks

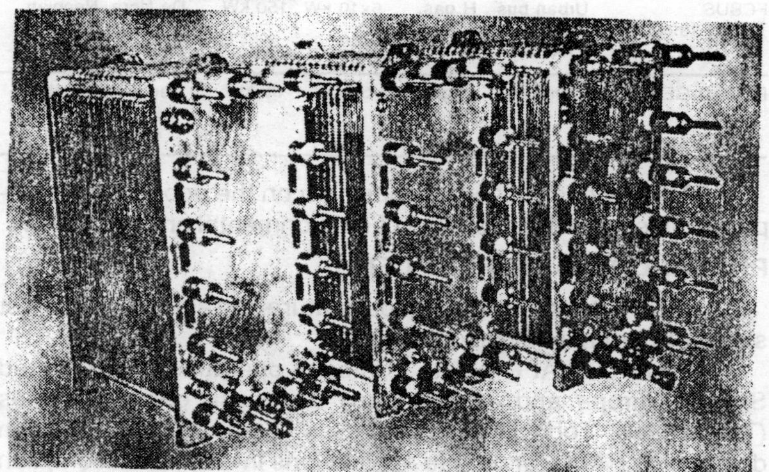

FUEL CELL SYSTEMS FOR TRANSPORTATION

Transportation in Italy

- accounts for almost 60 per cent of the oil consumption
- is the major contributor to air and noise pollution in urban areas

Fuel cell powered vehicles could represent over the long term an effective alternative to the ICE vehicles, with

- comparable performance
- greater energy efficiency
- reduced air and noise pollution

Solid Polymer Electrolyte Fuel Cells Development of Systems for Transportation

Project	*Vehicle Type*	*Fuel*	*FC Power*	*Total Power*	*Participants*
EQHHPP BUS Test from Jan. '98	Urban bus 12 m	H_2 liq.	8×5 kW	120 kW	Ansaldo, De Nora, Messer Griesheim, ASM Brescia
EQHHPP BOAT Test from April '98	Boat 20 m	H_2liq.	8×5 kW	120 kW	Ansaldo, De Nora, Messer Griesheim
FEVER Test from June '98	Passenger Car (Renault Laguna)	H_2liq.	3×10 kW	30 kW	Renault, Ansaldo, De Nora, Ecole des Mines, Volvo, Air Liquide

Contd...

FCBUS Test in 1999	Urban bus (Neoplan N4114)	H_2gas.	6×10 kW	150 kW	De Nora, Neoplan, Air Liquide, Univ. Genova, Ansaldo
HYDRO-GEN	Minivan	H_2gas (700 bar)	30 kW	30 kW	Peugeot, Ansaldo, De Nora, Solvay CEA, Renault

Market penetration depends both on the economical competitiveness and the evolution of environmental protection laws.

PHOSPHORIC ACID FUEL CELLS

PAFC technology is the most mature and already at the first stage of commercialisation.

Over 100 PAFC plants are today operating in the United States, Europe and Japan; many Japanese gas companies (Tokyo Gas, Toho Gas, Osaka Gas, Saiby gas) have been developing PAFC for on-site applications in order to accomplish early commercialisation.

PAFC electrical efficiency is from 40 to 45% lower heating value (LHV) and the total efficiency can be as high 75–80% LHV.

PHOSPHORIC ACID FUELS CELLS MAIN DEVELOPMENT PROGRAMS (1)

On-site Plants

- **Cogeneration Systems from 50–1000 kW.**
  In the 1990 International Fuel Cells and Toshiba set-up ONSI Corp. to produce small on site commercial cogeneration units (200–1000 kW).

World PC-25 (200 kW PAFC Power Plant) Fleet–January 1998

- n. 128 power plant in operation
- worldwide fleet cumulative operation exceeding 1,750,000 hours.
- Single unit longest cumulative operation over 40,000 hours
- longest continuous run over 9,5000 hours.

PHOSPHORIC ACID FUEL CELLS MAIN DEVELOPMENT PROGRAMMES (2)

Several systems ranging in size from few MW to 11 MW have been installed worldwide.

- an 11 MW distributed power plant built by IFC and Toshiba, at Tokyo Electric Power Company, in 1991.

- a 5 MW plant by Fuji Electric, for cogeneration, installed at Kansai Electric Power Company, in 1995.

PACKAGED FUEL CELL POWER PLANT
200 kW PC 25 UNIT/ONSI CORP.

- Rated electrical power output: 200 kW
- Fuel: natural gas
- Electric efficiency: 40%
- Total efficiency: 80–85%
- Thermal Energy Available: 700,000 Btu/h
- Mode of operation: grid connected; grid independent

FUEL CELL PROGRAMME
PHOSPHORIC ACID FUEL CELLS

The Italian industries are not involved in PAFC stack manufacturing, the national programme has been focussed on:

- development of adequate capability in designing, constructing and managing fuel cell power plants
- promotion of their introduction into the market
- commercialisation of plants, in cooperation with non-European suppliers.

Main projects:

- *1.3 MW power plant installed inside the Milan urban area (Aem, Ansaldo Ricerche and ENEA)*
- *200 kW CLC (A.Co.Ser., Bologna)*
- *25 kW Fuji-KTI (ENEA, Casaccia)*

- *1–5 kW (prototypes for military applications; Ansaldo Ricerche, ENEA)*

THE 1.3 MW POWER PLANT MAIN COMPONENTS

- **Electrochemical modules:** 2×670 kW stacks by International Fuel Cells Co.
- **Methane reformer and FPS:** Heat Exchange Reformer and other equipment by Haldor Topsøe AS
- **Turbocompressors:** RR 221 (l.p. stage) and TT 151 (h.p. stage) by Asea Brown Boveri TurboSystems
- **Inverter:** Self-commutated (GTO) by Ansaldo Ricerche
- **Control system:** INFI 90 products by Elsag-Bailey

1.3 MW P.P. Process Flow Schematic

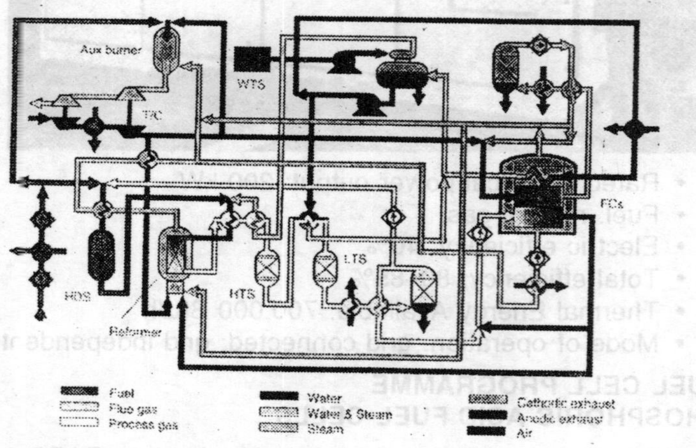

ATTIVITÀ CLC
SISTEMI "PACKAGE" PER COGENERAZIONE

CLC (joint venture Ansaldo-ONSI Corp) produces a **200 lW packaged cogeneration power plant** (PC25 power plant) equipped with a standard PC25 PAFC supplied by ONSI and balance of plant components manufactured in Europe

- **European fleet of PC25 (A and C models) fuel cell power plant—January 1998**

Total operating hours	296,704 h
Longest continuous run	5,729 h
Total electricity production	46,796 MWh

- CLC in studying **850 kW hydrogen fed modules**, for applications in the chemical industry (chlor-alkali plants, petrol-chemicals plants, ammonia plants).

MOLTEN CARBONATE FUEL CELLS

Electric power plant using MCFC technology could provide in the long term significant advantages over conventional co-generation and distributed generation systems, in terms of

- conversion efficiencies
- environmental benefits.

A considerable effort is still needed to solve technical problems and to bring cost down to acceptable levels.

For this purpose, extensive R, D & D programs are being carried out in USA, Japan and Europe.

MOLTEN CARBONATE FUEL CELLS PRESENT STATUS

MCFC technology is a present in the phase of full-scale testing, after the progress made in the last 20 years in the development of components, cells and stacks:

- 100–300 kW stacks have been tested in Japan and USA
- the construction and testing of large pilot plants are in progress (2 mW in USA, 1 MW in Japan).

Full-size plants are expected to be commercially available by 2002.

MOLTEN CARBONATE FUEL CELLS MAIN R&D PROGRAMMES(1)

United States

Technology development undertaken primarily by:

- **Energy Research Corp.**
  - 2 MW plant (sixteen 125 kW stacks) installed at S. Clara, CA; Demonstration terminated on March 1997. The plant operated for 4,100 hours in a grid-connected mode and generated over 2,500 MWh power.
- **MC-Power**
  - 250 kW plant at the Miramar Naval Air Station in San Diego, CA. The power plant, started up in February 1997, produced 158 MWh of electricity.

MOLTEN CARBONATE FUEL CELLS
MAIN R&D PROGRAMMES (2)

Japan

IHI, Hitachi, Mitsubishi, Sanyo, Toshiba and other developers have interest in MCFC field.

NEDO funds MCFC R&D programmes through the Molten Carbonate Fuel Cell Research Association.

- an 1 MW pilot plant is now being installed at Chubu Electric Kawagoe Power Station. (250 kW stacks supplied by IHI and Hitachi).

MOLTEN CARBONATE FUEL CELLS
MAIN R&D PROGRAMMES (3)

Europe

- **ARGE MCFC Development Consortium** [MTU, RWE and Ruhrgas (D); Haldor TopsØe and Elkraft (DK)]
  - 280 kW prototype ("hot module") in August 1997; stack supplied by ERC
- **Advanced DIR-MCFC Development:** BCN and ECN (NL), in cooperation with Stork and De Schelde companies
  - 2 kW stack, in 1998.
- **MOLCARE Project:** Ansaldo Ricerche, FN, ENEL/CISE and ENEA (I); Babcock & Wilcox Espanola and Iberdrola (E)
  - 100 kW cogeneration plant, in 1998

MOLTEN CARBONATE FUEL CELLS

The main mid-term objective of the programme is the construction and testing of a 100 kW congeneration plant, to:

- Develop a proprietary Italian technology for stack and systems
- Demonstrate on an intermediate scale the concept viability of a MCFC on-site congeneration plant by using the "Compact Unit" concept.
- Assess the technical problems involved

MOLTEN CARBONATE FUEL CELLS

R&D activities are carried out in the attempt to overcome the critical problems that still prevent the commercial exploitation of MCFC technology

The main topics are:

- alternative or improved materials for electrodes
- stack engineering (stack-to-mainfold seals, pressure plates design, manifold design, stack assembly)
- manufacturing technologies
- cost reduction
- conceptual and detailed design of MW plants

1994–1998 MCFC PROGRAMME ROLES OF ITALIAN PARTICIPANTS

Ansaldo Ricerche

- design and construction of cells and stacks
- cell and stack testing
- design and construction of 100 kW prototype plant
  Cooperation with BWE (E) and IFC (USA)

FN–Nuove Technologie e Servizi Avanzati (subsidiary company of ENEA)

- development of fabrication processes of cell active components
- production of cell active components (up to 1 m^2)

CISE (Research Company Subsidiary of ENEL)

- testing and characterisation of cells and stacks
- testing of 100 kW prototype plant

ENEA

- R&D activities on materials and components
- testing and characterisation of cells (100 × 100 mm)

MOLTEN CARBONATE FUEL CELL 1994–1998 PROGRAMME

- 5 kW stack with 700 cm^2 cells (1994)
- Prototype manufacturing line for components up to 1m^2
- a second 5 kW stack with improved design (1996)
- design, construction and test of 13 kW stack (active area 0.74 m^2) (1996)
- design and construction of 100 kW stack (early 1998)
- design and construction of 100 kW plant (1998)
- R&D activities on cell materials and components (new material for cathode, internal reforming, etc.) (1994–1998).

SOLID OXIDE FUEL CELLS

SOFC are currently being demonstrated in a 100 kW plant.

High temperature operating (up to 1000 °C) allows more flexibility in fuel choice and offers better performances in combined-cycle application.

SOFC approach 60 per cent electrical efficiency (4LHV) in simple cycles and over 70 per cent efficiency (LHV) in combined cycles. Total thermal efficiency can reach 85 per cent (LHV).

SOFC plants should enter the commercial marketplace by the year 2002.

SOLID OXIDE FUEL CELLS MAIN R&D PROGRAMMES (1)

Tabular Design

Westinghouse Electric Corp. has been developing tubular SOFC from more than 30 years.

- demonstration of 25 kW systems, in cooperation with utilities in the US and Japan; the most recent 25 kW plants completed over 13,000 hours of operation with an availability of 92 *per cent*.
- 100 kW cogeneration prototype installed inthe Netherlands, in the frame of a collaboration with a Dutch and Danish utilities.

Westinghouse has plan to demonstrate SOFC-GT systems ranging in size from 250 kW to 2.5 MW before 1999.

SOLID OXIDE FUEL CELLS MAIN R&D PROGRAMMES (2)

Planar Design

Many research organisations and industries, mainly in Japan and Europe, are working on planar design:

- small size cells (100 × 100 mm) and 10–15 kW stacks
- 100–300 kW stacks are expected after 2000.

Organisations involved:

Ceramatec and Ztek (USA), Murata, Fuji Electric, Sanyo Electric and Tonem (Japan), Siemens and Dornier (D), Einricerche (I), ECN (NL), Sulzer (CH), British Gas (UK), Statoil (N), Ris∅ (DK).

SOLID OXIDE FUEL CELLS PROGRAMMES IN ITALY

Enricerche

Development and testing of materials and components for planar SOFC in the framework in European and national collaborations.

Research activities on materials in CNR, ENEA, and Universities.

FUEL CELL SYSTEMS POTENTIAL MARKET SEGMENTS

Market Segment	*Capacity*	*Cell Type*
Commercial and residential cogeneration	<500 kW 200 kw–2MW	SPFC PAFC
Industrial cogeneration	5–200 kW	MCFC, SOFC
Distributed power	5-20 mW	PAFC, MCFC, SOFC
Central Station	50-150 MW	MCFC, SOFC
Transportation		SPFC

FUEL CELL MARKET DISTRIBUTION BY RATING–2020 STATIONARY APPLICATIONS

Rating	*Capacity*		*Volume*	
	(MW)	*(%)*	*(unit)*	*(%)*
10 kW ÷ 100 kW	2,100	18.3	68,900	85.5
100 KW ÷ 1 MW	3,100	27.7	10,500	12.9
1 MW ÷ 10 MW	3,200	28.6	1,100	1.3
10 MW ÷ 50 MW	2,900	25.4	140	0.2
Total	11,300	100.0	80,600	100.0

ITALIAN FUEL CELL MARKET DISTRIBUTION—2020 STATIONARY APPLICATIONS

Rating	*Capacity*		*Volume*	
	(MW)	*(%)*	*(unit)*	*(%)*
10 kW ÷ 100 kW	38	13.8	12,00	81.2
100 KW ÷ 1 MW	78	28.5	260	16.7
1 MW ÷ 10 MW	85	30.9	28	1.8
10 MW ÷ 50 MW	75	26.7	4	0.3
Total	275	100.0	1,550	100.0

FUEL CELL MARKET FORECAST BY APPLICATION TRANSPORTATION

Application		*YEAR*	
	2000 (MW)	*2010 (MW)*	*2020 (MW)*
Rail and Marine	—	25	200
Commercial Vehicles	<5	475	14,500
Passenger car	—	625	41,500
Other	—	15	875
Total	<5	1,140	57,075

FUEL CELL SYSTEMS ESTIMATED PRICES

	Current Price	*2005 Goal*
SPCF	14,00 – 30,000 \$/kW (THERMIE Programme)	1,500 \$/kW
PAFC	3,000 \$/kW	<1,000 \$/kW
MXFC	25,000 \$/kW (Total cost 2MW ERC plant)	1,200–1,500 \$/kW
SOFC	30,000 \$/kW	1,500–1,700 \$/kW

26

Perspectives of Fuel Cells Stationary Power Plants in Europe: from Present Status to Commercial Power Plants, based on Molten Carbonate Fuel Cell 500 kW Standard Modules

A. DUFOUR
Italy

1. FUEL CELLS TECHNOLOGIES

i) General

The large use of electrical energy, facilitated in the past by its low cost, has determined the improvement of life quality for people living in the industrialised world. As in the past decade, however, the real price of energy has increased rapidly, big efforts are being done to develop new, more efficient technologies. From this point of view, fuel cells represent a valid way for reducing the amount of the primary energy necessary to produce electrical energy.

Fuel cells are electrochemical generators which directly transform into electrical energy the free-energy variation, associated to an overall chemical reaction running through two electro-chemical processes taking place on separated anodic and cathodic regions. Their efficiency is higher than the traditional steam-water cycle because intermediate conversion of heat into mechanical energy is not needed.

FC concept was born 150 years ago, and has been developed in the industrial world in the 50's, firstly for the Gemini and Apollo NASA missions.

Today the space application is becoming a minor one, and the major efforts are concentrated in the stationary applications, for their overall performances, and in the mobile ones, for the direct feeding of the power trains in the electric vehicles. In this presentation we will stay on the first ones.

It seems very likely to many developers that, at the turning of this century, fuel cell power generation plants will be a convincing alternative to the traditional electric power generation plants, and, among the spread of different evaluations, the total installed power at the year 2010 will be given as a 5 digits value in MW (60,000 MW).

Ansaldo shares this expectations and is developing the FC technology since the early 80's.

2. FUEL CELL TYPES

FC technologies are normally defined from the type of electrolyte, that is:

- liquid for alkaline fuel cells (AFC), for phosphoric acid fuel cells (PAFC), and for molten carbonate fuel cells (MCFC)
- solid for solid oxide fuel cells (SOFC) and polymeric electrolyte fuel cells (PEMFC).

Another way of fuel cells first classification can be based on the operating temperature: PEMFCs and PAFCs (80–200 °C) are to be considered as low temperature fuel cells, if compared to MCFCs and SOFCs (650–1000 °C), which operate at high temperature. By the consequence, in the co-generation applications, for example, PAFCs and PEMFCs are to be preferred only if no high-temperature heat is required such as in domestic applications (present low operation temperature of PEMFCs can strongly limit their use also in this field). MCFCs and SOFCs do not suffer from these limitations and their high operating temperature allows to exploit part of the waste heat for sustaining the reforming processes (namely when natural gas is used) as well as to be suitably matched with a gas turbine for a high efficiency combined cycle.

Fuel cell plant electrical efficiency is higher for MCFC and SOFC (50–55%, potentially reaching 65% in a combined cycle) than for PAFC and PEMFC (40% for PAFC and, in the future, also for PEMFC) thus favouring the first ones when wider power ranges are required.

3. FUEL CELLS KEY ISSUES

The interest in using fuel cells to produce electric energy comes from the advantages that fuel cells offer in terms of high efficiency, good performance at base and partial load, very low emissions, modularity (easy adjustment of plant capacity to power-demand increase), and reduced plant erection time.

As a general statement, FC power plants, if compared to conventional power plants, are delivering to atmosphere almost no SOX, NOX and 1/3 of the standard CO_2. A direct consequence of this is the possibility of citing this type of plants almost everywhere, as the PRODE power plant in Milano has completely demonstrated.

4. DIFFERENT STAGES OF DEVELOPMENT

General

From a general point of view all the fuel cell technologies show, at various extents, the above listed advantages. Nevertheless specific features of each fuel cell type suggest to identify a specific field of application for each type of solution, in order to stress the potential advantages of any technology and minimise its possible drawbacks.

On the other end, the different level of maturity for the various fuel cell technologies does not allow an homogeneous comparison of technical and economical key parameters, and the subject is provoking a wide discussion among the developers, due to the fact that each one of them is generally directly involved in one single technology, and someone is also evaluating the gradient of the development, and not only the absolute levels already reached.

Let's take a judgement given by A.D. Little, a "super partes" body, who has defined a "Development Status of Fuel Cell Technologies", by taking for every type four development phases (technology development, field demonstrators, first commercialisation, "cost effective commercialisation").

PAFCs, due to their present commercial availability and operation experience as packaged cogeneration systems (PCS), are positively evaluated in terms of performance and costs.

On the other end, with regard to the other technologies—PEMFC, MCFC and SOFC—which are still under development, their commercialisation is expected within a period of 4 to 13 years according to single technology maturity level, and the level of public or private investments allocated.

Both MCFC and SOFC are considered in the field testing phase, even if the size reached is spanned by an order of magnitude, in favour of MCFC.

Finally, no doubt that the PEMFC technology is showing a clear acceleration in the development, due to the fact of excellent results acquired with new high current density membranes, and, particularly, to the highly concentrated effort in terms of financial investments, mainly related to their interest for mobile applications.

In the following table N.1 the main fuel cell installed stationary plants, rated 500 kW or above, are summarised:

Table 1 : The Main Fuel Cell Installed Plants

Plant Site	*Main Contractor*	*FC Power*	*Technology*	*Operating Pressure (bar)*	*Reformer Manufacturer*	*Testing Period*
1 MW banco	IFC	4×250 kW	PAFC	4	IFC	1976–78
New York 4.8 MW	IFC	20×250 kW	PAFC	4	IFC	1982–83
Tokyo 4,8 MW	IFC	20–250 kW	PAFC	4	IFC	1983–84
Chubu 1 MW	Fuji Mitsubishi	2×250 kW 2×250 kW	PAFC	4 4	Mitsubishi	1987–88
Chita 1 MW	Toshiba Hitachi	2×250 kW 2×250 kW	PAFC	6 6	Hitachi	1987–88
Tokyo 11 MW	IFC Toshiba	18x670 kW	PAFC	7	IFC	1991–93
Milano 1.3 MW	ANSALDO RICERCHE	2x670 kW	PAFC	7	H. Topsoe	1992–94
Kawagoe Power Plant	MCFC Res. Ass.	4x250 kW	MCFC			1997–98
San Diego, Calif	MC Power		MCFC			1995–98
Santa Clara, Calif.	ERC	1800 kW	MCFC		ERC	1995–97
Kansai Electric Power.Co	Fuji	5000 kW	PAFC			1992–94
Osaka Gas	Fuji	500 kW	PAFC		Fuji	1992–94

PAFC

The lowest viable size presently considered is around 100 kW, but smaller sizes could be produced as a consequence of the development of small reformers. In any case, the PC 25, developed by UTC/ONSI(US), Toshiba (Japan) and CLC/

Ansaldo(Italy), a 200 kW packaged system largely exceeding 150 units in operation, covers the largest part of the installed power capacity in the world, confirming that probably such a size is the most suitable to satisfy any present market requirement.

Nevertheless many larger plants, equipped with multiple stacks of about 500 kW each, are already operating namely in Japan (1, 5 and 11 MW) and also in Europe (PRODE plant in Milan, 1.3 MW). These arrangements are generally too expensive and they probably need to be re-designed in order to reduce costs and enhance their commercial competitiveness.

As concerns the PRODE, that has been built on the basis of International Fuel Cells' stacks by Ansaldo Ricerche in co-operation with AEM and ENEA, has been operated since 1995 by the Customer supported by the Ansaldo Ricerche project team. During 1996 and 1997 some tests have been carried out mainly to improve the plant behaviour for specific operational status changes and maintenance sequences. Particular attention has been also paid in checking stacks' characteristics during the time.

MCFC and SOFC

As above said, both high temperature technologies (MCFC and SOFC) are still in progress in the world, according to different development plans. The main efforts are aimed at solving some technological problems, moving towards industrial manufacturing processes, reducing costs, defining and developing suitable market-entry units. In particular MCFCs can be considered the most promising solution for a faster commercialisation, not only for their technical advantages which will be discussed later, but also because several demonstrations have been already or are being carried out in the range 100–2000 kW.

5. THE FC POWER PLANT

i) The Process

The core of the process takes place in the electrolyte, which can be solid or liquid, and is characterised by the operating temperatures and pressures, and by the overall conversion efficiency.

The different operating temperatures, that directly affect the available heat temperatures, range most favourably the

Table 2 : General Data for FC Types that could be Operated in the Electrical Energy Stationary Generation

Technology	*Operating Temperature*	*Ion Type*	*Electrolite Status*	*Present Efficiency %*	*Future Efficiency %*
Phosphoric Acid PAFC	200ºC	hydrogen	liquid	40–42	45–55(1)
Molten Carbonate MCFC	650ºC	carbonate	liquid	45–55	55–60 (1)
Solid Oxide SOFC	1000ºC	oxygen	solid	45–55	55–60 (1)

(1) Values for natural gas; for coal gas the values could be reduced by 5%.

technologies for the combined cycles applications, with respect to similar amounts of available heat.

For this reason, the high temperature fuel cells (specifically SOFC and MCFC) show a very good appeal, in spite of their still open issues, like thermo-mechanical stress and corrosion resistance of materials.

A very interesting development trend is under way for SOFC, by decreasing their operating temperature to the one valid for MCFC (650–700 ∫C).

The last two columns in the table, related to the present and future efficiencies, are specifically connected to an improvement of the process integration, as far as heat and mass flows are concerned. A general diagram flow for a FC power plant is shown in Fig. N.3 "Block Diagram for a Fuel Cell Power Generation Plant".

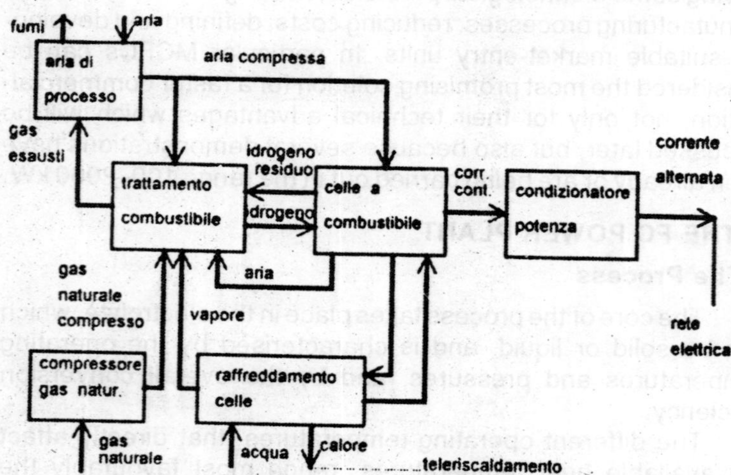

Fig. 3 : Diagramma a Blocchi di un impianto di generazione con celle a combustibile

As it could be easily seen, three main subsystems are present: fuel reformer, fuel cells electrochemical stacks, and power conditioning, and a few auxiliary sub-systems, as process air feeding subsystem, natural gas compressor, fuel cells heat removal subsystem.

Very interesting developments are underway for the so called "internal reforming" configurations, where specific stack cells are devoted to the natural gas reforming, instead of performing it in a separate subsystem. It is possible to take advantage of the exothermic heat flow in the cells, to compensate the endothermic heat flow in the reforming reaction. The only unsatisfactory aspect is related to the decrease in the life behaviour, which is inevitably linked to the poisoning of the catalysts, due to the impurities of the NG.

ii) The Reformer

The external reformer converts hydrocarbons rich fuel into hydrogen through a natural gas steam reforming or through an adiabatic reforming of liquid fuels, or through coal gasification processes.

In general terms, the different fuels acceptability of FC is very wide. Just as an example, PAFC have been operated on natural gas, methanol, propane, city gas, biogas and landfill gas.

iii) The Electro-chemical Section

Fuel Cells stack is composed by an arrangement of single cells of relevant surface (of the order of a square meter) to reach adequate current figures, starting from the typical current density levels (of the order of 200-400 mA/sqcm). These cells are overlapped and electrically linked in series to get a total voltage of the order of hundreds of volts, as the single cell voltage is usually comprised in the range 0.5–1 V. For most types of FC the development path is following the two lines:

- to maximise cell surface, by facing the technological limits due to the fabrication means physical limitations (mainly tape casting apparatus and continuous sintering ovens) and to the needs of having across the cell plane reagents temperature and, possibly, a uniform composition.
- to maximise the number of cells in series, to optimise the inverter technical and economical sizing, without entering into a materials creep risk, due to the high temperature of operation.

iv) The Inverter

The power conditioning section converts DC electrical energy into AC electrical energy. According to the different applications, the inverter is able to work connected to an electrical grid or "island operated", that's disconnected to any electric grid.

v) Auxiliary Systems

Among the many, it is worthwhile mentioning:

- control system
- fuel pre-treatment (as an example natural gas desulphuriser)
- exhausts recirculation system
- stack temperature control system
- thermal energy transfer systems: from stack to reformer, and, possibly to the bottoming cycle.

6. APPLICATION PERSPECTIVES

Every fuel cell developer has been giving or is due to give an answer to the question, which are the application perspectives of the potential product he is developing, and, of the same importance, in which time frame.

We believe that, as for any highly innovative product, the market approach develops through early application niches first, to face the classical chicken and egg problem, that is high costs related to low fabrication volumes.

Anyway, the rough table, giving for each technology the specific market segment and fuel of choice, is a first evaluation for sizes and fuels.

Table 4 : Applications for Fuel Cells and Fuel Types vs Power Range

Plant Typology	*Cell Type*	*Fuel*	*Power Range*
"Package" Cogeneration	PAFC MCFC SOFC	Natural Gas	200–1.000 kW
"Skid mounted Cogeneration	PAFC	Natural Gas Biogas	1–5 MW
Primary Substations	PAFC MCFC SOFC	Natural Gas Biogas	10–20 MW
Generation Stations	MCFC	Coal Gas	100–500 MW
Electrical or Hybrid Transportation	SPFC	Methanol Hydrogen	10–80 kW

"Packaged" power plants can normally be shipped completely assembled, with sizes within the limits of a standard container. The ones indicated as "pre-assembled" are skid-mounted, with a very poor additional assembling activities at the site. The ones for primary substations could be sited in an environment already "licensed" and close to the center of electrical use, being very promising for the short erection times.

The reduction in the electrical transmission and distribution losses, and the full value given to the active and reactive power control in the grid knots—allowed through the inverter operation, which is present in any case because of its primary function of converting DC current into AC current—are the prevailing aspects to be underlined. Another techno-economical key factor to be boosted is the possibility of taking local advantage from the available heat.

The different technologies have different degrees of maturity, and PAFC are considered today not only as market openers, but also as having the potential for taking a good share of the new installations after the year 2000.

Many business evaluations are available to determine the fuel cells market segmentation in the future. Just to mention a few:

Barnett (Ref. n. 7) evaluates in the year 2000 for the PAFC with 1500–1700 $/kW and efficiencies around 40 per cent a market articulated as follows:

Table 5 : PAFC Installed Power Around the Year 2000

Market Segment	*Total Installed power (MW/year)*	*FC installed power (MW/year)*	*Fuel Cell installed power (%)*
Centralised Generation	44300	400	0.9
Distributed Generation	9700	1400	14.0
Existing Power Plant Repowering	11700	900	8.0
Cogeneration	10200	1400	14.0
Total	75900	4100	5.4

Let's comment not the total installed power, which is very optimistic, but the shares in relative terms.

A big share is taken by distributed generation and cogeneration, that account for a 14 per cent of the total allowable market; the centralised generation does not appear to be a favourable application for fuel cells.

The same source, working with A.D. Little, has evidenced a market penetration for those countries where environmental and law conditions exist to accept more expensive technologies than conventional, as Japan, US, Germany and Italy; and similar evaluations are made by different sources. The installed power for country and around the year 2000 is evaluated as follows:

Table 6 : FC Installed Power Around the Year 2000 for Various Regions of the World

Geographical Area	*Total Installed (MW/year)*	*Fuel Cell Installed Power (MW/year)*
North America	21900	1800
Sourth America	6700	200
Western Europe	11300	800
Eastern Europe	3900	100
Japan	4600	400
Asia	25300	700
Others	2200	100
Total	**75900**	**4100**

Staying specifically on MCFC, a very recent market study (Ref. N. 4) gives at the year 2020:

(a) Table 7 : The MCFC Market Distribution by Sector

Sector	*MCFC Market (MW pa)*	*Distribution (%)*	*MCFC as % of Total FC*
Island and remote site power	160	5.3%	29%
Cogeneration and district heating	1.400	48.5%	35%
Power only DG-Baseload	1.100	37.9%	29%
Power only DG-intermediate	130	4.6%	6%
Power only DG-intermittent	—	—	—
Renewable energy systems	40	1.3%	12%
Other applications	70	2.5%	18%
TOTAL	2.900	100%	26%

(b) Table 8 : MCFC Market Distribution by Rating (Capacity and Volume)

Rating	*CAPACITY*		*VOLUME*	
	MW	*%*	*Units*	*%*
10 kW to 100 kW	—	—		
100 kW to 1 MW	670	23.1%	1.050	71.4%
1 MW to 10 MW	1.050	36.7%	360	24.6%
10 MW to 50 MW	1.150	40.2%	60	4.1%
TOTAL	**2.900**	**100%**	**1.450**	**100%**

(b) Table 9 : MCFC Market Distribution by Rating (Capacity and Volume) Specifically for Italy

Rating	*CAPACITY*		*VOLUME*	
	MW	*%*	*Units*	*%*
10 kW to 100 kW	—	—	—	—
100 kW to 1 MW	17	22.7%	25	72%
1 MW to 10 MW	28	37.3%	10	24%
10 MW to 50 MW	30	40.0%	1	4%
TOTAL	**75**	**100.0%**	**35**	**100%**

7. STATIONARY MCFC

i) Introduction

As we have seen, at least four types of fuel cells can be considered suitable for stationary applications:

- Polymeric Electrolyte Membrane Fuel Cells (PEMFC),
- Phosphoric Acid Fuel Cells (PAFC),
- Molten Carbonate Fuel Cells (MCFC), and
- Solid Oxide Fuel Cells (SOFC).

From now on, this presentation will only focus on MCFC. The reason is that Molten Carbonate Fuel Cells are among the most promising technologies, because of their environmental friendly operation for various fuels, their perspective low cost and high plant efficiency at full and partial load. In fact, the overall efficiency is typically some 50 per cent and can reach, as a consequence of their high operating temperature, 65 per cent if a bottoming cycle is added.

ii) Main Features of Molten Carbonate Fuel Cells

In the MCFCs the electrolyte is a molten salt mixture of lithium and potassium carbonates at about 650°C (recently sodium carbonate has been also considered for a binary mixture or as an addendum in a ternary mixture); it is retained in a porous matrix of lithium aluminate, properly wetting both the anode and the cathode. To this aim and to guarantee a perfect separation between fuel and oxidant, porosity and wettability of matrix, anode and cathode must be carefully selected and maintained during operation.

The anode is a porous sintered nickel plate, having a microstructure properly stabilised against creep. It is placed between the matrix and the current collector-gas distributor at the fuel side. Small quantities of chromium are generally employed as a structure stabiliser, but more recent studies showed that aluminium can secure even a better stability. The cathode is usually a porous plate whose micro-structure and composition are formed *in situ* after the first start-up of the cell which induces the formation of a lithiated nickel oxide sheet. Alternative materials or special coatings are under development to prevent or reduce the effects of NiO dissolution which can reduce the cell useful life. Like the anode, the cathode is placed between the matrix and the current collector-gas distributor at the oxidant side.

The current transport mechanism inside the cell matrix is based on the motion of carbonate ions from the cathode to the anode. This mechanism needs a continuous formation of carbonate ions at the cathode side which must be guaranteed by supplying proper flow rates of carbon dioxide to the cathode side or recycling CO_2 from the anode to the cathode. These two options allow to choose a suitable plant configuration according to the specific application and give the opportunity for an easy recovery of CO_2 from the anode exhaust for subsequent fixation processes if a gas turbine retrofits the fuel cell system.

The cells, each separated by the other by means of metallic separator plates, are piled in a stack where they are in series from the electrical point of view and in parallel with reference to the process gas flow path. Different technological solutions have been developed to distribute the gases to the cells and manage the gas flows inside them. To this purpose the separator plates can be constructed in various shapes, typically flat or corrugated, mainly depending on the gas manifold solution that has been adopted (internal or external).

Three main approaches are usually considered for managing the gases inside the cells: parallel co-flow, parallel counter-flow and cross-flow. They are aimed at optimising, with acceptable costs, some specific aspects such as cell performance, pressure control and temperature distributions. The two first ones require the internal manifold solution which needs a more complicated separator plate and generally induces higher pressure drops; the third arrangement can be used with the external manifold solution, which is the simplest one from the construction and the fluid dynamic point of view.

The operating temperatures of MCFCs (about 650°C) enable a strong thermal integration between the cell stack and the fuel reformer and are suitable for some co-generation industrial applications.

The solution of fuel internal (or partially internal) reforming is quite striking due to some advantages such as the possibility of enhancing the stack cooling and the simplification of the balance-of-plant system. Nevertheless it requires fuel consistency (only natural gas is suitable) and careful management of gas flow and temperature profiles in the stack to avoid the catalyst deactivation. Moreover it augments costs and makes worse the maintainability of the stack.

Some approaches have been considered for the fuel external reforming to get the main advantages of the internal one without its possible drawbacks. Several solutions have been studied having a strong thermal integration with the stack and the same modularity without affecting its constructions and, as a consequence, its costs and maintainability. By this way also a greater flexibility as concerns the fuel choice is achievable with no stack changes.

The Sensible Heat Reformer (SHR), which is the base of the "Compact Unit" concept developed by Ansaldo for the 100 kW plant under erection at ENEL/CISE and uses the heat transported by the gases leaving the anode, represents an interesting integrated modular option among external reformers.

By this solution, the deactivation of catalyst due to molten carbonate is strongly reduced because the catalyst is contained in a separate system. Moreover, if necessary, the catalyst reactivation or, in case, its substitution is feasible in a very simple way. An improved solution of the SHR, which will be discussed later, is now under development for the 500 kW market-entry units.

MCFCs can also be used with steam turbine bottoming cycles or integrated with gas turbine, boosting efficiencies up to 15 per cent over their own electrical efficiency (from 50 – 55% up to 65%) and keeping the temperatures very close to their optimum values foreseen for these arrangements. Besides, MCFC operating temperatures seem to be more appropriate than SOFC ones for fuel cell integration into gasification systems.

The technological level presently reached by MCFCs allows the construction of stacks characterised by a cell area up

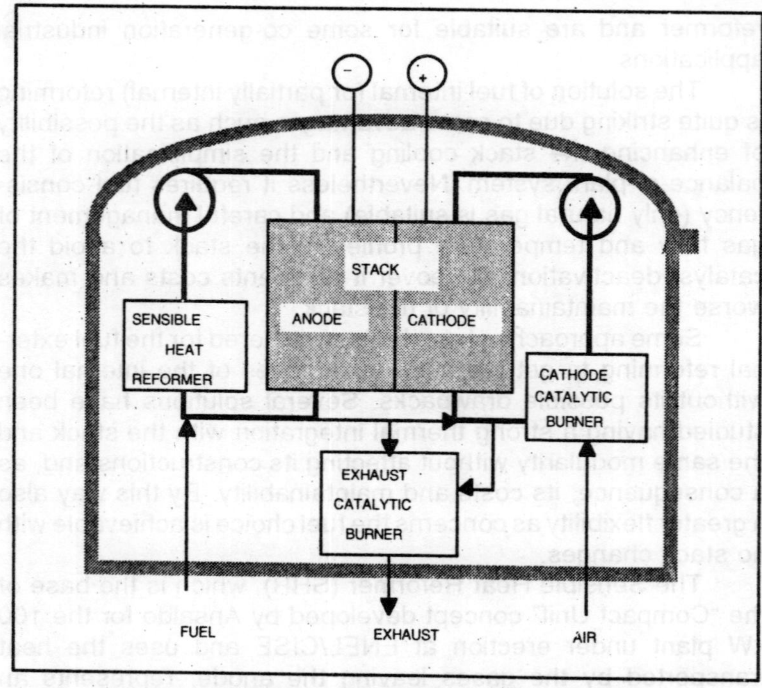

Fig. 10 : Schematic of the "Compact Unit", developed in the frame of the MOICARE programme

to 1 m², without those scale-up problems still affecting other high temperature fuel cell technologies. For this reason MCFCs can be regarded as the lowest cost fuel cell at MW-class level once the ongoing development on porous components stability and metallic part corrosion will confirm the envisaged solutions of these critical problems still preventing the achievement of the life targets.

Furthermore, the growing tendency towards dispersed generation, makes MCFCs particularly attractive for plants of MW size located at user's sites.

An additional advantages of MCFCs has already been mentioned: their good flexibility as concerns the use of various fuels. Besides pure hydrogen, in principle every hydrogen rich gas, which can be obtained by conversion of currently available primary sources—such as natural gas, oil, biogas and coal-gas—is a suitable fuel. In particular the use of coal is very attractive for this technology, being carbon monoxide a possible fuel too.

At present, MCFCs have reached the stage of scaling up to commercial size stacks with a power rate in the range of 100–400 kW. Some developers (as in Italy Ansaldo and FN) have already installed pilot fabrication facilities, which have an overall production capacity in the range of a new MW per year. Moreover, many progresses have been made in the design of the balance of plant and in the choice of ancillary equipment.

8. INTERNATIONAL AND EUROPEAN INITIATIVES

The great interest in the MCFC technology is confirmed by the relevant projects going on in the U.S., in Japan and in Europe. Just to mention the most important companies' involvement:

in the U.S.

Energy Research Corporation, cooperating with Mitsubishi, Sanyo and MTU, has installed a unpressurised 2 MW nominal unit at Santa Clara, CA, reaching a 44 per cent efficiency and an operation time of 4000 hours.

MC Power Corporation, cooperating with IGT, Bechtel, Stewart & Stevenson, and IHI, has installed at Miramar Naval Air Station in San Diego, CA, a 250 kW plant. Again, an overall efficiency of 44 per cent has been demonstrated along 2000 hours of pressurised operation.

in Japan

The Molten Carbonate Fuel Cell Research Association, funded by NEDO, is installing a 1 MW non-integrated external reforming pilot plant at Chubu Electric Kawagoe Power station, with stacks made by IHI and Hitachi. The goals are 5000 hours of operation and overall efficiency of 45 per cent.

The Mitsubishi Electric Corporation is developing a 200 kW Advanced Internal Reforming plant, an hybrid of the Direct Internal Reforming and of the Indirect Internal Reforming.

The Toshiba Corp., initially working on IFC technology, as Ansaldo did, is now implementing on a full scale stack 91.2 sq m) all the improvements already testing separately, like new electrolyte, new electrode configurations and an advanced type separator, internally manifolded and flexible.

in Europe

The ARGE DFC consortium, led by MTU, Haldor Topsoe, Elkraft, Ruhrgas and RWE, which is operating a 280 kW very compact unit (6m by 2.5 m). At the moment, no official information about the performances are available. The final goal of the MTU project are: expected life of 40,000 hours; 50 per cent overall efficiency at the sub-MW level and up to 65 per cent in larger configurations, with a bottoming cycle.

The BCN company (with investments of Stork and Royal Schelde) is now developing an internally manifolded direct internal reforming 2 kW test stack, after having cancelled a 250 kW co-flow internally manifolded demonstration plant.

The MOLCARE consortium, grouping in various ways Ansaldo(I), BWE(E), ENEL/CISE(I), Iberdrola(E), Endesa(E), ENEA(I), CNR(I), FN(I), which is now testing a 100 kW unit. Further details of the complete MOLCARE project will be given in the following.

9. SPANISH ITALIAN MOLCARE PROGRAMME

i) The Team

For the above mentioned reasons, Ansaldo has been active in the MCFC so promising technology since the early 80's, but the efforts of the companies involved and the level of their expenditures, have pushed towards the idea of forming a R&D consortium like organisation, together with Babcock and Wilcox Española (previously TGI), with the overall goal of running a common R&D programme and boosting the development of the MCFC technology towards an industrial maturity and a commercial exploitation.

The acronym MOLCARE stands for MOlten CARbonate Europe, and around the basic agreement a set of interested partners, with various roles, has been established:

in Italy: ENEL, the national utility
CISE, a development company owned by ENEL
ENEA, the national body for alternative energies
FN, a ceramic components manufacturer

in Spain: IBERDROLA, the most important Spanish private Utility
ENDESA, the most important Spanish public Utility, now undergoing a privatisation process

in the U.S.: International Fuel Cell Corp. (IFC), through a co-operative research and development agreement, which today is expired.

The MOLCARE programme covers a wide range of activities, such as basic material development, fabrication processes scale up for the main stack active components, proto-type stack manufacturing and testing, detailed design of all specific plant components and engineering design of MW size power plants (Fig. 11).

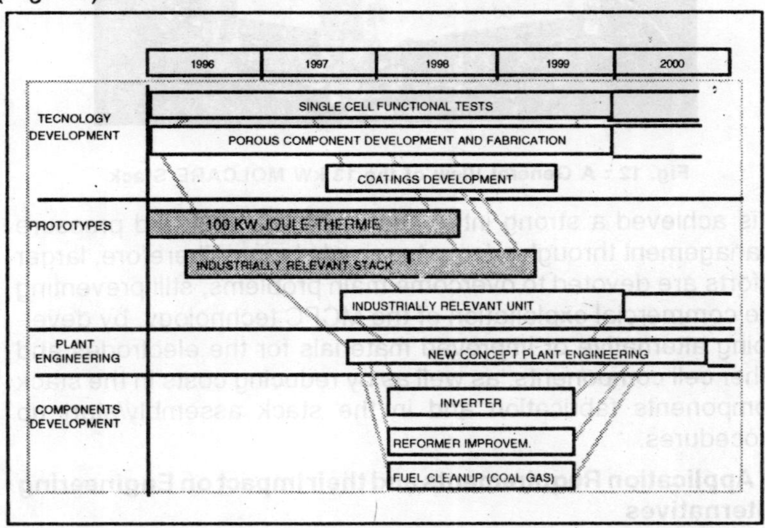

Fig. 11 : The General Plan of the MOLCARE Phase 1 Project

At the beginning the relevant fuel will be natural gas, but it is intended to explore also different types of fuels, like coal gasification products, low BTU gases, and so on. After the testing of a 13 kW, 20 cells, 0.75 m^2 stack successfully carried out in the second part of 1996 (*see Fig. 12 for a general view of the stack*), the MOLCARE programme main objective is the realisation of a 100 kW stack to be first tested in Spain, and then installed in a demonstrative cogeneration plant at the CISE premises near Milan (Italy).

The heart of the plant is an electro-chemical module, the "Compact Unit", that is a pressure vessel which contains the fuel cells stack and its closer ancillary devices (Fig. 10). By this way

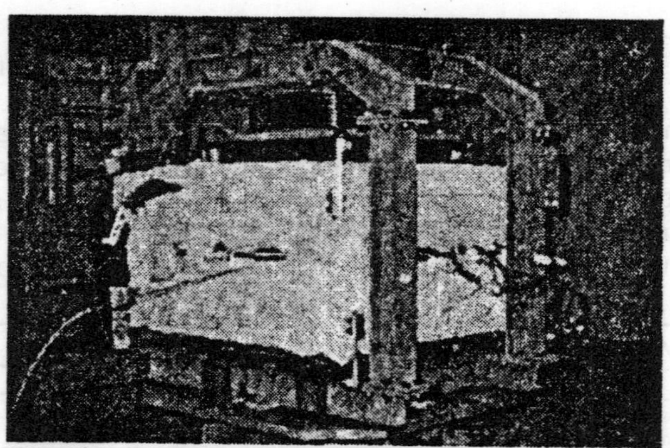

Fig. 12 : A General View of the 13 kW MOLCARE Stack

it is achieved a strong integration both for heat and pressure management through adequate recycle loops. Therefore, larger efforts are devoted to overcome main problems, still preventing the commercial exploitation of the MCFC technology, by developing alternative or improved materials for the electrodes and other cell components, as well as by reducing costs in the stack components fabrication and in the stack assembly/start up procedures.

ii) Application Requirements and their Impact on Engineering Alternatives

When moving a new technology from laboratory towards pre-commercial-scale production, particular attention has to be paid to requirements coming from expected applications, as some of them can significantly affect design and manufacturing alternatives. This warning is very important in the case of MCFC stacks, the core components of MCFC power plants, where inter-dependencies between materials, manufacturing processes, technology alternatives, system configuration and cost targets are very tight.

As regards the application scenarios, MCFCs seem a promising generating system, eventually in connection with a bottoming cycle, in the following typical fields:

- 250 kW– 50 MW: cogeneration

- 1 – 100 MW: dispersed generation
- 50 – 150 MW: repowering of existing power plants, as alternative to the addition of a gas turbine cycle.

Their perspectives for high power generation (100–600 MW range) could also be considered, but they are subject to main general constraints existing for other types of power plants.

The internal reforming option, not developed in the MOLCARE Project, is especially tailored on natural gas and in principle appears very attractive for the stack thermal management because uses directly on the spot the heat generated by electrochemical reactions inside the stack. Nevertheless it strongly affects stack technology and design because the catalyst, according to a variety of specific solutions, must be placed inside the stack by adding appropriate compartments or modifying the anode gas ducts. In general it is also necessary to introduce special alkali shields to prevent accessibility of cell electrolyte to the reform catalyst, resulting in a catalyst deactivation.

All these arrangements require to considerably modify the cell and stack design and to introduce additional components, thus increasing the stack height and complexity. Moreover, due to the deactivation of the catalyst by possible content of sulphur in the fuel, also reliability appears to be reduced and maintenance needs increased. If one considers that internal reforming choice is not useful with fuels different from natural gas, this alternative, in spite of its promising advantages, seems also unfavourable to stack standardisation processes and cost reduction.

With the external reforming option, which has been chosen by Ansaldo as technical basis for the MOLCARE project, the process is carried out, at some extent, independently from the stack, which arrangement is therefore unaffected by the fuel type thus giving the opportunity for a strong fuel cells standardisation and for a potential cost reduction. Various solutions are possible with external reforming systems aimed at meeting specific requirements. In general they can be integrated with suitable fuel treatment sections and tailored to accept any kind fo raw fuel without significantly affecting stack design, as the only aspect to be carefully considered is the gas pressure drop in the stack channels in case of high recycling rations. Among the various external reforming approaches, some solutions such as the

Sensible Heat Reformer (SHR), seem suitable to match main advantages of internal reforming without its problems, with simpler control criteria and easier maintenance or replacement.

iii) The Engineering Alternatives and their Impact on Fabrication Technologies

The main parameters derived from system analysis and cost constraints, affecting fabrication technology are typically cell area and shape (square or rectangular), cell number in the stack (or in the submodule), gas manifolding to the cells and fuel reforming arrangements.

It is well known that, in principle, a larger cell area reduces the cell cost per surface unit, at least up to about 1 m^2. Two main reasons, in fact, limit the reduction of the cost per m^2 when cell area is further increased.

The first one comes from the manufacturing costs of the repeat parts which rapidly rise with the component area due both to the bigger number of scraps and to the increase of investment for larger facilities. For the latter aspect, a rectangular shape of the cells could significantly mitigate the cost rises versus the cell area increase, thanks to the width reduction for some continuous manufacturing equipment (namely the tape casting machine and the furnaces). Nevertheless, a number of other constraints significantly limits the choice of rectangular shapes having a great ratio between the two side dimensions. Many of them come from system configuration and namely from anode and cathode recycle ratios. In fact, higher gas flow rates in the gas ducts inside the cells result in bigger pressure drops, thus producing a differential pressure potentially dangerous for the matrix integrity. On the other hand, a big increase of the cell area results in higher currents, thus increasing, at the same power capability, the cost of the electric power conditioning system (namely the inverter).

From previous considerations, it appears that it is not useful to greatly stress cell area increase or rectangular shapes: a better solution could be obtained by increasing the cell number in the stack. By this way, in fact, no prohibitive cost rise comes from component manufacturing processes and not too high currents are needed. Nevertheless, also the cell number in the stack must be limited as some other specific problems can arise.

One of the most relevant of them, namely when an external gas manifolding system is used, is the compacting of the stack due to adjustments during start up as well as to creep phenomena and micro-structural instabilities of porous components during operation, which can produce a slide effect between the cell package and the manifolds, so reducing the gasket sealing capability possibly resulting in a gas leakage.

iv) The manufacturing process on semi-industrial facilities

The transfer of materials composition and fabrication technologies set up at laboratory level to industrial processes and lines, which has been started since the beginning of the 90's, is still on-going and represents a relevant effort and a key factor for the final technical and economical success of the whole project.

Generally, the process is based on the following steps:

- Powder, solvent and binder characterisation;
- Milling and mixing;
- Tape casting and drying;
- Heat treatments: debinding, sintering, pre-oxidation and pre-filling.

The first part of the manufacturing process is common to the three porous components: the anode, the cathode and the matrix. Of course ingredients and preparation times change, but essentially the process mainly consists of the dispersion of powders (ceramic powders in the case of matrices and metallic powders in the case of the electrodes) into a suitable solvent, in order to obtain a slurry with adequate viscosity. The whole material has to be dispersed and disagglomerated by milling, and then intimately mixed with the organic binders. The solvent and the binder have to be selected according to different criteria such as the subsequent drying and sintering processes requirements, the material toxicity, their stability and, finally, their costs.

After the dispersion, the homogeneous slurry is used in the so called "tape casting" process, in order to obtain "green" tapes, which are flexible when not yet sintered.

The tape casting technique enables the production of tapes with different areas and suitable thickness, being modelled by a blade (Doctor-Blade system) that spreads a slurry of the appropriate viscosity on a flat moving support.

The Doctor-Blade system generally consists of an open box with a vertically moving side, so that the slurry can flow out of a slit. The box is fixed on the tape casting machine, and the moving "mylar" support runs under the slit. The height of the blade states the thickness of the tape.

The ratio between the thicknesses of the green tape and of the sintered product is about 2:1. Several factors affect the thickness of the tape like the stability of concentration and viscosity of the slurry and the speed of the tape casting transporting film.

In the tape casting technique, the solvent is eliminated by evaporation, in various phases by different mechanisms. When the tape is still liquid, the solvent is easily transported by capillarity to the surface where its evaporation is facilitated by an air stream at adequate temperature.

In general, a controlled solvent evaporation has to be obtained to avoid the formation of a surface film that may inhibit the evaporation step and generate defects and cracks. Therefore, a careful balance of the whole process is very important to prevent possible defects on the green tape. After the drying step, the tape is ready to be stripped away from the "mylar".

A subsequent debinding treatment is aimed at the complete elimination of the binder and must be performed to wholly eliminate ashes or chemical impurities from the tape.

The following sintering treatment has the purpose of obtaining porous components with the right porosity and medium diameter of pores, which are the most important features for the correct operation of the cell. The definition of the process parameters, such as the maximum temperature, the heating cycle and the atmosphere is also very important.

After sintering, only anodic components undergo the pre-oxidation and the prefilling treatments: in this case it is necessary to optimise respectively the quantity of water and carbonates, in order to assure a good final quality of the product.

Each step of the manufacturing process is controlled by the collection of several data and by carrying out all the planned analyses. Different types of analytical controls are required, namely on the precursors, on the intermediate products and on the final components. Consequently, powders are characterised for their chemical and physical properties (chemical composition and impurities, density, surface area and granulometry); the green tapes are controlled for their integrity and thickness, and the sintered electrodes are controlled for dimensions (thickness, in particular), density, porosity, pore sizes, morphology and, in the case of the anode, alligation. Other data must be controlled and collected: the preparation and the aeration times for the slurries, the parameters of the casting step and the heat treatment cycle.

All the data are then summarised in a "identity card" which follows each component all through the transfer from the production site to the MCFC stack assembly site. At the moment, tests are being performed on the whole production as this seems to be the best way to fully understand and keep the process under control, in order to get elements to transform it into a standardised one. The present requirements, in fact, are still very tight, so that it is very difficult to identify reasonable tolerance ranges for the quality parameters. As an example, the anode is subjected to a longer and more complex process so that the risk of manufacturing rejections increase. In general, the present experience shows a percentage of rejection in the order of 30–35 per cent.

v) Guidelines for Cost Reduction

As a consequence of the aspects and processes previously discussed, it appears that MCFCs have a high potential for strong cost reductions. In general the main items to be addressed for such cost reduction are manufacturing facilities of porous components, cell and stack assembly, and stack conditioning. A few considerations on each of them.

The very high total cost of porous components manufacturing is mainly dependent on present production volumes and quality. The first one is connected with the manufacturing facilities, which cannot be optimised on the present low production volumes; the second one is connected with the tight requirements which are still stated for raw material purity and quality as well as for final products tolerance ranges. From this point of view many opportunities for significant cost reduction can be taken:

- the use of materials with lower purity could reduce the supplying cost, even if this does not seem the most important factor;

- a careful definition of wider tolerance ranges of the different properties, giving the right emphasis to the characteristics that really influence the quality of the components and their performance in the cell;
- the highest number of rejections is, at present, generated during the heat treatment; in fact, about the 50 per cent of the scraps is due to the use of batch furnaces that cannot assure thermal cycle reproducibility. Of course, a continuous furnace would give a more repetitive production, but the great investments associated with this tool would be justified only by a production carried out on a large industrial scale;
- the necessity to check several properties during the manufacturing process for the whole production increases the costs. It is expected that the future standardisation of the process will give elements to make simpler but more selective control plans;
- the use of an automatic production line, provided with the proper instrumentation to minimise the man power, would represent an optimisation of the manufacturing processes and would decrease the total costs.

At present, batch-type furnaces are preferred for the initial limited production volumes and for the need to clearly define all the process parameters. It is under evaluation the possible threshold for the economic utilisation of a continuous furnace that could range in the 20,000 to 30,000 m^2/y. The different rejection rates have also to be considered.

A particular attention has to be paid to the cell and stack assembly processes, looking at the specific and tailored operations which are involved. Present needs for stack assembly can be covered by controlled atmosphere rooms equipped with specific tools to make easier the most delicate operations.

Figure 13 gives an overview of the Ansaldo assembly room while assembling the 100 kW plant of the MOLCARE project.

The final solution for a mass production will be necessarily a fully automated line. A better definition of sizes, procedures, inspection needs is required to move from present tool-aided process to semi-automated and, then, to fully automated systems. Again, the optimum level of automation is strongly dependent on the actual production volumes. A quantitative evaluation

Fig. 13 : The Ansaldo Assembly Room, during the Assembly of the 100 Plant

of the cost reduction really achievable step by step is quite difficult. In any case it is evident that a fully automated production plant offers the opportunity for greatly lowering costs and improving quality and repeatability. Looking at the high number of the elementary components needed for the cell and, subsequently, the stack assembly (several thousand pieces for a 100 kW stack) and at the wide variety of technologies which are involved, a strategy, aiming at committing the preparation of selected pre-assembled parts to qualified vendors for homogeneous technologies, has to be pursued.

At present, the first conditioning of stack generally represents the final phase of the manufacturing process. This phase is quite expensive and critical due to its duration (typically some hundreds hours per stack), the utilisation of complex, high-cost facilities, the consumption of specific reactant gas, the close tolerances on both temperature cycles and gas compositions, the control of effluent gas and products coming from burn-out or decomposition of organic binders and carriers. On the other hand, a correct conditioning cycle is essential to get the design performance. The idea of carrying out this treatment at the developer premises is now prevailing, but some examples exist where the conditioning is planned and performed directly on the plant, which must be specifically arranged to meet the first-start-up requirements. In the perspective of an industrial production, however, the conditioning cycle should be performed in the factory, both as the final phase of the manufacturing process and as the checkout before the stack delivery.

To this end, simpler standard cycles with less tight requirements and suitable automated facilities will be identified. By this, a strong reduction of costs will be achievable for this part of the manufacturing process.

vi) Preliminary Comparison with Competing Systems

On the basis of the gained experience and of previous analyses and discussions dealing with main items to be addressed for a cost reduction goal, a possible "learning curve" of costs versus production volumes can be stated.

It takes into account the main factors which can affect costs such as: improved expertise, less tight constraints on component tolerances and materials purity, simplification of design and fabrication processes, automation of manufacturing and inspection procedures, introduction of new cost-effective materials/ technologies.

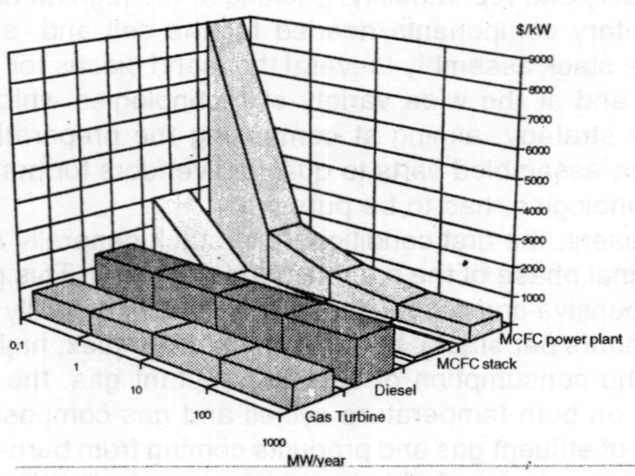

Fig. 14 : MCFC and Competing Systems' Costs versus Production Volumes

In spite this qualitative analysis and assuming that the stack cost is about 30 per cent of the whole plant, (Fig. 14) shows a preliminary comparison of MCFCs with main competitor system, typically up to 5 MW diesel generators, having a price of about 1500$/kW and over 5 MW gas turbine, having a price not exceeding 800$/kW. The prices of both these generator systems are assumed independent from production volumes, being based on mature technologies.

From this analysis it appears that a production volume of MCFC power plants of a few MW/y should make such technology competitive with diesel systems, also without taking into account some specific advantages of MCFC systems such as higher efficiency and lower emissions. With respect fo gas turbine and to combined cycles power plants, the competitiveness should be reached just over a standard production volume of about 10 MW/y, provided that the ongoing MCFC demonstrations have shown a suitable reliability level.

10. FROM DEMONSTRATORS TO COMMERCIAL PLANTS, BASED ON 500 kW STANDARD MODULES

i) Updating the MOLCARE Programme

As it has already been illustrated, the MOLCARE programme covers a wide range of activities including material and technology development, manufacturing processes, prototype realisation and testing, system engineering, plant design and demonstration actions. During the first part of the programme (Phase 1) the attention has been focussed on the development of base technologies in order to improve, validate and consolidate, at single cell and sub-scale stack level, materials and solutions for scaling-up to larger stacks and then to plants. In particular, some manufacturing processes and suitable facilities have been developed and set-up both for porous and metallic components of stacks including coatings and relevant treatments on separator plates to reduce or prevent corrosion.

In the subsequent phase (see Fig. 15), more attention will be paid to demonstrative plant design also including an experimental activity carried out by means of suitably sized prototypes. In addition to some special works performed in parallel to solve the problems identified by experimental activities, a big effort has been devoted to the development, design and testing of stacks which are needing further improvement. A scale-up process related to the area of cell as well as to the number of

Molcare 2

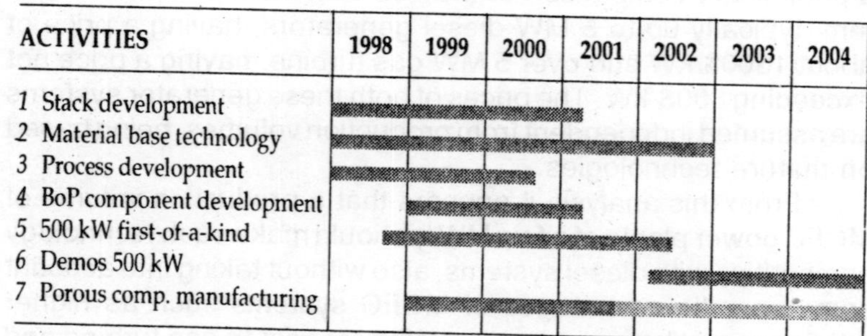

Fig. 15 : General Structure of the MOLCARE Phase 2 Programme

cells in a stack has been identified in order to reach, through intermediate steps, the target of a "full-area" 20 cell stack.

In this phase of MOLCARE programme, the main mid-term target is the erection and operation of a MCFC demonstrative cogeneration plant having an electrical output of 100 kW, including design, development and construction of the cell stack. Both these activities have been partially funded by the EU under the JOULE II and THERMIE programmes.

As regards MCFC plant engineering, Ansaldo Ricerche has taken great advantages from its know-how mainly based on the experience gained by designing, building and operating the 1.3 MWe PRODE PAFC power plant in Milan realised in co-operation with ENEA and AEM (Milan Municipal Energy Authority).

In parallel with the ongoing last phases for the realisation of the 100 kW plant, the last part of the MOLCARE programme is addressed to the definition of market-entry and pre-commercial units. To this aim, particular attention has been and is still devoted to cost analysis, selected solutions, industrial viability, plant sizes and configurations.

A 500 kWe output system has been selected as modular unit, defining its configuration and general layout, as consequence of the careful evaluation phase now in progress. This choice has been done taking into account the modularity features of fuel cells whose plant size easy fit the increasing power demand (at least by multiplying the base units up to a reasonable number), the economical parameters and the present level of MCFC technology. Starting from the intermediate results of the ongoing activities, it has been planned to improve the present design by adopting some new technological solutions without any remarkable modification neither of the previous stack arrangement, nor of the original Balance of Plant (BoP). In particular, with reference to the ongoing project, the shape will be rectangular instead of square consistently with a modified process which should improve some features of the present sensible heat reformer.

A lot of attention will be paid to the system compactness by optimising size and location of all the components; in such a way thermal integration as well as power density will be improved without penalising maintainability. Besides, a suitable layout will allow the unit to be really modular and thus to be used as a building block for MW class larger power systems. Consistently with these targets, a parallel technological programme based on selected support activities has been defined. Looking towards an industrial application, these activities mainly aim at improving both materials and manufacturing technologies in terms of cost, lifetime, performance, reliability and environmental impact.

ii) The 100 kW Prototype Power Plant

a) General

The 100 kW MCFC co-generative power plant, whose commissioning is ongoing at the CISE site in Milan, is the main mid-term goal of the MOLCARE programme. This plant can be regarded as the first rendezvous" between stack and BoP technologies. The main objective of this project, partially funded by EU, is the experimental validation at a significant power size of the "*Compact Unit*" (CU) concept that will generally characterise the next market entry units.

The present CU arrangement can withstand an absolute pressure of 3.5 bar. The heat management is guaranteed by proper anode and cathode recycles which allow to provide the desired gas temperatures, flow rates, compositions and pressures at the anode and cathode inlets. In particular the anode recycle is able to supply all the heat required for the fuel reforming process that occurs into the sensible heat reformer.

As the auto-thermal running is a key point of the CU concept, particular attention has been paid to the heat loss analysis by comparing different insulation materials and their configurations in order to limit costs without affecting performance. Suitable process conditions have been defined for the CU starting procedure; the slope of pressure and temperature from room conditions up to operating ones has been split into several steps, specifying the relevant key parameters for each of them with reference to any part composing the CU itself. A dynamic simulation of the main operational and single failure transients has been performed by means of a suitable model starting from the available steady state analysis. The main results of this activity has been used as additional specifications for the Application Software Requirement.

The CU hardware has been designed to be operated with many different fuels, such as natural gas (the first option) coal gas and biomass derived gases. CU configuration can be rearranged for each fuel simply by modifying the management procedures of the gas streams, both internal (such as anode and cathode recycles) and external (such as fuel inlet, oxidant inlet and exhausts) without affecting the cell stack. Although the project is essentially aimed at validating the CU operational features working with natural gas, some tests with other fuels namely coal gas, are planned to verify the suitablility to use alternative sources.

b) The 100 kW Stack

The CU is powered by a 100 kW MCFC stack consisting of two modules composed of 75 cells each. The geometrical area of a single cell is about 0.75 m^2. This stack is the final result of a scale-up procedure based on development, design, construction and testing of several sub-scale stacks. Different cell areas (100, 700 and 7500 cm^2) and cell numbers (10, 15, 20 and 50 cells) have been investigated to focus and solve specific aspects of the stack technology from the fabrication, assembly, first conditioning and operation points of view. In particular the final steps of this process have been addressed to check separately a first "full-height" stack whose cell number was very close to that one of the 100 kW single module and a second "full-area" stack where the area of the single cell was the same as the 100 kW one. Table 16 shows the main features of these final steps compared with those one related to the module of the 100 kW stack.

Table 16: Main Features of a Single Module of the 100 kW Stack Compared with the Two Final Sub-scale Stacks in the Scale-up Process

100 kW stack,	*Sub-scale stacks*					
Main features	***single module***		***"full-height" stack***		***"full-area" stack***	
General configuration	*• external manifolds • Cross flow*		*• External manifolds • Cross flow*		*• External manifolds • Cross flow*	
Cell number (#)	75		50		20	
Geometric cell area (cm²)	7500		900		7500	
Active cell area (cm²)	6700		700		6700	
Rated power (kW)	50	(a)	4	(b)	13	(b)
Operating Pressure (bara)	3.5	(a)	1.5	(b)	3.5	(b)
Power density (kW/m²)	1.00	(a)	1.14	(b)	0.97	(b)
Test duration (hours)	up to 8000		900	(b)	1300	(b)
Thermal cycles (#)	≥1		1	(b)	3	(b)

Notes: (a) design data
(b) experimental data

The experimental results coming from these sub-scale stacks, the knowledge gained during their construction as well as the set-up of the manufacturing and assembly facilities allowed in 1997 the construction of two modules forming the 100 kW stack (Fig. 13) at Ansaldo Ricerche laboratories. At the beginning of 1998 the two modules have been sent to Spain for start-up and first test campaign at the test stand facility of IBERDROLA in Guadalix. The aim of this experimental phase is to evaluate, after stack conditioning, its behaviour without any influence from the other BoP components.

c) The Guadalix Test Facility

The 100 kW MCFC stack test facility has been built by PEPIBERDROLA at the Training and Dissemination Centre located in San Augustin de Gaudalix, near Madrid. Its main functions are:

I. to permit a semiautomatic stack conditioning which mainly consists in burning out organic products and oxidising the cathode nickel;

II) to characterise the MCFC stacks under various working conditions.

The facility is equipped with a control system with pre-programmed functions for its operation under different power states and consists of the following major components and sections: vessel, purge gas and mixing gas systems, heaters and off-gas coolers, gas recycle loops, gas storage systems, water treatment and steam boiler, control room and clean room. The general layout of this installation (Fig. 17) includes:

- *Vessel:* the stack to be tested is placed inside a vessel which can be pressurised by means of inert gases;
- *Mixing gas stations, heaters and off-gas coolers:* they allow a suitable gas management by regulating compositions, pressures and temperatures of the stack inlet and exit gases;
- *Gas recycle loops:* they allow to simulate the operating condition of the stack when it is integrated in particular power plant configurations;
- *Gas storage systems:* includes liquid carbon dioxide and nitrogen tanks and a high pressure gaseous hydrogen storage on mobile platforms;
- *Water treatment and boiler:* has the function to provide steam to the plant at a pressure up to 12 bar.
- *Control room:* it hosts the plant control system, based on a Remote Input Output (RIO) type architecture processor and an "on-line" high resolution Mass Spectrometry gas analyser;
- *Clean room:* to guarantee the quality of atmosphere during checks and/or final preparation of the stacks before testing and overhauling during or after tests.

Fig. 17 : View of the Stack Test Plant of S. Augustin de Guadalix (Madrid)

d) CISE Prototype Plant

Unlike the Guadalix Test Facility, that is mainly oriented to stack conditioning and characterisation, the facility at CISE is a real MCFC plant aiming at proofing the CU concept integrated with the Balance of Plant systems. Therefore the site is provided with all the equipment forming the BoP including infrastructures and services which are necessary for the plant operation such as natural gas pipeline, steam, site air for controls, cooling water, and a.c. power for auxiliaries. The main BoP systems supporting the 100 kW CU are arranged on an outdoor site; they are:

- *Power Conditioner* to convert the dc power produced by fuel cells to ac power for stand-alone and grid-connected operation and provide electrical insulation and reverse current protection.
- *Centralised Control System* to provide process and operational control and allow the monitoring of the whole plant.
- *Fuel Clean-up System* to condition (namely desulphurise) the natural gas for the use in the CU;
- *Process Air System* to provide air to the CU at suitable pressure for fuel cells; the system also includes an heater for start-up.
- *Steam Supply System* to control the available site steam to be used together with the fuel in the CU.
- *Auxiliary Power System* to distribute power to blowers and heaters in the CU and in the BoP.
- *Co-generation System* to recover the waste heat in the CU exhausts.
- *Nitrogen and CO_2 Supply* Systems to provide purge gas for the vessel.
- *Hydrogen Supply System* to provide hydrogen for start-up, fuel clean-up and coal gas simulation.
- *CO Supply System* to provide CO for coal gas simulation.

Some standard equipment used in MW class systems is not useful for this plant. Thus, simpler solutions such as e.g. motor-driven air compressor, steam site supply and absence of bottoming cycle turbo expander powered generator have been preferred. Nevertheless, during the result analysis the effects of such simplifications will be taken into account.

iii) Next Development of the Market Entry Units

a) General

Distributed generation in the power range 100 kW – 50 MW, including co-generation and district heating, is generally targeted as the principal market area for MCFCs. This one is often considered as an emerging market which could influence the future of power generation business promoting plants located near load centres instead of centralised systems and their relevant transmission and distribution networks.

Even if traditional large size systems will continue to cover by far the greatest part of power demand, distributed generation share is expected to increase significantly in the next decades up to 15-20 per cent reaching 100 GW per year and, probably, over.

A great potential for distributed generation is offered by the emerging countries having not yet strongly invested in the transmission and distribution systems associated with centralised generation.

A significant share of this market sector, up to about 10 per cent, seems to be approachable for fuel cells even if their utilisation in larger size power plants will be feasible later, when fuel cell technology will reach a higher level of maturity. In fact in the medium term MCFCs will be suitable for plant size in the range 500 kW - 5 MW, the upper on being easily obtainable exploiting the fuel cell modularity. In particular, the expertise already gained through the 100 kW project design, shows that a 500 kW co-generative power unit is the most suitable base to match technical, economical and market requirements.

This last arrangement, which must be regarded as first-of-a-kind plant, is based on the present 100 kW CU concept with only few improvements mainly related to the stack design, the fuel reforming system and relevant recycles in order to improve heat management, efficiency, flexibility and compactness. The goals of this last development are:

- to be competitive with traditional power plants of the same size, also providing heat for co-generation;
- to limit the overall dimensions by locating the cell stack and its modular integrated reformer, that is linked with the catalytic burner, inside a suitable, truck transportable vessel;

- to simplify the maintenance procedures and reduce the servicing needs by a suitable compact arrangement and an enhanced stack life;
- to reduce the number of components and optimise their layout to place the whole plant over a skid;
- to be suitable for use as a "building block" for larger sizes by some small modification of the BoP.

The plant is expected to reach an electrical efficiency of about 54 per cent and an overall efficiency (power plus heat) of about 80 per cent.

b) The 500 kW Base Unit

According to the above mentioned criteria, the 500 kW plant is based on an improved *Compact Unit* which hosts two 250 kW modular stacks. Each stack is integrated with a reformer that is linked to a relevant catalytic burner, both having the same modularity of the stack. In such a way the anode recycle is avoided and only a blower is needed to move the hot gas from the reformer flue gas side to the cathode inlet. All these components are set inside an internally insulated pressure vessel designed and built to ensure a correct gas management and limit the constraints in terms of sealing between gas manifolds and stack.

Key feature of the 250 kW stack configuration, split into two 125 kW "*Twin Stacks*" with common cathode inlet and cathode outlet chambers, is its capability of assuring a very effective thermal management without the negative effect of the differential pressure due to high flow rates.

This solution allows a slight increase (from 7500 to about 8000 cm^2) of the cell area presently used for the 100 kW plant, a better distribution of temperatures over the cell area, as smaller differential pressure across the cell matrix. The configuration of the 125 kW stack is quite completely derived from the design of a "technological" stack (this is a part of a project partially funded by EU), which will allow to validate such a solution at a sub-scale size.

The fuel reforming system is based on a Modular Integrated Reformer forming a unique body with the catalytic burner to exploit the heat generated by the complete combustion of the anode exhaust gases. These ones are cooled down by the endothermic CH_4 conversion and then are recycled to the cathode by a blower whose impeller is located inside the CU. The oxidant leaving the cathode is partially mixed with the anode exhaust gases before entering the catalytic burner and is partially used to feed a gas turbine driving the air compressor and also an electrical generator. The other components of BoP are:

- *Fuel Processing System:* its goal is to purify and desulphurise the natural gas and mix it with steam before feeding the reforming section.
- *Air Compression System:* it is designed to supply suitably pressurised air to the cathode inlet after filtering and desulphurising stages.
- *Co-generation System:* it allows the utilisation of the hot gases leaving the turbine to supply heat for district heating and/or air conditioning.
- *Power Conditioning System:* it converts dc current generated by the stacks into ac current at 380 V and 50 Hz for stand alone or grid connected operation.
- *Control data and acquisition system:* it will permit the automatic control of the plant with the final goal of an unmanned or remote operation and monitoring.

The experimental operation of such a plant will be addressed to verify the behaviour of the whole system under real operating conditions with reference to a.c. output, fuel consumption, environmental impact, cell stack and ancillaries behaviour for a comparison with competitive traditional plants of the same size and power range. The testing will also allow to analyse the behaviour of the plant during unmanned and/or remote modem control operation. Moreover, with reference to industries and utilities requirements, the present suitability of the plant to be used as a building block for larger power production systems will be evaluated.

c) The MW Class Plants

In the medium term, the modularity of MCFCs will allow to satisfy the requirements of a power market ranging up to a few MWs which is becoming of interest for electrical utilities.

From the cost point of view, it is preferable to focus the attention on actual potentialities and possible limits of configurations that foresee several standard units in parallel (i.e. the above mentioned 500 kW plant) with as small as possible adjustments. Present modules are expected to show this feature. In any case, if application requirements will suggest to review BoP components, whose modularity is not mandatory, the CU will represent the actual building block for MW class plants thus giving a strong opportunity for an effective cost reduction.

From the efficiency point of view, the recovery of exhaust gases enthalpy contents by proper turbo-expander-powered-generators can be quite attractive also for few MW plants even if the integration of fuel cells with bottoming cycles is traditionally a feature of largest size plants.

Both these technical and market aspect analysis are now in progress in order to develop the detailed design of the 500 kW plant and get elements for the evaluation of the costs. From this point of view, the target is 1500$/kW by 2005 which can be regarded as really competitive if also the environmental impact is taken into consideration; the fuel cell emissions are negligible if compared with those ones of traditional plants.

11. CONCLUSIONS

After an overview of fuel cell technologies status and prospects and main features, focus has been kept on stationary power plants, mainly in relation to high temperature technology of MCFC.

This technology is very promising because of a unique set of advantages, mainly efficiency, life, fuel flexibility and low fabrication costs prospects. As a direct consequence, many developers in US, Japan and Europe are involved in different teams, to demonstrate the technology potential.

As an example of development programmes, a diffused description of the two phases of the MOLCARE project has been given, trying to explain how the present achievements should demonstrate that Molten Carbonate Fuel Cells technology, except for a few technological improvements under development, is ready for moving from prototypes and proof-of-concept solutions towards first-of-a-kind pre-commercial and market entry units.

In fact among the other fuel cell types, MCFCs—in addition to some technical advantages—show the best potential for a faster commercialisation, at least in a power range between few hundreds kW to few MW.

The technology developed in carrying out the first part of the MOLCARE Programme, thanks to the joint efforts of several qualified partners, has allowed the construction of a 100 kW prototype co-generation plant powered by the larger stack ever designed and completely realised in Europe.

Even if further helpful elements are expected to arise from the testing activities, the experience already gained in performing this work in terms of base technology, manufacturing processes, sub-scale stack testing and plant engineering has enabled the Spanish and Italian partners to define the ways towards commercialisation. It has been possible to identify the size and the general configuration of systems, which will be suitably used as connected to the grid or stand alone market-entry unit or as a building block for bigger plants.

Some evaluations are also in progress to define more in detail the expected allowable cost for MCFC power plants in comparison with other competitive systems, as well as to state the relevant targets and identify the viable ways to achieve them. The preliminary results already achieved are very promising.

Acknowledgements

Part of the work described in this paper for the MOLCARE project has been performed under the financial support of the European Commission in the framework of the JOULE and THERMIE programmes and of the ENEA/Ansaldo Italian Association Contract.

References

1. M. Brossa, A. Dufour, E. Hermana, J.F. Jimenez, F. Sanson, “The European MOLCARE Programme: a 100 kW Demonstrative Plant and Engineering Development of the MCFC Technology”, Fuel Cell Seminar ’94 San Diego, California, Nov. 28–Dec. 1, 1994.
2. A. Torazza, A. Dufour, L. Giorgi, J.G. Martin, G. Rocchini, F.J. Simon, ‘Guidelines for Design and Development of Industrially Relevant MCFC Stacks”, Fuel Cell Seminar ’96, Orlando, Florida, Nov. 17–20, 1996.
3. A. Dufour, M. Campagna, E. Luchi, A. Torazza “The Molten Carbonate Fuel Cell Manufacturing Technology under Development in the Frame of the MOLCARE Project” Energy Week

February 1997, Houston, Texas U.S.A.
4. Esco Vale "Analysis of the prospects for fuel cell applications" Report 2051 August 1997.
5. Kuehn Fuel cells near commercial reality. Power Engineering International Feb. 1994.
6. European Commission DG XII e DG XVII "A ten year fuel cell research, development and demonstration strategy for Europe" November, 1997.
7. Barnett & al. Journal of Power Sources, 1–2–1990.
8. A. Dufour, A. Torazza, V. Regis, G. Rocchini, E. Hermana: "MCFC Compact Unit and System improvements for 100 kW–2 MW Dual-Fuel Demonstration Plant", POWER-GEN Europe '95, 16–18 May, 1995 RAI Amsterdam, The Netherlands, Book II, Volume 5, pages 859–874.
9. J. Garcia Martin, A. Dufour: 'Present Status of Fuel Cell Activities in Spain. The Spanish-Italian MOLCARE Programme" POWER-GEN Europe '97, 17–19 June, 1997, Madrid Spain, Volume IV, pages 451–461.
10. F. Parodi, A. Perfumo, G. Rocchini, M. Scagliotti, A. Moreno, J.F. Simon: "Testing of a Molten Carbonate Fuel Cell Stack of 13 kW", 5th Grove Fuel Cell Symposium, Poster Session, 22–25 September 1997, London, UK.
11. A. Torazza, A. Dufour et al. "MOLCARE development towards MCFC commercial power plants based on 500 kW standard modules" To be published at Power—Gen. 98 Europe Milan, June 9–11, 1998.

27

Technical and Economical Aspects of Fuel Cell Systems for Residential Applications

P. SILBERG
Germany

1. INTRODUCTION

Preservation of resources, environmental and climate protection are the outstanding energy political requirements at present. According to the Federal Government resolution of 1990, the CO_2 emissions in Germany are to be reduced by 25 per cent in 2005 (in comparison to 1987). At the conference in Kyoto 1997, it was decided to reduce the global emissions of the most relevant greenhouse gases significantly until 2008–2012. The future power generation plays an important role in achieving these goals. Using fossil fuels as primary energy source, mostly internal combustion engines are used for energy conversion at present. These thermal engines convert the energy contained in the fuel via thermal and mechanical energy into electrical energy. In order to reduce CO_2--emissions, new innovative technologies are required to enhance the efficiency of power generation. Due to the principle of electrochemical energy conversion, which implicates the direct conversion of the fuel without using a thermal cycle, the fuel cell technology is a high promising technology.

This direct energy conversion offers several advantages. In contrast to conventional thermal engines (e.g. gas engines, diesel engines), the electrical efficiency of fuel cells is not limited to the carnot-factor. Fuel cells show not only a high but also an almost load independent efficiency over a wide range of the power output. Due to their low emission level of atmospheric pollutants (CO, NO_x, SO_2 and hydrocarbons), fuel cell power plants can be installed on-site, where energy is consumed, even

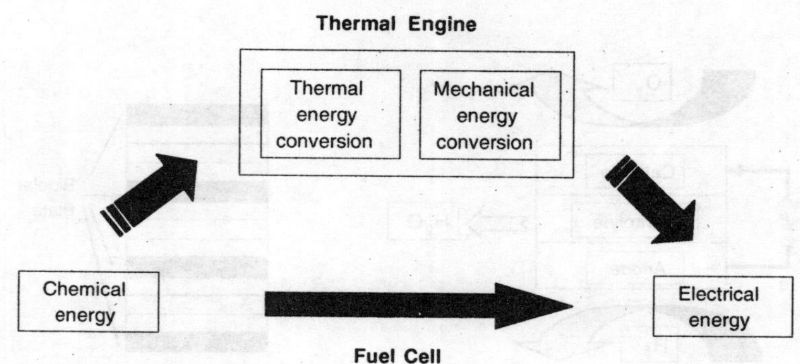

Fig. 1 : Direct Energy Conversion with Fuel Cells in Comparison to Conventional Indirect Technology

inn densely populated areas. Fuel cell systems have also the advantage to be modular and can therefore be built in a wide range of power requirements from a few hundred Watts up to megawatt sizes.

2. FUEL CELL DESIGN

In principle a fuel cell works like a battery. The reactants—e.g. hydrogen and oxygen (from air)—are fed continuously to the electrodes. Electrical current and heat are generated, and the only product of the cold combustion of hydrogen is pure water.

The key-component of any fuel cell is the assembly of *p*ositive electrode, *e*lectrolyte and *n*egative electrode (PEN). The electrolyte, covered with the two electrodes, must be gas-light to separate the feed gases and therefore to enable the oxidation of the fuel gas at the anode side and the reduction of oxygen at the cathode side. Furthermore, the electrolyte must be highly conductive for ions (H^+ or O^{--}) to ensure the transport of the ions between the electrodes. The electrons released at the anode are transported back to the cathode via an external load i.e. electric charge is transported.

For achieving a technically suitable voltage level, single cells must be assembled in series. So-called bipolar plates have

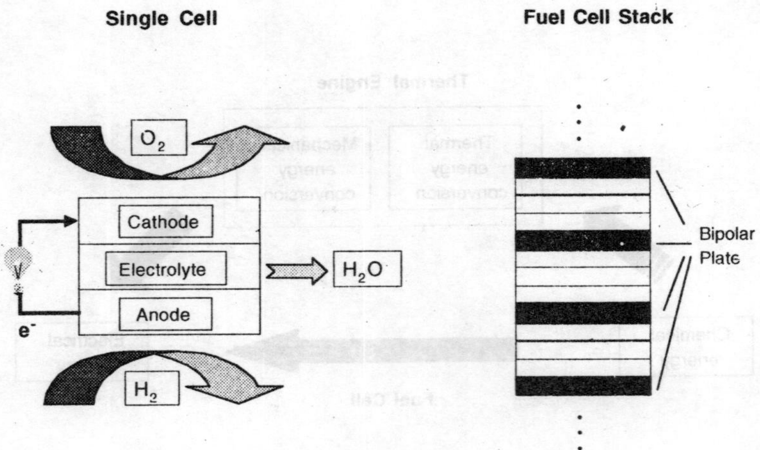

Fig. 2 : Single Cell and Fuel Cell Stack Design

to be integrated between the single cells which facilitate the gas supply to the electrodes and electrically connect the single cells (see figure 2). Usually cells and bipolar plates are piled up one onto the other in the form of sandwiches, creating the fuel cell stack. To avoid an increase in temperature, surplus heat must be extracted by cooling with air or water, depending on operating temperature.

3. FUEL CELL SYSTEMS

The fuel cell principle, the "cold combustion" of hydrogen and oxygen, was discovered by Sir William Grove in 1839, but the technical development of fuel cells did not start before the end of World War II. During the last decades low-temperature fuel cells (alkaline fuel cells (AFC) and polymer electrolyte membrane fuel cells (PEMFC) were developed for space programmes (Gemini, Apollo, Spacelab) and submarine applications. In recent years good progress has been achieved in the development of electric vehicles powered by PEMFC. For combined heat and power (CHP) applications 200 kW phosphoric acid fuel cells (PAFC) produced by ONSI/USA are commercially available.

Usually fuel cells are classified according to their operating temperatures into low, medium and high temperature fuel cells (see Table 1). The different operating temperature has its roots in the electrolyte materials used in the different fuel cells systems.

Tabel 1 : Overview of Different Fuel Cell Systems

Fuel Cell	*Operating System*	*Electrolyte Temperature*	*Fuel*	*Oxidant*
Alkaline Fuel Cell (AFC)	60–90°C	alkaline lye (30% KOH)	Pure H_2	Pure O_2
Polymer Electrolyte Membrane Fuel Cell (PEMFC)	50–80°C	Polymer membrane	Pure H_2 and reformed H_2	Air
Phosphoric Acid Fuel Cell (PAFC)	160–220°C	immobilizised phosphoric acid	reformed H_2	Air
Molten Carbonate Fuel Cell (MCFC)	650°C	immobilised molten carbonate (Li_2CO_3/K_2CO_3)	reformed H_2 and CO	Air/CO_2
Solid Oxide Fuel Cell (SOFC)	900–1000°C	ceramic oxygen ion conductor (ZrO_2/Y_2O_3)	reformed H_2 and CO	Air

It must be considered that complete fuel cell cogeneration systems not only exist of a fuel cell stack. Several peripheral compounds are needed for safe and efficient generation of electrical power and users heat. Using natural gas as fuel, this must be converted in a chemical reactor—the so-called reformer—to generate a hydrogen-rich gas which is delivered to the fuel cell stack. As sulfur compounds would destroy the reforming catalyst, the natural gas must be desulfurised first. The electrochemical energy conversion inside the fuel cell stack delivers direct current, which has to be transformed to alternate current by means of a DC/AC inverter. Furthermore a complete power plant needs pumps, gas or/and air compressors, heat exchangers, additional burner etc.

Low temperature PEMFC and AFC show a very good dynamic behaviour and a very high volumetric power density. In contrast to the high temperature systems, their start-up time is very short. It is rule that increasing operating temperature decreases the sensitivity towards impurities in the feedgas of the cell. Therefore MCFC and SOFC are less sensitive towards sulfur compounds than the other fuel cell systems, and they are able to run with carbon monoxide as well as with hydrogen. This is a great advantage compared to PAFC and particularly PEMFC and AFC, where the feedgas must carefully be cleaned from carbon monoxide. High temperature fuel cell systems can also reach higher net electrical efficiency, even if they are coupled with gas and steam turbines. Due to their high operating temperature, the natural gas reforming in MCFC and SOFC can be carried out directly at the anode side. This internal reforming enhances the electrical efficiency by approx. 5–10 per cent in comparison to the low and middle temperature systems, where the hydrogen production must be carried out in an external reformer. On the other hand material problems occur at elevated temperatures. MCFC causes problems because of the corrosive electrolyte, and the main disadvantage of SOFC is the brittleness of the ceramic electrolyte, which is therefore very sensitive to thermal stresses.

4. FUEL CELL APPLICATIONS

The field of application of a fuel cell system depends on its specific characteristics. Mobile application as well as stationary application are currently under development.

i) Mobile Application

Fuel cells have very special characteristics for transport duties. There are few alternatives to the fuel cell if a pollution-free vehicle with adequate performance and range is desired. The PEMFC technology seems to be the most suitable fuel cell technology for traction systems and is being pursued by numerous companies (Ballard Power Systems, BMW, Chrysler, Daimler-Benz, Energy Partners, Ford, General Motors, Honda, H. Power, Intl. Fuel Cells, MAN, Neoplan, De Nora, PSA, Renault, Siemens, Toyota, Volvo, VW). Although PEMFC are still very expensive, the cost reduction potentials are huge and very promising. Specific costs of US$ 50–100/kW$_e$ for complete PEMFC systems including fuel reformer and gas purification technology are needed for large-scale market penetration. It is very difficult to predict, if this goal can ever be achieved.

sii) Stationary Application

The high efficiency at full and partial load as well as the low atmospheric emissions make fuel cells very attractive for stationary application. Fuel cells in large-scale as well as in small-scale application seem to be ideally suited for combined heat and power production, but it must be considered that fuel cells have to face severe competition from still developing technologies. The types of competing technologies depend on the field of application, In conventional large-scale power plants, energy efficiencies up to 58 per cent are achieved using a combined gas and steam turbine (GUD) generation scheme. However, the efficiency can further be enhanced up to 60–70 per cent if a GUD process is used as a topping cycle for SOFC power plants. Due to the low specific investment costs of conventional GUD power plants it will be difficult for large-scale fuel cell systems to be economical competitive.

Furthermore, middle-sized cogeneration systems (100 kW–1 MW) for the use in offices, restaurants, hotels, hospitals and housing estates already exist in form of well-established and low cost engines and gas turbines in packaged cogeneration plants. Although fuel cells show higher electrical efficiencies and ecological advantages it will not be easy to penetrate this market.

a) Fuel Cell for Residential Applications

In contrast to large- and middle-sized cogeneration systems, no well-established conventional cogeneration technology competes with fuel cell technology for residential application. Furthermore, natural gas is available from a dense distribution grid in many European countries. This well developed natural gas infrastructure will push the introduction of small scale fuel cell systems into the energy market.

In the well developed countries, the household demand in electricity and heat requires a great part of the primary energy consumed in these countries. That is why this field of application offers a great potential for saving primary energy sources and reducing CO_2– as well as air pollutant emissions. In Europe an obvious trend in residential electricity and heat supply in new buildings points towards significant energy savings. In Germany a new law expected for 1999 will require that so called low energy houses with a very low heating demand be built. Therefore, the operation of small cogeneration systems with a high quotient of electrical energy and thermal energy will best fit, so that the operation of fuel cells with an electrical power output below 25 kW will become very attractive in the near future.

Among the five important fuel cell systems, PEMFC and SOFC are suitable for small scale cogeneration. Both systems show advantages as well as disadvantages, and it is difficult to say which system will be best suitable for residential application. PEMFC presently require very clean hydrogen gas as fuel. That is why the reformed gas mixture must be purified in a costly process. Especially carbon monoxide contents must be very low (<10 ppm). As PEMFC typically operate at 40–80ºC, heat recovery of the system is difficult. Furthermore, the PEMFC must be operated under pressure at the anode and at the cathode side to achieve a high electric efficiency of the stack. The use of compressors can result in a paratsitic load of up to 20 per cent of fuel cell output.

The exhaust temperature of the SOFC is > 200 ºC and heat recovery is easy. The SOFC is operated at ambient pressure, but the very high operating temperature causes material problems inside the stack as well as using heat exchangers for air-pre-heating. Therefore, high temperature corrosion will probably reduce the lifetime of the SOFC in a non-tolerable manner.

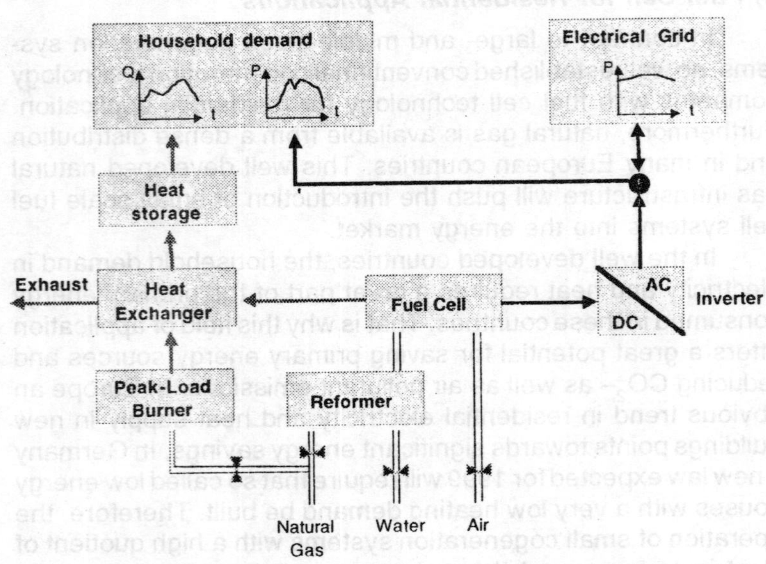

Fig. 4 : Block Diagram of a Fuel Cell Cogeneration Unit

Figure 4 shows a block diagram of a fuel cell system for residential application. In order to fulfill the household demand in heat and electric power, the fuel cell system will best be grid connected. This net-parallel operation facilitates fuel cell sizing independent of the peaks in electrical demand. An additional peak-load burner allows the decoupling of actual heat demand and fuel cell heat production. In this way, the sizing of the fuel cell not depends on the maximum heat demand, because peaks in heat demand can be delivered by the additional burner. A heat storage allows the decoupling of heat and power production over a wide range and in summer the operation of the peak load burner will not be necessary.

In Switzerland, the company Sulzer HEXIS is developing small scale cogeneration SOFC systems. A prototype with an electrical power output of about 1 kW is still operating in the gas laboratory of DEW (Dortmunder Energie and Wasser), the local utility of Dortmund (see figure 5). In a common project between Sulzer, DEW and our Institution of Electric Energy Systems at the University of Dortmund the behaviour of SOFC stacks under real field conditions with natural gas was investigated.

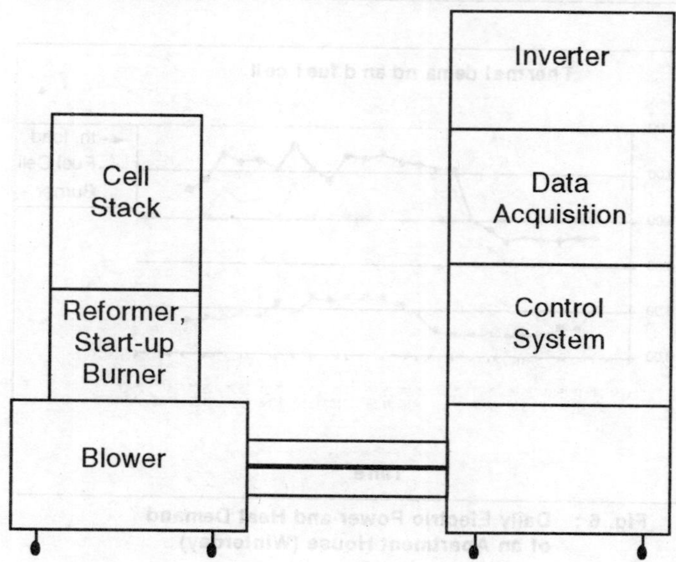

Fig. 5 : Layout of the Sulzer HEXIS Field Test at DEW in Dortmund

In addition to the technical work, detailed measurements of the heat demand and the electrical power demand in a single family residence and an apartment house were carried out. In figure 6 a typical load profile for an apartment house—consisting of 6 flats with 15 persons—during winter-time is shown.

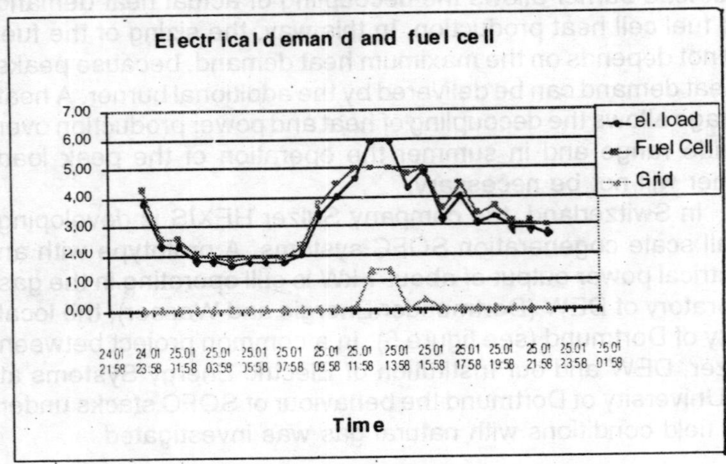

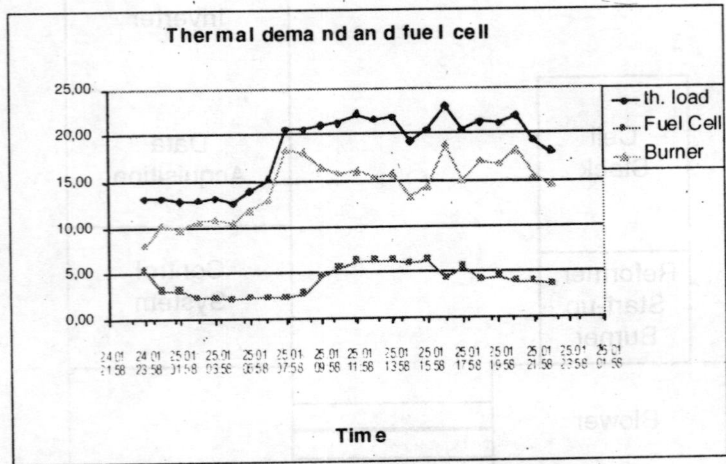

Fig. 6 : Daily Electric Power and Heat Demand of an Apartment House (Winterday)

The operation of a 5 kW_{el} SOFC fuel cell system, assumed to operate in an electricity-oriented mode, is simulated (see figure 6). The electricity demand curve shows a typical electrical peak load at lunch-time. For about one hour surplus electricity must be received from the grid. The heat demand is considerably higher than the electricity demand and the difference between fuel cell heat production and heat demand is delivered by the peak load burner.

In the summer season the electrical demand is not quite different, but the heat demand is significantly lower and no heat from the additional burner is required.

Crucial for the market penetration of fuel cell systems are specific investment costs and operating costs. The competitive specific investment costs vary with the type of application. For residential application a fuel cell system must be competitive with the conventional heat and power supply, mostly consisting of a gas or oil burner and the public grid. In this connection it must be considered that the fuel cell stack costs are about 1/3 and the peripheral costs are about 2/3 of the whole system costs. The average lifetime of a fuel cell stack is expected to be five years in contrast to the lifetime of the peripheral compounds which is expected to be about 15 years. In view of competitiveness the stack costs play a dominant role, because the reinvestment-costs of the stack must be considered.

In principle, cogeneration systems can be operated in different modes. Conventional systems, based on internal combustion engines, are often operated in a heat oriented mode, where the heat demand of the user is fulfilled all the time, and the electricity is only "by-product". On the other side an electricity-oriented mode may be economically more attractive, because of the high electricity costs private customers have to pay to the utility. Therefore, the substitution of electrical energy is most cost-reducing, and it must be the goal to receive as little surplus electricity from the grid as possible.

The operation of fuel cells in residential buildings may also be attractive for local utilities. Though the electricity prices for local utilities are lower in comparison to private users and consequently the reducing effect of energy substitution is not so high, operating a fuel cell offers the chance to replace peak power. The utility option currently seems to be most promising for small natural gas and electricity supplying communities, as only then the operational flexibility of small decentralised cogeneration systems can be utilised.

6. SUMMARY

Fuel cell technology is a very promising technology in view of saving primary energy sources and reducing the emissions of CO_2 and several atmospheric pollutants. The role fuel cells will play in the future will be determined by the prevailing economical and environmental regulations. The most important applications will be found where avoiding pollution is a crucial supplementary condition. Nevertheless fuel cells must be economically competitive with still existing and established conventional technologies. Fuel cells will have a good chance to compete with other technologies in mobile applications as well as in decentralised combined heat and power production. The operation of small-scale fuel cell cogeneration systems for residential application offers a great potential for saving primary energy sources and for reducing the local emissions to a minimum. This application will also be economically attractive for private customers as well as for local utilities. However, a number of technical issues have to be solved and the investment costs as well as the maintenance costs must be lowered. If these requirements are fulfilled, fuel cells will play an important role in the future energy system.

28

Methodological Approach to the Energetic Problem of the "Isolated Consumption"

A. LANTIERI
Italy

SUMMARY

This report describe a methodological approach to solve the energetic problems of the "Isolated Consumption". In this context, the definition of "Isolated Consumption" cover the villages, the rural, industrial and mining productive district that are distant from the energetic grid of the country.

This approach consider relevant that the selected technological solutions enable a "sustainable developing" of the considered reality.

In fact it is, now, international accepted conviction that the environmental tenability (in terms of resources consumed and solid, liquid and gaseous wastes produced), economical (the total costs must be acceptable for all the interested subjects) and social (to satisfy the needs of the communities respecting their values and priorities) is pursued.

GENERAL BACKGROUND

In the world, many countries have an energetic situation (production and distribution) that not satisfy the thermal and electric needs of the countries.

In particular many people that live in places not served by the principal energetic network have precarious electric and thermal sources supplied.

The problem of this "Isolated Consumption" is considerable in the countries which have a extension, a particular typology of the territory and are in political and economical transition.

In this context, the definition of "Isolated Consumption" cover the villages, the rural, industrial and mining productive district that are distant from the energetic grid of the country.

The process of political and economical transition of the countries makes more difficult the adequate energetic problems solution of these lands, because new technology, methodology and innovative professional figures, not always available in the interested lands, are needed.

METHODOLOGICAL APPROACH

We consider relevant to promote in these countries stable and qualified organisations with the aim to give solution to these problems.

These Organisations will be permanent centres to analyse the needs of these lands, to propose appropriate solutions; finally these centres will be the place where the needs of the lands meet the scientific and technological solutions.

These centres will be also the place where the local executive cadres will be trained in the appropriate technologies and methodologies to define and manage programmes for the solution of the energetic problems in these lands, utilising alternative energies in line with their resources and needs.

It is very important that the selected technological solutions enable a "sustainable developing" of the considered reality.

In fact it is, now, an international accepted conviction that the environmental tenability (in terms of resources consumed and solid, liquid and gaseous wastes produced), economical (the total costs must be acceptable for all the interested subjects) and social (to satisfy the needs of the communities respecting their values and priorities) is pursued.

On this problem Italy and in general, the western countries have much experience in the development, construction and conduction of thermal and electric plants for "Isolated Consumption".

The Italian Universities, industries and research centres develop advanced activities of study, experimentation and realisation of plants in the aim of the combustion, cogeneration, accumulation and combined cycles.

Activities in this field are developed in the framework of programmes financed by the UE, but, considering the extent of this problem, it is important that bilateral and multilateral cooperation, and joint venture with partners of western countries are implemented.

The "Energetic Centres for the solution of the Isolated Consumption problems" located in regional zones are fundamental to train the local executive cadres, and to give them instruments (instruments for the needs analysis, for the planning and the estimation between different solutions, technologies, and methodologies to execute the detailed project of the intervention). We consider that the more effective intervention is jointly the know-how transfer with the training on the job.

In Russia, for example, for some time the scientific community studied the problem to supply thermal and electric energy to the "Isolated consumption", but this problem, during the period of planned economy, was minimised with systems that did not take in to account the economical and the consolidation of the new market logic in the society, as priority problem.

The scientific community effected several analysis about the entity of the problem, and it emerged that several millions of people have precarious electric and thermal sources supplied.

The scientific community considers that the solution of these problems needs much time, a large quantity of financing resources and especially a different methodology approach to the problems.

So it is considered necessary to find a methodological effort to approach and solve the problems as well as supplying to the regional communities the instruments and the methodologies to study the problems in a more profitable manner.

This approach is more fit to the new demand of the country reality that needs training for new executive cadres who implement the development of the alternative energies which, in many cases, are the only or the most qualified to solve, adequately, the problems of the energetic supplying to these lands.

The Russian Ministry for Fuel and Energy, for example, has developed and is carrying out programmes to implement the Russian Energetic System, developing the utilisation of the alternative resources and of the technologies for the energy saving. More and more often the Russian scientists join the energetic problem with the environmental capability.

The European programmes are devoting many funds to

resolve the problems concerning the improvement of the Russian energetic systems and the development of the alternative energy in Russia, but there is a lack about the specific problems of these lands that we define "insulating consumption".

We consider relevant this particular segment of the energetic problem in several countries and important that the western countries give a contribution in this to extend the ongoing general process of the modernisation of the energetic system in these countries.

MAIN OBJECTIVE

The main objective of the western countries is to support the other countries to analyse the situation of the places with these problems and to constitute specific centres supplying them the necessary instruments to start the Centre's activities and to research the conditions and the aggregations for its future conduction and implementation.

These centres will become the national and international reference point for the training of the local executive cadres and for the planning and definition of programmes and projects for the solution of the energetic problems of the isolated lands of the country.

The Centres should have at least the following aims:

- training for the operators in this field;
- monitoring and definition of typical places representatives of the interested local reality for the solution of the energetic problems;
- analysis of the available resources in the typical places;
- våluation of the environmental tenability of the technological proposal;
- definition of the appropriate technologies for the thermal and electric energy supply and for the energetic saving;
- making and management of data base to meet the territorial demands and the national and international offer;
- monitoring, analysis, design and realisation of demonstrative plants.

DATA BANK ORGANISATION

In order to endow the Centre with instruments able to facilitate the information, methodologies and technologies transfer, data banks will be organised. The data banks structure will be conceived to permit its progressive implementation.

This data bank will contain at least the information to manage:

- the energetic supply and demand of the interested places;
- the potential energetic of the interested places;
- the potential energetic technologies able to solute the energetic problems and their tenability;
- the potential national and international suppliers of these technologies.

On the basis of the data reported on the data bank, studies will be effected in the places considered, for the energetic solution of these places.

The study will include all the analysis necessary to define the place energetic plan that will be the starting point for the intervention detailed Project.

EXPECTED RESULTS

The minimum results expected concerning the installation of these centres are:

- Training of central and local experts for the definition of energetic plan of areas utilising alternative technologies;
- the implementation in the countries of the instruments and methodologies to effect analysis for the solution of the energetic problems of the lands defined "Isolated Consumption";
- individualisation of the productive districts, other economical and productive subjects and their involvement in the Centres management as utiliser or as operative and financiers of the Centre;
- monitoring and definition of the typical places representative of the lands with energetic problems and individualisation of the available energetic resources;
- individualisation of the manufacturers of energetic technologies suitable for the necessity and conditions of the lands take to reference;
- organisation of data banks to meet the lands necessity and the disposable technologies;
- definition of the scenarios for the possible energetic solutions of the lands;
- identification of reference points for the solutions of the

energetic problems utilising alternative technologies, for the training of the sector operators and for the monitoring, analysis, design and realisation of pilot plants.

29

Nuclear Power Technology for Sustainable Development

V.M. MOUROGOV
and
V. KAGRAMANIAN
Austria

INTRODUCTION

In the second half of the 20th century nuclear power has evolved from the research and development environment to an industry that supplies 17 per cent of the world's electricity. In these 50 years of nuclear development a great deal has been achieved and many lessons have been learned. By the end of 1997, over eight thousand five hundred reactor-years of operating experience had been accumulated.

The past decade, however, has seen stagnation in nuclear power plant construction in the Western industrialised world, slow nuclear power growth in Eastern Europe and expansion only in East Asia. The prospects for nuclear energy have been affected by a number of factors: slower economic development and general reductions in the rate of increase in energy demand, coupled with over-supply in some countries; the TMI and the Chernobyl accidents with their effort on public confidence in nuclear power; slow progress in properly implementing nuclear waste disposal; difficulties, for utilities in some countries, in transforming from a rapidly growing industry to routine operation of ageing facilities; electricity supply deregulation; and increased competition from natural gas.

The turn of the century is potentially a turning point for nuclear power prospects because of (a) increasing world energy consumption, nuclear power's contribution to reducing greenhouse gas emission, nuclear fuel resources sustainability, and improvements in operation of current nuclear power plants; (b) advanced reactor designs that will improve economics and availability, and further enhance safety; and (c) continued strengthening of the nuclear power safeguards system.

This presentation focuses on general nuclear energy issues, the status of the development of advanced reactor designs and the IAEA activities supporting this development.

THE GLOBAL ISSUE OF ENERGY SUPPLY

Today's global pattern of energy supply is not sustainable. The provision of affordable energy services is a fundamental prerequisite for economic growth and development. The plentiful energy resources in the past and the enormous efforts in research, development and engineering have produced the high living standards, enjoyed today, by the industrialised countries. For these countries achieving economically, environmentally, and socially sustainable development is a top priority.

Today and in the near future, better living standards and increased employment opportunities for the developing countries are inevitably linked to the provision of substantially more energy services. Taking into account the population growth, the anticipated increase in energy services will require more than twice as much energy production over the next half century. Over the next two decades India plans to triple and China to double the combustion of coal for electricity generation alone. Where transmission and distribution infrastructures are already in place, natural gas will be the preferred fuel for electricity and heat generation and for households. With the increase in the income per capita, and with growing trade volumes in a global market place, the demand for oil product fuelled transportation will expand rapidly.

There is an international consensus that heavy dependency on fossil fuels—which today account for more than 85 per cent of the total energy supply—must be controlled. Their use adversely affects the atmosphere through emissions of greenhouse gases along with noxious gases and toxic pollutants thus becoming an obstacle to sustainable development both on a regional and on a global scale. A specific feature of fossil resources is their uneven distribution around the globe. For example 60 per cent of the proven oil reserves are in the Middle

East. 40 per cent of the gas resources are in the countries of the former Soviet Union (FSU) and 40 per cent are in the Middle East. Coal is also very unevenly distributed with more than 80 per cent of the proven reserves being concentrated in three regions: North America, FSU and China. The uneven distribution of fossil fuel resources and the high cost of transport systems and infrastructures will be an additional issue to be taken into account when deciding on future energy supply.

While energy efficiency—in generation, transmission and end use, and the new renewable technologies are an essential element of the sustainable energy policy of the industrialised countries, they may be far from sufficient and in many cases even inadequate to compensate for the expected increase in the demand for energy in the rest of the world.

The global challenge is to develop strategies that foster a sustainable energy future, less dependent on fossil fuels.

THE NUCLEAR POWER POTENTIAL

Though it is not problem free, nuclear power is recognised as having an advantage in contributing to the goals of sustainable development. It has been deployed in the industrialised countries when energy supplies have been insecure and has largely contributed to the stable and predictable energy supply necessary for their economic growth.

From today's point of view of sustainability criteria, the entire energy chain from nuclear fuel production to radioactive waste disposal has very limited emissions of greenhouse gases and other pollutants and practically does no harm to the environment.

Further it is an established technology and commercially available—currently 473 nuclear plants are operating or being built in 32 countries. It can also be reasonably competitive, the resources it uses are plentiful and have no other useful application.

Nuclear power is unevenly used in the world. More than 95 per cent of it is deployed in the industrialised countries and the countries of Central and Eastern Europe and Russia. But the contribution of nuclear energy in the energy mix of the rest of the world and in particular the developing countries, is very small. While nuclear power has reached a level of saturation in several European countries and North America, it continues to expand in Asia. At the same time countries in Eastern Europe and the FSU, heavily dependent on nuclear power, are experiencing serious difficulties due to a breakdown in the economies and the infrastructure necessary to keep the nuclear power plants operational and to further expand their nuclear power programmes.

The future will see a mix of energy sources. The make-up of this mix cannot be precisely defined—it will depend not only on environmental considerations, but also on technological, political, and market factors. The experience to date shows that in most of the countries which have reached a quasi-sustainable level of development, nuclear energy has played an important role in supplying a part of the required energy. Most of these countries will try to preserve their nuclear energy generation and capability and probably will seek to renew it in the future when the life of the current plants is exhausted. Inevitably, the countries whose economies will continue to grow rapidly, will be better placed to include nuclear power in their energy supply system for meeting their energy needs but also for security of supply, environmental awareness and access to high technology.

THE KEY NUCLEAR POWER ISSUES

In an increasingly competitive and international global energy market a number of issues will affect not only the choice of nuclear power, but also the extent and manner in which it will be used in a sustainable mix of energy sources:

- Enhancing reactor safety
- Improving nuclear power generation economics
- Minimising environmental impact
- Improving resource utilisation

Let me take these issues one by one.

Enhancing reactor safety: With over 8500 reactor year of operation world wide, nuclear power generally has an excellent safety record. But the Chernobyl accident demonstrated that one very severe nuclear accident has a potential to cause national and regional radioactive contamination. Although safety and environmental impacts are becoming a key issue for all energy sources, many in the general public perceive nuclear power as particularly and intrinsically unsafe. In order to reduce the risk of accidents a number of approaches are used:

- International collaboration to promote internationally accepted safety and engineering standards.

- Enhancement of the integrity of the reactor vessel and reactor systems (such as double containment)
- Development of advanced reactor designs with enhanced safety systems

Unquestionably, the most convincing demonstration of safety will be through the safe performance of existing plants and the avoidance of any major incident in the future.

Improving nuclear power generation economics: The present challenge for nuclear power is economics—Can nuclear power in private ownership in a deregulated industry remain competitive? Success in meeting this challenge is critical to maintaining a role for nuclear power as a viable energy potion. Without getting the economics right, its potential environmental benefits may well become irrelevant. Nuclear power plants will increasingly have to compete directly, in an open energy market, with other suppliers of electricity. This competitive environment has significant implications for plant operations including among others, the need for efficient use of all resources, including personnel; more effective management of plant activities, such as outages and maintenance; and sharing of resources, facilities and services among utilities. The ultimate objective is to provide electricity services at competitive costs without compromising operational safety.

As fuel costs are relatively low, reduction of overall costs by decreasing construction, operation and initial financing expenses for new reactor designs is essential to the overall economic viability of nuclear energy.

Nuclear energy also has the potential to provide an economic source of heat for non-electric applications including district heating, desalination and high temperature process heat especially through development and application of small and medium sized reactors.

Minimising environmental impact: Although nuclear energy has distinct advantages over today's fossil burning systems—in terms of fuel consumed, pollutants emitted and waste produced—a further reduction in environmental concerns can positively influence public attitudes. As the overall health and environmental impact of the reactor and nuclear fuel cycle is small, attention is directed at techniques to deal with spent fuel, accumulated plutonium and radioactive waste. Reprocessing of spent fuel and recycling of most dangerous actinides in future fast reactor is analysed in some countries as a solution for the fuel cycle back-end issues.

Improving resource utilisation: Known and likely resources of uranium should assure a sufficient nuclear fuel supply in short and medium term even with reactors operating primarily on once through cycles with disposal of spent fuel. However, as uranium demand increases and reserves are decreased to meet the requirements of increased nuclear capacity, there will be economic pressure for the optimal use of uranium in a manner that utilises its total potential energy content per unit quantity of ore. Recycling of generated plutonium in thermal reactors and introduction of fast reactors in the longer term is considered in some countries as a solution.

HIGHLIGHTS OF GLOBAL DEVELOPMENT OF ADVANCED NUCLEAR POWER PLANTS

World-wide, considerable efforts are being made to develop advanced nuclear power plants with the aim to address the above mentioned issues. Various organisations are involved, including governments, industries, utilities, universities, national laboratories, and research institutes. Expenditures of development of new designs, technology improvements, and the related research of the major reactor types combined is estimated to exceed US$ 2 billion per year.

The full spectrum of advanced nuclear power plant designs of concepts covers different types of designs—evolutionary ones, as well as innovative designs that require substantial development efforts. A natural dividing line between these two categories arises from the necessity of having to build and operate a prototype or demonstration plant to bring an innovative concept to commercial maturity, since such a plant represents the major part of the resources needed. Designs in both categories need engineering, and may also need research and development (R&D) and confirmatory testing prior to freezing the design of either the first plant of a given line in the evolutionary category, or of the prototype and/or demonstration plant for the second category. The amount of such R&D and confirmatory testing depends on the degree of both the innovation to be introduced and the related work already done, or the experience that can be built upon. This is particularly true for designs in the second category where it is entirely possible that all a concept needs is a demonstration plant, if development and confirmatory is essentially completed. At the other extreme, R&D, feasibility tests, confirmatory testing, and a prototype and/or demonstration plant are needed in addition to engineering. Different tasks have to be accomplished and their corresponding costs in qualitative terms are a function of the degree of departure from existing designs. In particular, a step increase in cost arises from the need to build a reactor as part of the development programme.

Through activities within the framework of its nuclear power programme, the IAEA is serving as an international source of objective reference information about the different concepts being developed and the project status, as well as typical development trends throughout the world.

OVERVIEW OF WATER-COOLED REACTOR DEVELOPMENT PROGRAMMES

To assure that nuclear power remains a viable option in meeting energy demands in near and medium terms, evolutionary water cooled reactor designs, aimed at achieving certain improvements over existing designs through small to moderate modifications, are being developed in a number of countries. Utility requirement documents have been formulated to guide these activities by incorporating experience from current plants with the aim of reducing costs and licensing uncertainties by establishing a technical foundation for the advanced designs. Large evolutionary units with power outputs of 1300 MWe and above, which incorporate proven, active engineered systems to accomplish safety functions, and mid-size evolutionary design, which place more emphasis on utilisation of passive safety systems, are being developed. Common goals for these new designs are high availability, user-friendly features, competitive economics and compliance with internationally recognised safety objectives.

ADVANCED LIGHT-WATER REACTORS (ALWRs):

Designs for ALWRs are being developed in a number of countries as summarised below.

United States: Important programmes in development of ALWRs were initiated in the mid-1980s in the United States. In 1984, the Electric Power Research Institute (EPRI), in co-operation with the US Department of Energy and participation of US nuclear plant designers, initiated a programme to develop utility requirements fro ALWRs to guide their design and development. Several foreign Companies have also participated in, and contributed funding to, the programme. Utility requirements were established for large boiling-water reactors (BWRs) and pressurised-water reactors (PWRs) having power ratings of 1200 to 1300 MWe, and for mid-sized BWRs and PWRs having power ratings of about 600 MWe.

In 1986, the US Department of Energy, in co-operation with EPRI and reactor design organisations, initiated a design certification programme for evolutionary plants based on a new licensing process, followed in 1990 with a design certification programme for mid-size plants with passive safety systems.

In May of 1997 the System 80 + plant of ABB Combustion Engineering and the ABWR plant of General Electric received Design Certification from the U.S. Nuclear Regulatory Commission highlighting the success of the U.S. Department of Energy's design certification programme for large evolutionary plants. The Westinghouse AP-600 design, which has been submitted to the U.S. NRC under the DOE's design certification programme for mid-size plants with passive safety systems, which was initiated by DOE in 1990, is expected to receive Final Design Approval in 1998. Also the first-of-a-kind-engineering, FOAKE, programme has been completed for the ABWR in September 1996 and similar work on the AP-600 is scheduled for completion in 1998. The power company in Taiwan, China, recently selected the USA's ABWR design for two new units planned to go into operation in 2004.

France and Germany: In Europe, Framatome and Siemens with their joint company, Nuclear Power International, together with Electricité de France and the group of "nuclear" German utilities have worked out the design of a new advanced reactor, the European pressurised-water reactor (EPR), a 1525-MWe plant with enhanced safety features. This reactor is designed to meet the common set of utility requirements endorsed by the major utilities of the European countries as expressed in the European Utilities Requirements (EUR) document. The "Basic Design Report" was issued in May 1997 and marked the successful completion of a key milestone, and the design is being reviewed jointly by the French and German safety authorities. This procedure will provide strong motivation for the practical harmonisation of the safety requirements of two major countries, which could later be enlarged on a broader basis. Siemens is also, together with German utilities, engaged in the development of an advanced BWR design, the SWR-1000, which will incorporated a number of passive safety features, for initiation of safety functions, for residual heat removal, and for containment heat removal. This effort is currently in the basic design stage.

Sweden and Finland: In Sweden, ABB Atom, with involvement of the utility Teollisuuden Voima Oy (TVO) of Finland, is developing the BWR-90 as an upgraded version of the BWRs operating in both countries.

Republic of Korea: In the Republic of Korea, an effort started in 1992 to develop an advanced design known as the Korean Next Generation Reactor (KNGR), a 4000-MWth PWR design. The basic design is currently being developed by the Korea Electric Power Corporation (KEPCO) with support of the Korean nuclear industry. The goal is to complete a detailed standard design by the year 2000.

Russian Federation: In the Russian Federation, design work is under way on the evolutionary V-392, an upgraded version of the VVER-1000, and another design version is being developed in co-operation with the Finnish Company Imatran Voima Oy (IVO). Also being developed is a mid-sized plant, the VVER-640 (V-407), an evolutionary design which incorporates passive safety systems, and the VPBER-600, which is a more innovative, integral design. Construction of the first unit of the VVER-640 is planned in Russia and Kazakhistan. Construction of 1000-MWe VVERs is being planned in the People's Republic of China, India and Iran.

Japan: In Japan, the development of the ABWR started in 1978 as an international co-operation between five BWR vendors. The resulting conceptual design was received favourably by TEPCO and other Japanese utilities, and as a result, the ABWR was included in the third standardisation programme of

Japan from 1981. Preliminary design and numerous development and verification tests were carried out by Toshiba, Hitachi and GE together with six Japanese utilities and the Japanese Government. Two ABWRs, the Kashiwazaki-Kariwa units 6 and 7 were subsequently ordered by TEPCO and have been successfully taken into commercial operation in November 1996 and July 1997 respectively.

The Ministry of Trade and Industry is conducting an "LWR Technology Sophistication" programme focusing on development of future LWRs and including requirements and design objectives. A large, evolutionary 1350-MWe advanced PWR is being developed by Japanese utilities together with nuclear vendors, with construction of a twin unit being planned at the Tsuruga site. In addition, an advanced BWR Improvement and Evolution study was started in 1991. It involves development of a reference 1500-MWe BWR that reflects the accumulated experience in operation and maintenance of BWRs. Also in progress are development programmes for a Japanese simplified BWR (JSBWR) and PWR (JSBWR), projects which involve vendors and utilities. The Japan Atomic Energy Research Institute (JAERI) has been investigating conceptual designs of advanced water-cooler reactors with emphasis on passive safety systems. These are the JAERI Passive Safety Reactor (JPSR) and the System-Integrated PWR (SPWR).

China: In China, the Nuclear Power Institute (Chengdu) is developing the AC-600 advanced PWR, which incorporates passive safety systems for heat removal.

In all of these countries, the advanced LWRs under development incorporate significant design simplifications, increased design margins, and various technical and operational procedure improvements. These include better fuel performance and higher burn-up, a better man-machine interface using components and improved information displays, greater plant standardisation, improved constructability and maintainability, and better operator qualification and simulator training.

Heavy-water cooled reactors (HWRs): In addition to light water-cooled reactors, the technology for HWRs has also proven to be economic, safe, and reliable. Approximately 7 per cent of all current operating plants are HWRs. A mature infrastructure and regulatory base has been established in several countries, notably in Canada, the pioneer in the development of the HWR concept. Two types of commercial HWRs have been developed, the pressure tube and the pressure vessels versions, and both have been fully proven. HWRs with power ratings from a few hundred MWe up to approximately 900 MWe are available. The heavy-water moderation yields a good neutron economy and has made it possible to utilise natural uranium as fuel which leads to lower fuel costs compared with LWRs. The amount of fissile material is quite limited, however, and the pressure tube designs are therefore using on-load refuelling to achieve adequate reactivity for the plant operation. The effectiveness of this on-load refuelling has been successfully demonstrated; the annual and lifetime load factors of most of the pressure tube HWRs have been among the best of all commercial reactor types. Safety performance has also proven to be very good.

Canada: The continuing design and development programme for HWRs in Canada are primarily aimed at reduction of plant costs and at an evolutionary enhancement of plant performance and safety. Two new 715-MWe CANDU- 6 units with improvements over earlier versions of this model are under construction in Quishan, China. Up-front basic engineering continues on the 935-MWe CANDU-9 reactor, a single unit adaptation of reactor units operating in Darlington, Canada. The two year licensability review by the Canadian Nuclear Safety Commission was completed in January 1997, and found that the CANDU-9 meets the country's licensing requirements. Further studies are being carried out for advanced versions of these reactor models to incorporate further evolutionary improvements and to increase the output of the larger reactor up to 1300 MWe.

India: Also under development is an advanced 500-MWe HWR in India, and construction of such units is planned. This HWR design takes advantage of experience feedback from the 220-MWe HWR plants of indigenous design operating in India.

OVERVIEW OF GAS-COOLER REACTOR DEVELOPMENT PROGRAMMES

Significant activities are occurring in the development of high-temperature gas-cooled Reactors (HTGRs), particularly with regard to the utilisation of the gas-cooler reactor to achieve economically competitive systems for highly efficient generation of electricity and for process heat applications. Technological advances in component design and processes—coupled with the international capability to fabricate, test, and procure the components—provides and excellent opportunity for achieving HTGR commercialisation.

United Kingdom, Germany, and United States: Gas-cooler reactor have been in operation for many years. In the United Kingdom, nuclear electricity is mostly generated in CO_2-cooled Magnox and Advanced Gas-Cooled Reactors (AGRs). Other countries also have pursued development of high-temperature reactors (HTGRs) with helium as coolant, and graphite as moderator. The 13-MWe AVR reactor was successfully operated for 21 years in Germany demonstrating application of HTGR technology for electric power production. Other helium-cooled, graphite-moderated reactors have included the 300-MWe Thorium Rich Temperature Reactor in Germany, and the 40-MWe Peach Bottom and 330-MWe Fort St. Vrain plants in the United States.

Russia Federation, United States, France, Japan: MINATOM, General Atomics, Framatome and Fuji Electric have combined their efforts for the co-operative development of the gas turbine-modular helium reactor (GT-MHR). This plant features a 600 MWth helium cooled reactor as the energy source coupled to a closed cycle gas turbine power conversion system. The net efficiency of this advanced nuclear power concept is 47 per cent. Substantial progress in the development of components such as magnetic bearings and fin-plate recuperators make this type of HTGR plant a feasible alternative for future consideration in the production of electricity. This plant is under consideration for the purposes of burning plutonium and for commercial deployment.

South Africa: In South Africa, the large national utility, Eskom, which has an installed generation capacity of about 38,000 MWe, is in the process of performing a technical and economic evaluation of a helium-cooled pebble bed modular reactor. It would be directly coupled to a gas turbine power conversion system for consideration in increasing the capacity of the utility's electrical system.

China and Japan: In China a test reactor is under construction which will have the capability of achieving core outlet temperatures of 950ºC for the evaluation of nuclear process heat applications. Construction of China's High Temperature Reactor (HTR-10) at the Institute of Nuclear Energy Technology (INET) continues with initial criticality anticipated for 1999. This pebble bed reactor of 10 MWth will the utilised to test and demonstrate the technology and safety features of the HTGR. Development of the HTGR by INET is being undertaken to evaluate a wide range of applications. They include electricity generation, steam and district heat production, combined steam and gas turbine cycle operation, and the generation of process heat for methane reforming. The HTR-10 is the first HTGR to be licensed and constructed in China.

The principal focus of Japan's HTGR R&D programme is completion of the High Temperature Engineering Test Reactor (HTTR) at the Japan Atomic Energy Research Institute (JAERI) site in Oarai, Japan. This 30-MWth helium-cooled reactor will be utilised to establish and upgrade the technology for advanced HTGR development, and to demonstrate the effectiveness of selected high temperature heat utilisation systems. Initial criticality of the HTTR is anticipated by mid 1998. The start-up physics testing programme for the HTTR will then continue through the remainder of 1998.

OVERVIEW OF LIQUID METAL–COOLED REACTOR DEVELOPMENT PROGRAMME

The feasibility of a nuclear chain reaction involving only fast neutrons was recognised from the earliest days of nuclear power in the 1940s. It was realised that fast reactors would have potential advantages over thermal reactors because excess neutrons would be available, which could be used for breeding fissile material from fertile material. This provided the key to utilising the enormous world-wide energy reserve represented by uranium-238.

The development of civil fast reactors started in several countries, notably the USA, the USSR, the UK and France, in the late 1940s. This involved test reactors such as CLEMENTINE and EBR-1 in the USA and BR-2 in the USSR.

Subsequently demonstration reactors such as EBR-2 (USA), BOR-60 (USSR), Rapsodie (France) and DFR (UK) were constructed in the 1950s and 1960s, leading to prototype power reactors such Phenix (France), PFR (UK), and BN-350 (USSR-

Kazakhstan), and finally full-scale power plants SPX (France) and BN-600 (USSR-Russia).

In the earliest days great importance was attached to the breeding of fissile material, but the increasing availability of cheap fissile fuels from the 1980s onwards shifted the emphasis to include other uses for fast reactors both in the critical and subcritical mode, particularly for the control of plutonium stocks and the treatment of radioactive wastes. In spite of these additional functions the main long term importance of fast reactors as breeders, essential to world energy supplies, remains unchanged.

Significant technology development programmes for LMFRs are proceeding in several countries, notable France (in co-operation with smaller efforts in Germany, the United Kingdom, and other European countries), Japan, India, and the Russian Federation. Activities continue in some other countries at a lower level.

China: In China, the basic research work on LMFRs was started in 1964. Since then and up to 1987, the major work has been on neutronics, thermal hydraulic, and sodium technology. During 1991–92, the conceptual design of the 15-MWe Chinese experimental fast reactor (CEFR) was completed and during 1992–93, the conceptual design was confirmed and optimisation studies were carried out. Since 1993 onwards, major work has involved the preparation of a detailed design.

France: In France, the commercial introduction of LMFRs is being postponed. Meanwhile, the application of an additional important aspect of these reactors—to transmute long-lived nuclear waste and to burn plutonium—is being developed (CAPRA project). The current programmes on operation of the 350 MWth Phenix reactor reflect these requirements. One objective of extending the lifetime of the Phenix reactor by another 10 years is to perform the necessary irradiation experiments. The 1200 MWe Superphenix reactor has recently been shut down.

India: In India the fast-breeder test reactor (FBTR) is in operation. Fuel development, material irradiation, and sodium technology are the principal technical programmes. The introduction of FBRs is linked to their economic acceptability. The basic design features are now selected for the 500-MWe Prototype Fast Breeder Reactor (PFBR). The emphasis in 1997–98 is on detailed design, engineering development, sodium technology, and materials technology. Reduction in construction time is an important target.

Japan: In Japan, the prototype LMFR "Monju" with the capacity of 280-MWe reached initial criticality in April 1994 and was connected to the grid in August 1995. Reactor operation was interrupted in December 1995 due to a leak in the non-radioactive secondary cooling system. The design of a 660-MWe demonstration fast-breeder reactor (DFBR), which is expected to be constructed in the beginning of the next century, is in progress. In addition to this main stream of development work, studies are being performed regarding the development of technology capable of meeting the diverse needs of future society. These needs include the reduction of environmental impacts and the assurance of nuclear non-proliferation, demands that widen the technological options.

Republic of Korea: The Republic of Korea plans to develop the conceptual design of its first fast-breeder reactor, the 330-MWe Kalimer plant, by 2001. Construction is planned to enable achievement of criticality during 2011.

Russian Federation: Russia's experience in the operation of experimental and prototype fast reactor (the BR-10, BOR-60, and BN-600) has been very good. Efforts are directed towards improving safety and reliability and making the LMFRs economically competitive to other energy sources. While these efforts would take some time, the use of LMFRs over the near-term to brun plutonium and minor actinides is foreseen on the basis of new BN-800 reactors.

United States: A promising integral fast reactor (IFR) concept, comprising the LMFR and its entire fuel cycle, has been developed at the Argonne National Laboratory (ANL) and General Electric Company over the past two decades. A distinguishing element of the IFR concept is its unique fuel cycle based on metallic fuel and pyrometallurgical processing. At ANL, an effective fuel cycle technique has been developed whereby spent fuel is reprocessed and new fuel is fabricated at the reactor sites. The plutonium is not separated from the higher radioactive actinides; these are recycled together in the reactor and never leave the reactor site. Advantages of the IFR system are in areas of (i) fuel performance, (ii) passive safety, (iii) economics, (iv) waste management potential, and (v) proliferation resistance. When the first LMFRs were constructed in the 1950s, it was expected that the burn-up potential of metallic fuel would be limited to a brun-up of 3-5 per cent of heavy atoms (ha). In fact, by using an U-Pu-1- per cent Zr alloy and ferritic-martenistic HT9 cladding and duct, a burn-up of about 20 per cent has been achieved in the USA EBR-II. All irradiation results in the EBR-II and FFTF have demonstrated reliable performance of metallic fuel and the potential to achieve high burn-up in prototypical fuel elements.

SMALL AND MEDIUM SIZED REACTORS

Small and medium sized reactors (SMRs) are of particular interest for non-electrical applications of nuclear energy, such as desalination of seawater and district heating. But SMRs are also a suitable option for electricity generation in countries with small electricity grids or for remotely located areas such as for populated small islands.

In several countries there is an emerging interest in small and medium sized reactors which do not require on-site refuelling. These reactors may either have a long core life-time or are returned to the vendor for maintenance and refuelling.

Barge mounted reactors are a typical example of this SMR category. They could be supplied to countries having an immediate need for energy, which do not yet have a fully implemented nuclear infrastructure as needed for large sized plants. Barge mounted reactors could be operated under the auspices of the vendor and since no on-site refuelling is required they create a new approach to proliferation resistance. Transportable barge mounted SMRs offer at the same time the potential for the reduction of the financial risk compared to conventional nuclear power plants.

Currently a project with barge mounted reactors is being implemented in the Russian Federation to supply electricity and heat for the northern part of Siberia. The reactor considered in the frame of this project is a redesigned version of the proven KLT-40, which has been used for the propulsion of ice-breakers.

IAEA ACTIVITIES SUPPORTING DEVELOPMENT OF NUCLEAR POWER TECHNOLOGY

As an international forum for exchange of scientific and technical information, the IAEA plays a role in bringing together experts for a world-wide exchange of information about national programmes, trends in safety and user requirements, the impact of safety objectives on plant design, and the co-ordination of research programmes in advanced reactor technology.

Activities in areas of nuclear power technology development are based on the advice of International Working Groups (IWGs) consisting of leading representatives of national programmes and international organisations for each major type of reactor.

To support its information exchange function, and to provide balanced and objective information to all Member States on advances in reactor technology, the IAEA has recently prepared two technical documents—Status of Advanced Light Water Reactor Designs and Fast Reactor Database. Other recent documents address improvements in reactor system components and technologies for improving availability and reliability of current and future reactors.

Enhancing Communication: In view of the importance of communication to the public, and the technical community in general, consistency and international consensus are desirable with regard to the terms used to describe advanced reactor designs. In 1991, drawing on advice from reactor design organisations, research institutes, and government organisations, the IAEA issued a document entitled Safety Related Terms for Advanced Nuclear Power Plants that is being widely used. More recently, using the same approach to obtain advice from involved parties, the IAEA published Terms for Describing New Advanced Nuclear Power Plants. The document's purpose is to clarify the meaning of terms by drawing distinctions between design stages reflecting the maturities of designs.

Co-operative Research: The IWGs advised the IAEA to establish international co-operative research programmes in areas of common interest. These co-operative efforts are carried out through Co-ordinated Research Programmes (CRPs), which typically are three to six years in duration, and often involve experimental activities. Such CRPs allow a sharing of efforts on an international basis and benefit from the experience and expertise of researchers from the participating institutes.

As an example, the IAEA has co-ordinated work to collect and systematise a database of thermophysical properties for a broad spectrum of light- and heavy-water reactor materials over a wide range of temperatures; the database has been published. For liquid metal-cooled reactors, the results of co-operative activities on material behaviour also have been published recently. In other co-operative programmes, the IAEA is establishing sets of thermohydraulic relationships for water-cooled reactors and liquid metal-cooler reactors appropriate for use in analysing reactor performance and safety. In the field of gas-cooled reactors, the principle focus has been on the four specific technical areas which are predicted to provide advanced HTGRs with a high degree of safety, but which must be proven. These technical areas are: the safe neutron physics behaviour of the reactor core; reliance on ceramic coated fuel particles to retain fission products even under extreme accident conditions; the ability of the designs to dissipate decay heat by natural heat transport mechanisms, and the safe behaviour of the fuel and reactor core under chemical attack (air or water ingress). Activities in HTGR applications focus on design and evaluation of heat utilisation systems for the Japanese HTTR.

All these activities are indicative of global co-operation for the development of advanced types of nuclear power reactors. As countries move ahead with plans, and new plants are introduced, further enhancements can be expected in areas of plant economics, reliability, and safety. Through its International Working Groups on advanced reactors, the IAEA will be encouraging the international exchange of information on non-commercial technology and co-operative research. It will also assist countries in harmonising user requirements, and in the preservation of key technological data on advanced nuclear power systems.

References

1. Sustainable development and nuclear power, IAEA, Vienna, 1997.
2. Status of advanced light water cooled reactor designs: 1993, IAEA-TECDOC-968 (IAEA, Vienna, September 1997).
3. Progress in liquid metal fast reactor technology: 1996, IAEA-TECDOC-876 (IAEA, Vienna, April 1996).
4. Design and development of gas cooler reactors with closed cycle gas turbines, IAEA-TECDOC-899 (IAEA, Vienna, August 1996).
5. Advances in HWR technology, IAEA-TECDOC-984(IAEA, Vienna, November 1997).
6. Design and development status of small and medium reactor systems in 1995, IAEA-TACDOC-881 (IAEA, Vienna, May 1996).

30

Towards Modular, Inherently Safe, Passive Reactors

M.CUMO
Italy

THE SEARCH FOR NEW DESIGN CRITERIA

In the field of nuclear power plants, unlike what happened in other industrial fields, the technological evolution has rarely led to better performances with lower costs.

The evolution of the design criteria and of projects' characteristics in the nuclear field has led so far to more advanced performances at the cost of an increasing complexity of the systems. Plant costs, consequently, increased dramatically. The evolution of safety requirements contributed substantially to the increase of complexity and costs.

While the first-generation nuclear plants (50–60s) were mainly based on traditional codes and regulations, presently special nuclear codes and regulations are available (10 CFR 50, etc.) and the design and construction of a new nuclear plant must obey the following "new" conditions concerning the accident analysis and defining substantial inputs to the emergency core cooling systems' performance:

- compliance with developed codes for pressure components, specifically oriented to the nuclear field, including also the construction (ASME III) and operation phases (ASME XI: in-service inspection);
- application of the hypothesis of rupture of piping containing high-energy fluids, with the consequent protection of safety-relevant plant parts from the effects of pipe whips, jet impingement, flooding, etc.;
- design of the plant to face external human events (sabotage, aircraft crashes, explosions, etc.) and catastrophic natural events, with obvious consequences for the design

and qualification of components, systems, structures;
- design of plants and structures so as to make radiation doses to personnel during normal operation "as low as reasonably achievable";
- application of the post-Three-Mile-Island experience: review of the role of the plant operators, with consequent modification of the interface man/plant, adoption of instrumentation and control systems for preventive diagnostics, adoption of sampling systems even during accidents, etc.;
- adoption of probabilistic risk assessment methodologies and application of the assumption of combination of potential events able to lead to the core damage.

The Chernobyl accident has confirmed the importance and the dangerousness of the human factor; the need of being able to manage even severe accidents is felt by many people, and new tendencies in licensing processes are recognised in several countries.

During the last years the plants' complexity increased very much; nevertheless, the exasperation of the complexity of a plant system may not always be the best solution of the problem of the safety guarantee. In fact, the complexity may cause the invalidation of two aspects representing the pre-supposition of the "inherent safety" of a plant: its reliability and simplicity.

In the aim of researching the essential design criteria of a nuclear plant of small-medium size, equipped with an "inherently safe" nuclear reactor, to be used as a multi-purpose heat generator, a research group has operated since 1983 at the University of Rome "La Sapienza", later enforced by a research and design group of the Italian Agency for Energy, Environment and Innovations (ENEA). This team performed a critical review of the design criteria internationally adopted in the nuclear field, in order to find new plant solutions able to satisfy the requirements of high safety standards, reduced production costs, together with a high reliability and a substantial system simplicity.

In this context, new design criteria never adopted previously were considered, and some were adopted as main design criteria for the new plant (for instance, the criteria of the inherent capability of the plant to shut down the reactor in anomalous conditions of enthalpy for the primary coolant, or to cool down the reactor in faulted conditions for the main coolant circulation). Furthermore, considerable importance was given from the very first steps of the new design development to aspects that were previously given importance mainly at the latest phases of the plant design (as the criterion of the low doses to personnel).

The possibility of guaranteeing the residual-heat removal in emergency conditions by means of a completely passive system greatly increases the plant inherent safety degree, allowing the selection of a very simple (and, therefore, reliable) circuitry of the whole system containing the primary coolant.

The "new" residual-heat removal emergency system is based on natural circulation of fluids. Pumps, motors, other active electric components, emergency diesel generators with annexed auxiliaries are avoided. The availability of an adequate heat sink with an infinite capacity (external air) guarantees a theoretically never ending cooling, without the need of external water feed requirements.

Such a choice considerably affects the whole plant lay-out and design, which is substantially different from those of "traditional" big-size plants.

With the exclusion on accidents due to fuel element handling, the maximum number of potential nuclear accidents in traditional Light Water Reactor plants is originated by the coincidence of a high pressure and a high specific internal energy in the core cooling fluid. In the worst accidents (large and small loss of coolant accidents, control rod ejection, etc.), the origin of the deterioration of the core cooling is a consequence of ruptures in the primary-coolant pressure boundary.

In these plants, the possible accidental evolutions are different with respect to large or small ruptures, and the configuration of emergency systems called for the intervention is different as well.

One of the main goals considered is the removal of possible causes of rupture of the pressure boundary and, therefore, also the removal of causes of rupture of the emergency cooling system pressurised boundary. This has been achieved through the adoption of a null pressure difference between the primary coolant and the environment housing the primary coolant boundary.

This solution allows the operation of the primary system

boundary (including the reactor vessel) in the absence of primary stresses (ASME III), eliminating the possibility of ruptures.

The primary cooling system shall be wholly contained in a low-enthalpy fluid pressurised containment.

Even if, as a consequence of catastrophic events, the external containment should break, the loss of the low-enthalpy containment fluid would be very low, depending on the volume of a pressurised gas cushion; this loss would not be dangerous for the core and could permit a safe shutdown in an intact pressure boundary (the primary-coolant boundary).

On the other hand, losses from the primary cooling system into the pressurised external containment, would be extremely unprobable and without consequences, owing to the low flow rates.

The adoption of a low-enthalpy water pressurised containment, solving *a priori* the problem of ruptures in the primary-coolant pressure boundary, exalts the plant inherent safety already achieved by means of the completely passive emergency core-cooling system.

MARS, Multi-purpose Advanced Reactor inherently Safe, is the name of the design above described. Its purposes have been outlined regarding the passive and inherent safety characteristics. Let's consider now other, sometimes interconnected purposes, which may be of no lesser importance.

It is well known that nuclear electricity costs are capital intensive so that there is a strong convenience in the reduction of the construction period with the related scalar interests. Being extremely time consuming the constructions on the site, the solution envisaged is, with the obvious exclusion of the civil works, a construction entirely in factories. This requires components with flanged connections which may be easily assembled once transported to the site. In the same time the main controls and inspections may be performed within the factories, with greater easiness, better reliability and shorter items. The flanged connections represent one of the main characteristics of the MARS project and are possible for the external pressurised containment. Another important item is the modularisation of the MARS concept, so that many units become complementary for a given power station which may gradually increase its capacity, following the network requirements much more closely and reducing the monetary risks. Useless to say that modular plants are particularly fit for factory works and for chain productions with the special controls and quality assurance characteristics required by nuclear components. This shift from site to factories may be indeed a revolutionary change in the nuclear industry. The plant life extension, to maximise the utility profit, is easily achieved with the flanged connections of all the main components through the quick substitution, if necessary, of the single defective component without impairing the whole plant. The life extension may be so prolonged till extreme values.

In the same way the decommissioning operations, often financially uncertain and risky, may be promptly and easily performed. In the MARS design, for this reason all the parts within the reactor building are realised in assembled steel parts which may be disconnected. Even the biological screen around the core is composed by blocks of steel and borated paraffin.

A gold rule in nuclear engineering is to maximise the experience up to now matured: being the pressurised water reactors the most experienced ones, the MARS reactor is a pressurised reactor which, in normal operation, behaves like a traditional PWR. The added parts pertain only to abnormal and accidental situations with passive devices like the additive control rod bank for the scram, the special check valves for the emergency cooling loop, the natural convection loops for the residual heat removal for an indefinite time.

The physical elimination of the loss of coolant accidents and of the control rod ejection accident, due to the external pressurised containment, is a decisive step towards safety, which entails as well the saving of a reinforced reactor building against internal accidents. This building, which may be mainly underground, has the task to protect the plant against external dangerous events (by nature or man). Even a physically impossible accident, the core melt, has been studied. In this case the melted core would be frozen in the bottom of the internal pressure vessel, due to the presence of cooling water in the external shroud.

Due to the "traditional" components adopted in the MARS design, an evaluation of the cost of the plant is not difficult or uncertain. This has been performed as well, leading to encouraging results. It is well known that small plants suffer for a scale factor in the cost of the unit product (in this case electricity and heat). But this is true for fixed designs in which mainly the scale is changed. If design changes and substantial simplifications are operated, the final results may even reverse this trend.

In conclusion, the proposed solution seems to have many characteristics of interest for a market sector which comprises small utilities and reduced electricity networks, as well to increase, with co-generation, the overall efficiency. These small, modular, cogenerative plants in no way counteract the development of big plants which, in the western world, have reached outstanding safety and reliability records; they simply look to a different sector of the energy market, so to complement and strengthen the employment of the nuclear energy both for developed and developing countries.

Their fuel and fuel cycle facilities are the same; the same may be the plant builders. Their passive and inherent safety characteristics may enhance the public reliance in the only source of energy which presently may facilitate the solution of global problems.

THE MARS NUCLEAR POWER PLANT

The design was focused on a nuclear power generation capacity of 600 MWth, corresponding to about 150 MWe in the case of only electrical production, with a modular solution to satisfy progressively increasing power requirements from the station.

The design was carried out using a step-by-step approach, which addressed all major issues regarding plant safety, performance and cost.

Well-proven technological features of most experienced PWRs (geometry and materials of fuel assemblies; layout of the primary cooling system; geometry and materials of control rods and their drive mechanisms; core internals; etc.) have been chosen in the design of the primary coolant system.

This criterion has been settled together with the requirements imposed by a design that has its own characteristics, depending on the final use, size and destination market (lower operation pressure, reduced core thermal fluxes and, most of all, inherent safety characteristics of the safety decay heat removal system and of the additional scram system).

In order to make the MARS plant economically competitive with other nuclear plant solutions and with fossil-fueled thermal power plants, it was necessary to adopt a somewhat extreme simplification criterion in the design approach. This simplified design involves mainly in-shop construction with only a "few" operations of easy assembling and simple substitution of all mechanical components for maintenance purposes, and easy removal for the final, total decommissioning to be performed in the site.

The MARS reactor main data are shown in Table 1.

Table 1 : MARS Mod 6 Reactor Characteristic Data

Rated power	600	MWth
Inlet core coolant temperature	214	°C
Outlet core coolant temperature	254	°C
Rates pressure	75	bar
Fuel bundles	89	
Fuel rod array	17 × 17	
Fuel rods per fuel bundle	264	
Fuel rod external diameter	0.95	cm
Fuel rod active length	260	cm
Fuel rod cladding thickness	0.63	mm
Fuel rod pitch	1.26	cm
Control rod guide tube diameter	1.224	cm
Control rod guide tube thickness	0.4	mm
Control rod diameter	0.978	cm
Core heat transfer surface	1823	m^2
Core average linear power density	98.2	W/cm
Core average heat flux	32.9	W/cm^2
Core average volumetric power density	56.5	kW/litre

The MARS core is a heterogeneous one including 89 "standard" PWR fuel assemblies. The assemblies are zircalloy-cladded with rod array 17 × 17, including 264 fuel rods and 25 positions for zircalloy guide tubes for control rods (black, Ag-In-Cd; gray, stainless steel) or for burnable poisons (borosilicates).

The core has been optimised to limit the power peak factors (the ratios between local generated power and core average power) close to unity and to achieve criticality conditions with a little concentration of boron in the primary coolant (about one fourth than in traditional PWRs) in order to simplify the boron concentration control systems. This limitation also reduces the radioactive wastes produced by the plant.

The core is equipped with a standard shutdown system and with an additional, passive-type, shutdown system (ATSS) whose operation is based on the differential thermal expansion of a bimetallic sensor that causes, in case of an abnormal increase in the core temperature, the falling of neutron absorber rods in the core itself. (Fig. 1).

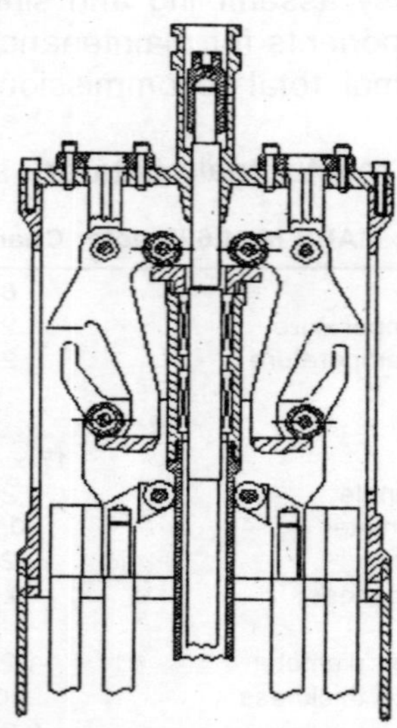

Fig. 1 : Operational Scheme and Self-Releasing Head of ATSS

The primary cooling system, shown in Fig. 2 and whose hydraulic scheme is shown in Fig. 3, includes one loop only, with 25" I.D. pipes and one vertical-axis U-tube steam generator.

Connected to the reactor vessel is the safety core cooling system (SCCS) whose function is to transfer the heat generated in the core to the external environment in case of unavailability of normal cooling systems, relaying only on natural circulation of fluids and using the external air as final heat sink.

The pressure inside the primary cooling system (75 bar) is controlled by a vapor-bubble pressuriser. On/off valves are installed in the primary cooling loop, in order to isolate, if necessary, the steam generator (S.G.) and the primary pump (i.e., in the event of a SG tube rupture).

The primary cooling system and the safety core cooling system are inside a pressurised containment (CPP), filled with

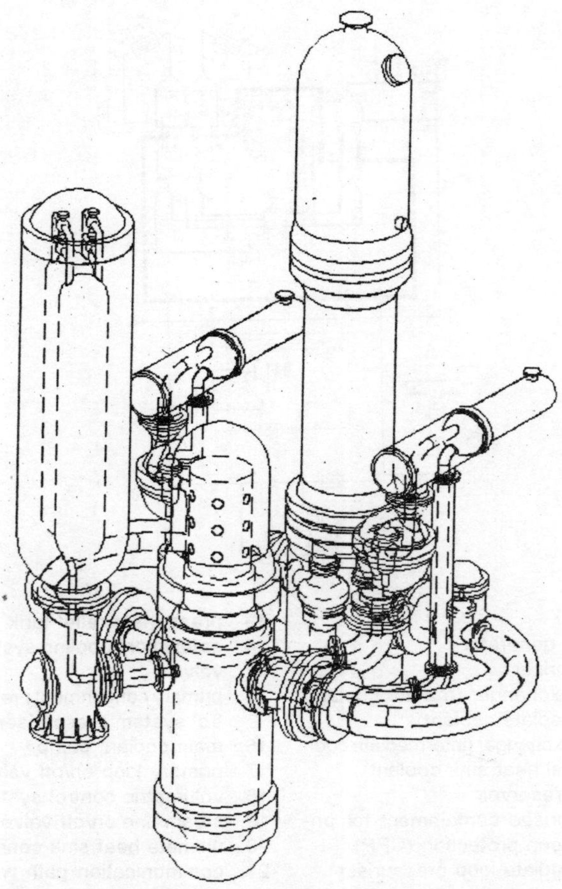

Fig. 2 : Primary Cooling System (enveloped by the pressurised containment)

water at the same pressure as the primary coolant, but at a lower temperature (about 70°C).

The function of the CPP is to avoid stresses of the primary loop due to pressure, thus makes impossible the rupture the primary loop itself for the propagation of material defects, if any. On the other hand, even if a crack in the primary loop should exist, no primary coolant leak takes place since no pressure difference exists between the internal and the external side of the loop.

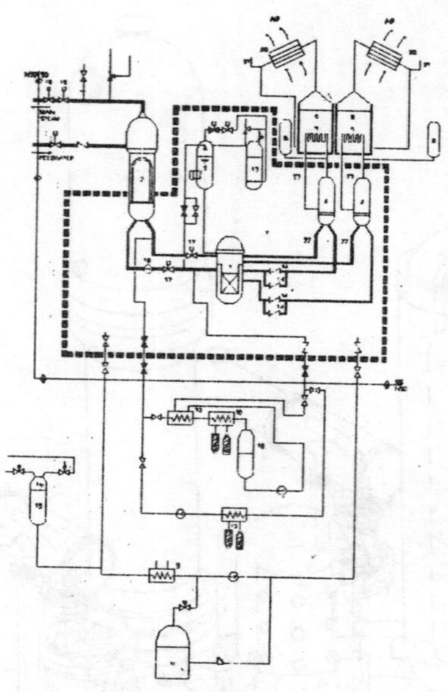

1. reactor
2. steam generator
3. pressurizer
4. heat-exchanger (reactor coolant/ intermediate coolant)
5. heat-exchanger (intermediate coolant/final heat sink coolant)
6. water reservoir
7. pressurised containment for primary loop protection (CPP)
8. intermediate loop pressuriser
9. heat-exchanger (primary containment water cooling system)
10. chemical and volumetric control system heat-exchangers
11. water storage tank
12. residual heat removal system heat-exchanger
13. pressuriser relief tank
14. safety core cooling system check valve
15. primary containment pressure control system pressuriser
16. main coolant pump
17. primary loop on/off valve
18. volumetric control system tank
19. steam line on/off valve
20. ultimate heat sink condenser
21. communication path with the atmosphere
22. safety core cooling system primary loop
23. safety core cooling system intermediate loop

Fig. 3 : Primary Cooling System and Main Auxilliaries Hydraulic Scheme

The presence of the CPP also makes possible the adoption of flanged connections in the primary loop (impossible in traditional PWRs just for leakage problems due to the pressure differences between the internal and the external side), greatly simplifying both the plant construction and maintenance and decommissioning phases.

The above mentioned design choice impose several limits to the design itself.

The first limit, due to the particular type of safety decay heat removal system (SCCS) adopted, is the thermal power design value, that cannot exceed approximately 1000 thermal MW per unit, and which in the solution herein described has been chosen equal to 600 MWth. Another characterising parameter is the pressure in the primary system, chosen equal to 75 bar, which is different from the pressure values usually adopted in PWRs for the production of electric power (150-170 bar). This choice, that leads to a loss in thermodynamic efficiency of the plant because of the limitation of the higher isotherm in the steam cycle, has nevertheless allowed the adoption of the pressurised Containment for Primary loop Protection (CPP, the pressurised boundary that envelopes the primary cooling system and the emergency core cooling system), substantially eliminating the possibility of loss-of-coolant accidents, the highest risk in these reactors.

The inclusion of the primary coolant system (average operating temperature: 234°C) inside the low-enthalpy-water-filled pressurised containment (CPP, at the temperature of 70°C) requires thermal insulation to reduce heat losses form the primary coolant system. An insulating system has been designed on the external side of the whole primary coolant boundary (only the lower head of the reactor vessel is thermally insulated in the internal part) through matrices of stainless steel wiring, that cause the presence of semi-stagnant water and can resist to high pressure and fast pressure gradients, with acceptable shape modifications. This system causes losses of about 03 per cent of the reactor thermal power.

The main auxiliaries of the Nuclear Steam Supply System, which are not relevant to safety, are:

- the Volumetric Control System (VCS) which allows the volumetric control of the primary coolant.
- the primary Coolant Cleaning System (CCS) and the chemical Additive Control System (ACS) which allow to control the chemical characteristics of the primary coolant
- the Residual Heat Removal System (RHR) which allows the primary coolant cooling during shutdown conditions.

The low temperature of the fuel rods in the MARS core (consequence both of the low temperature of the core coolant required by the low operating pressure and of the low power density of the core) allows a strong retention of fission products within the rods themselves (due to the drastic reduction of the fission products diffusion coefficients); also the low operating temperature of primary coolant allows to limit the oxidation of all materials which physically interface the reactor coolant.

In addition, the absence of possibility of fast nuclear transients, together with the considerable thermal inertia of the reactor coolant system makes thermal transients slow, thus limiting thermal stresses.

These features allow a drastic reduction of the radioactive fluids flow rate and of their contamination, as well as of the amount of solid wastes produced.

The contamination of the fluid stream to be treated is reduced to 5 per cent only of the value of streams in traditional PWRs withe the same thermal power. Additionally, low activity solid waste produced have a volume reduced by a factor 10 with respect to traditional PWRs (obviously, referred to the same thermal power).

Even if the engineered safeguards of the MARS plant are able to guarantee core coolability up to (and including) severe accident conditions, the adoption of a Reactor Containment Building has been incorporated into the design as an additional barrier against radiological releases and to perform a special protection against external events (Figs. 4 and 5).

The reactor containment building is made of reinforced concrete to face the impact of the reference missile (aircraft) on the external surface as required by the Italian nuclear regulations for external events, while all the internal events (including severe accidents) might be faced even adopting a small-thickness-steel-made building.

A relevant portion of the building is located under the ground level to simplify the building protection against both external impacts (reducing the exposed surface) and earthquake effects (increasing the overall building stiffness).

All components and structures located inside the building, including the structures supporting the reactor vessel and the steam generator as well as the working floor and service floor

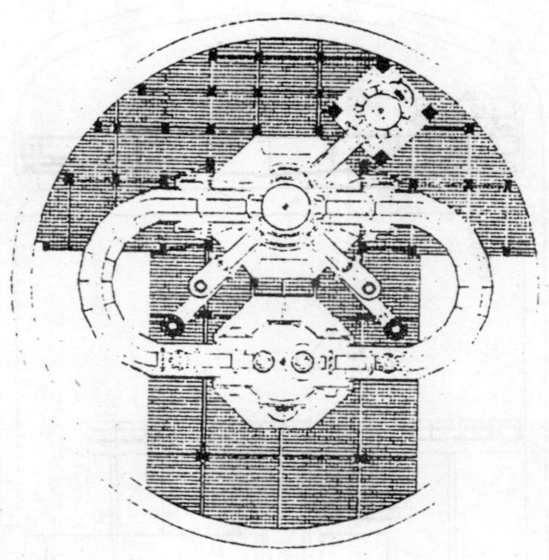

Fig. 4 : Reactor Building Plan at Primary Loop Nozzles Elevation

structures are made of steel beams, mainly bolted. This allows the easy assembly and removal of any component (even the largest ones), together with their support structure. Also the radiological shielding around the reactor is made of steel boxes filled with borated paraffin.

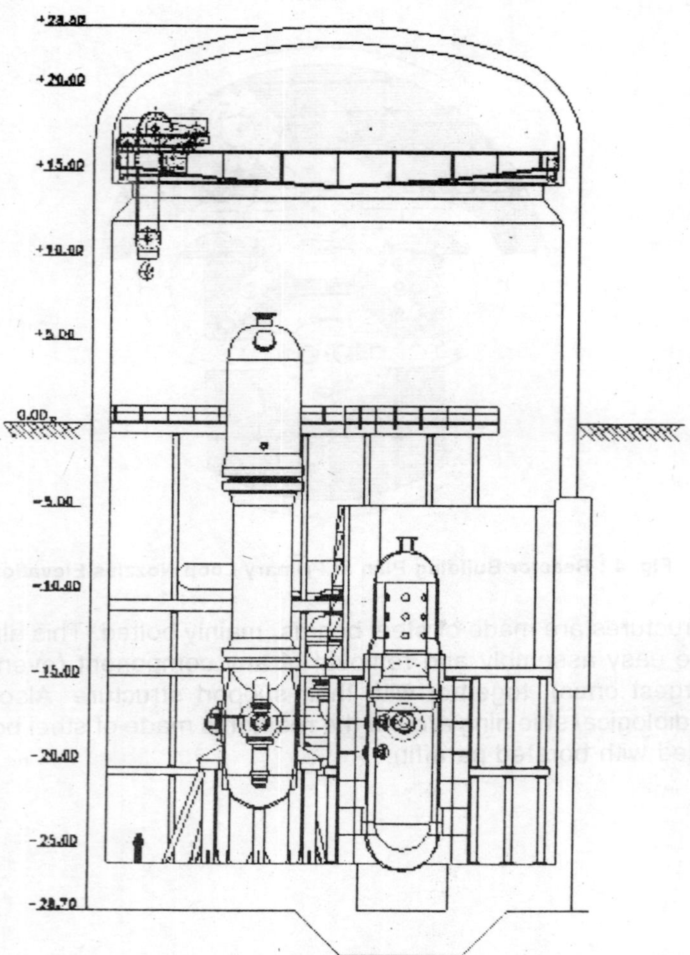

Fig. 5 : Reactor Building Section

PART — 6

BIOCLIMATE ARCHITECTURE

31

Passive Energy Systems and Architectural Expressiveness

MANFREDI NICOLETTI
Italy

ABSTRACT

The passive approach to the themes of energy savings is essentially based on the morphologic articulations of the constructions. The experience of Studio Nicoletti in low energy building is exemplified by five groups of designs based on different typologies. Cool air stored in the heart of buildings. Tall buildings: energy savings based on minimisation of the structural materials and bioclimatic systems. Double skin for cooling the external envelope of the buildings. Urban climate: climatisation of large urban fragments, protected either by a glass envelope or a bioclimatic pergola. Integration of various methods including extensive use of natural lighting, like in the design for the New Acropolis Museum in Athens.

FORM AS A TOOL FOR ENERGY CONTROL

There is no doubt that the passive energy approach is certainly the one that, being supported by the material shape of the buildings has a direct influence on architectural language and most greatly influences architectural expressiveness. Terms like "passive cooling" or "passive solar use" mean that the cooling of a building or the exploitation of the energy of the sun are achieved not by machines but by the building's particular morphological organisation.

Form is our main tool. To give form to the material things that we produce is an ineluctable necessity. In architecture form in fact summarises and gives concreteness to its every value: to economy, to aesthetics, to functionality, and therefore also to energy efficiency.

Our target is to enrich our expressive message with forms producing an advantage energy-wise.

Many concrete examples and a whole literature have recently grcwn up around these subjects and the wisdom of forms and expedients that belong to millennia-old traditions has been "rediscovered". Such a revisiting however is only, or most especially, conceptual, since it must be filtered through today's technology and needs, both being almost irreconcilable with those of the past.

Among the "historical" concepts two are of special importance.

One consists in the effort to establish rational and friendly strategic relations with the physical environment. The other recognises the interactions between the psyche and physical perception in the creation of the feeling of "comfort".

The former, which may be defined as an "alliance with the environment", deals with the physical parameters involving a mixture of natural and artificial ingredients such as soil and vegetation, urban fabrics, pollution. The dominant outside parameter is of course the sun's irradiation, our planet's primary energy source. All these elements can be measured in physical terms and are therefore the subject of science.

Within the second concept we consider our emotional and intellectual energies, which are our prime inexhaustible source of renewable power. In this case "cultural" paratmeters are involved, which are not exactly measurable. However, they represent the very essence of the architectural quality.

Objective, scientific measurements parameters tell us very little about our emotional way of perceiving which is actually influencing the messages of our physical sensorial organs. Our perceptual reality arises from a multitude of sensorial components—visual, thermal, acoustic, olfactory, kinaesthetics—and also from the organisational quality of the space in which different parameters come together, like the sense of "order" or of "serenity". But such practical evaluations as that of usefulness can be involved too. The evaluation is a wholly subjective one but it can also be shared by a set of experiencing persons. Therefore these "cultural" parameters are different in different contexts in spite of the inexorable levelling on a planet-wide scale. But the parameters change not only with the cultural environment, in the anthropological sense, but also in relation to function. What is suited to a gallery of contemporary art may not be for an archaeological museum, or for a concert hall. The scientific—measurable—parameters can thus have their meanings very profoundly altered by the non-measurable but describable cultural parameters.

ENERGY EFFICIENCY AND ARCHITECTURAL EXPRESSION

Our efforts must not stop at the banal technical solution of problems, forgetting the complexity of our intellectual and emotional life, or the symbolic and more generally psychological values of which architecture is the vehicle.

The designs that are presented here summarise some of the experience our team gained in the attempt to create architectures that are energy-efficient in a very broad sense.

The passive approach to the themes of energy savings is essentially based on the morphologic articulations of the constructions. The form, in its geometric and material sense, conditions the energy efficiency of a building in its interaction with the environment. It is then very hard to extract and separate the parameters and the elements relative to this efficiency from the expressive unit to which they belong. By analysing energy issues and strategies by means of the designs, of which they are an integral part, we will more easily focus the attention on the relationship between these themes, their specific context and their architectural expressiveness. For ease of reading they have been clustered in five groups, on the basis of affinities of the means employed for energy purposes.

The low energy target means also to eliminate any excess in the quantities of material and in the manufacturing process necessary for the construction of our built environment. That claims for a more sober, elegant, essential expression, which is not jeopardising at all, but instead enhancing, the richness and preciousness of architecture, while contributing to a better environment also from an aesthetic view point.

STORING COOL AIR

The use of great shaded spaces in the center or on the periphery of buildings is an "historically" efficient method. The purpose is to create reservoirs of air cooler than that outdoors, both to set in motion movements of air in the very heart of the building organisms and to constitute zones acting as filters to the atmosphere and to the sunlight. Obviously, the strong contrast in thermal and visual perceptions when passing from the outdoors to the interior is the main psychological effect, of quiet protection and well-being.

In the 300m. long departmental organism of **Udine University City** the air-cooler space is the central distribution spine of the building wings defining a series of inner square courts. The three-levelled sequence of rooms for teaching administration and research are located at both sides of the inverted pyramid shaped spine which receives natural light from inclined skylights. They have four different orientations rotated by about 45° in respect to the cardinal points depending on their location in the organism.

For thermal reasons it should be avoided any direct penetration of sun-beams, particularly in the hottest months. For the design of the skylight inclination we simulated "how the sun sees" the opening of the skylights.

As a result those spaces enjoy a very pleasant temperature and natural homogeneously diffused light (Fig. 1). This kind of

Fig. 1 : University City of Udine—General View

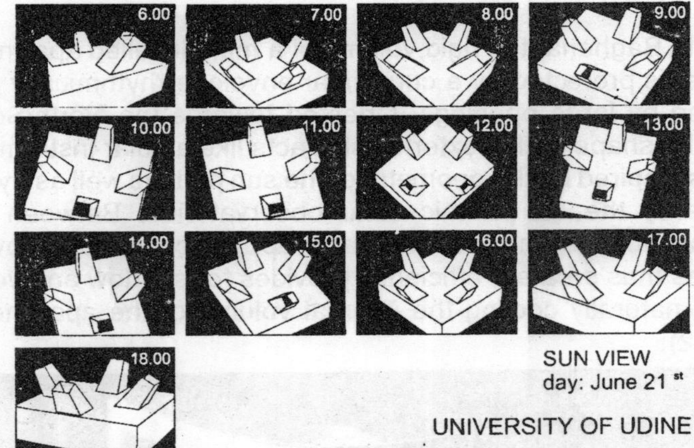

Fig. 1 : University City of Udine—Sun-views of the Sylight System

Fig. 1 : University City of Udine—Section Sun-views of the Main Building

illumination, reflected by the soft lawn-green vinilic floors, gives a feeling of calm and serenity, coherent with the university intellectual life-style. The cooled air reservoir characteristics are to be found in several buildings I have designed: the **Pisa Hall of Justice**, and in Rome, the University's **Department of Computer Science** and the **Social Security Agency's** headquarters.

DOUBLE SKIN

Another group of designs is based on a double skin typology. "This system is screening the interior spaces from the violence of the sun's rays with opaque or semi-opaque surfaces and air spaces.

The double skin system is the main architectural feature of the **Casa Moncada** in Bagheria, near Palermo, and of the awarded competition design entry for the **New Bibliotheca Alexandrina** in Egypt.

In Bagheria, the rigid volume of a multi-levelled apartment house, is protected by a curvilinear envelope rhythmically perforated by large openings. Oriented towards the North/South axis the shape of this external skin acts like a solar instrument. It was inspired by the geometry of the sun path as well as by the shape of the old Islamic solar observatories. Between the internal volume and the external wrap, the continuous flow of loggias thus created, efficiently provides for shadow and ventilation naturally cooling the internal volume of the apartments (Fig. 2).

Fig. 2 : Bageria (PA), Casa Moncada—frontview

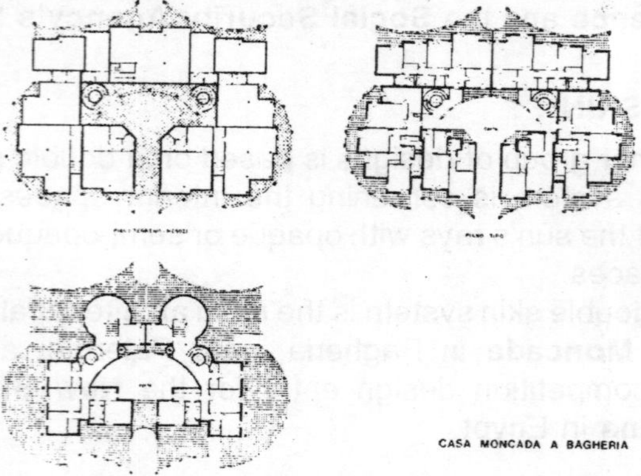

Fig. 2 : Bagheria (PA), Casa Moncada—plans

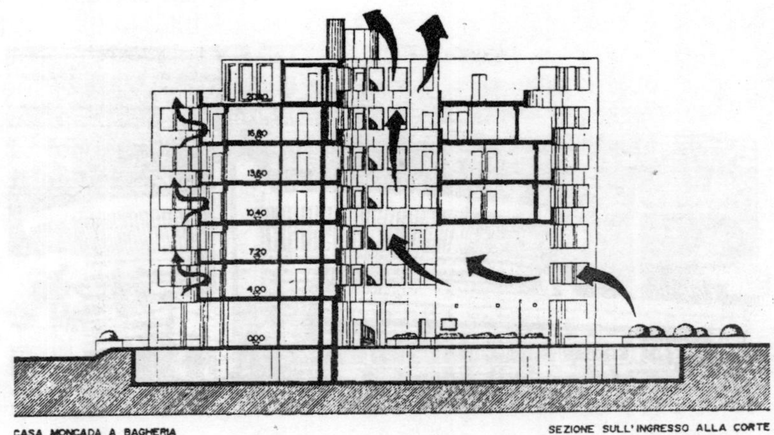

Fig. 2 : Bagheria (PA), Casa Moncada—Section

In Alexandria, the enclosure of the Reading Hall, similar to a gigantic masharabijia, has two concentric walls with two different types of openings. Only the internal ones are protected by glazing. The external walls open up views into the surroundings, offering protection against direct sunlight. The inner envelope—in contact with the climatised area—is constantly in the shade and cooled by natural air circulation. Furthermore, the inverted cone-shaped volume—with its 71° inclination—is completely shaded during the hottest period of the summer solstice, when the sun is about 82° over the horizon. The 110m. diameter lens-like air-space covering structure and the underground book storage area for 7 million volumes (for which darkness and low temperatures are vital) also optimise energy balance using natural methods (Fig. 3).

In the Hall of Sport of Palermo, the idea of the dynamic of sports is joined to that of lightness. An efficient system of protection against the solar radiation was built by super posing on the two end windows sunshade screens formed by large stainless steel pipes that describe warped surfaces in space. In addition, to create its 60m. span tensostructure of the roof, a moderate amount of steel was needed, its exceptional properties under traction being exploited (Fig. 4).

Fig. 3 : Biblioteca Alexandrina, Egypt—General View

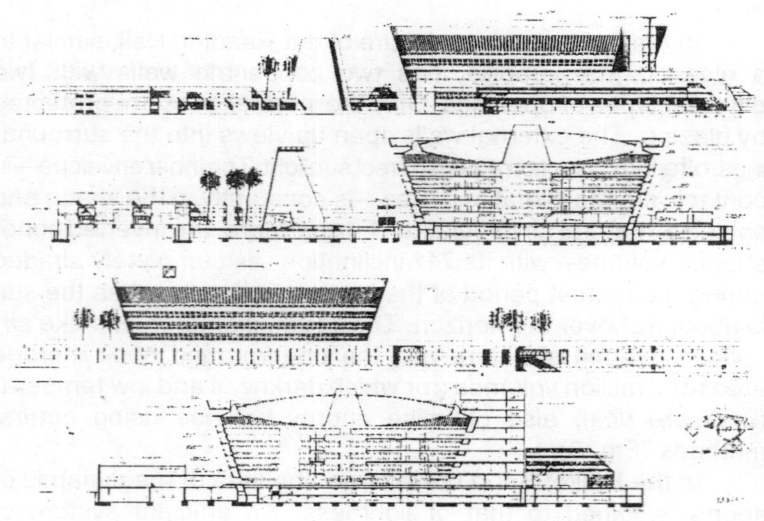

Fig. 3 : Biblioteca Alexandrina, Egypt—Sections

Fig. 3 : Biblioteca Alexandrina, Egypt—Detail View of Reading Room

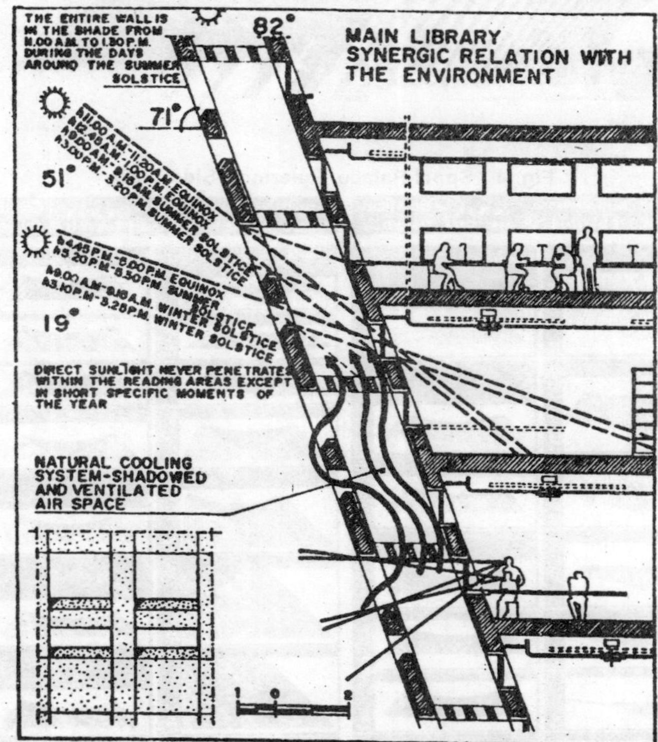

Fig. 3 : Biblioteca Alexandrina, Egypt—Solar Scheme of the Double Skin of Reading Room

Fig. 4 : Sport Palace, Palermo—Side View

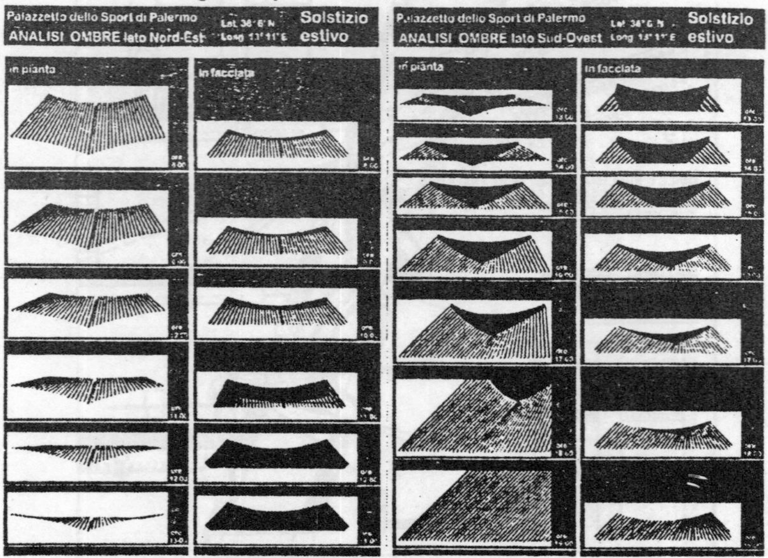

Fig. 4 : Sport Palace, Palermo—Shadow Projected by the Sun-screens on June 21

THE CRYSTAL PALACE

The great glass shell that protects an entire "citadel" is a theme of great fascination that is developed too in the designs for the **Rome Bank Headquarters** on the Ostiense and the **Cardiff Opera House** complex.

Similar to a large boat emerging from the earth to float on the River Tiber, the new Bank Headquarters has a triangular plan and a triangular longitudinal section. The building is situated near to a monumental group of redundant gasholders.

Their exceptional profiles erupt into a 15,000 m^2 glazed interior plaza, giving life to all the bank activities.

The control of the bioclimatic of the plaza is determined by the north facing roof. Because of its inclination, the angles of incidence of the sun's rays on it are low. The modular shading system integrated into the roof space-frame permits a view of the gasholders without the sun penetrating into it during the months between the equinoxes and the summer solstice.

The system's geometry provides a foliage effect, recalling the experience one has under a tree.

The volume envelope is clad in blue-black granite with blue glass panels, following the rhythm of the two bands of windows of the offices on each floor which, with the help of light-shelves, permit a very deep penetration of natural light into the working spaces. Warmth roman travertine dominates the interior of the plaza (Fig. 5).

Aided by the climate, the glass shell typology has a radical solution at Cardiff and one that is much more measured in Rome. Both design summarize the conflict between transparency, absolute visual permeability and the violence of the sun's radiation that represents, especially in Mediterranean countries, one of the

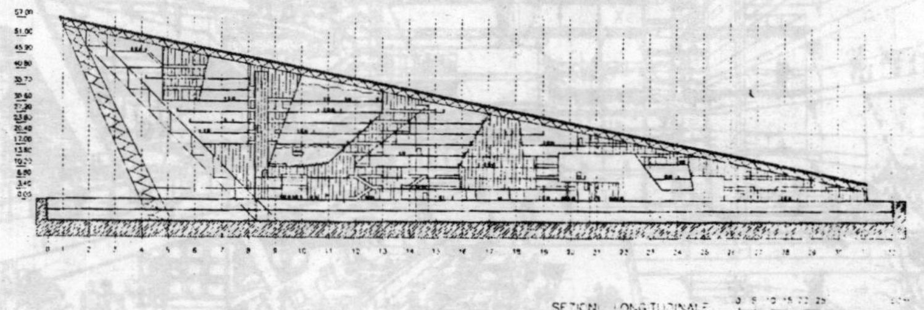

Fig. 5 : Headquarters of a Bank, Rome—Section

Fig. 5 : Headquarters of a Bank, Rome—General View

Fig. 5 : Headquarters of a Bank, Rome—Interior View

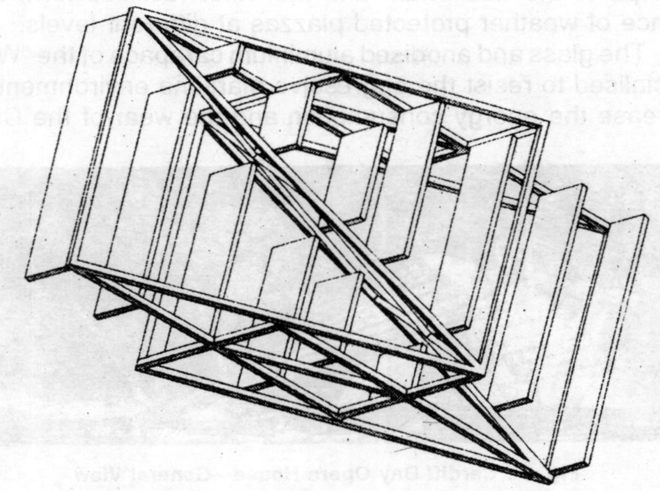

Fig. 5 : Headquarters of a Bank, Rome—Detail of Single Element of Foilage-Sunshades to the Space Frame Structure of the Glass Roof

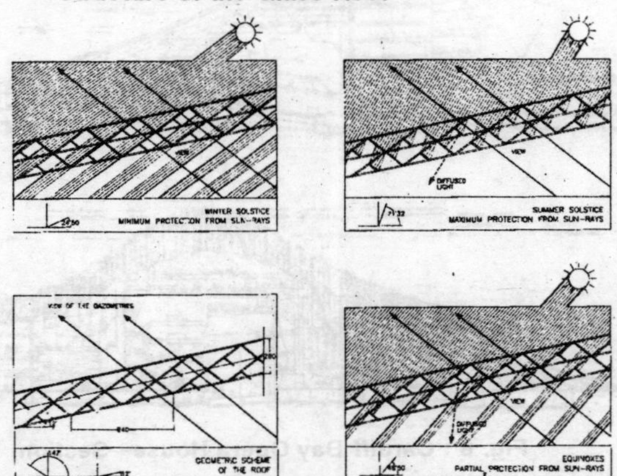

Fig. 5 : Headquarters of a Bank, Rome—Functional Schemes of the Foilage-Sunshades

most arduous challenges to the contemporary architecture that relies essentially on passive systems for climate control.

In Cardiff, the crystalline warped geometry of a huge glass "Wave" envelops the entire "City of music". Under the Wave, foyer and terrace restaurants, bar and exhibition areas on the rooftops of the auditorium and the other facilities form a sequence of weather protected piazzas at different levels.

The glass and anodised aluminium carapace of the "Wave" specialised to resist the aggressive maritime environment, will decrease the energy consumption and the wear of the City of

Fig. 6 : Cardiff Bay Opera House—General View

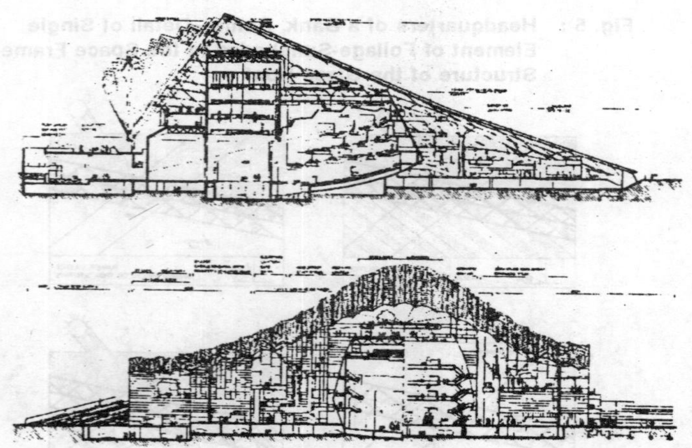

Fig. 6 : Cardiff Bay Opera House—Section

Music and will allow an extremely compact and flexible scheme.

This 37,000 cubic meters of this public space is never artificially refreshed or directly heated.

For most of the summer, the Atrium space will be controlled at an optimum level by adjusting the amount of the ventilation using the openings on the lower edge and on the upper crest of the Wave.

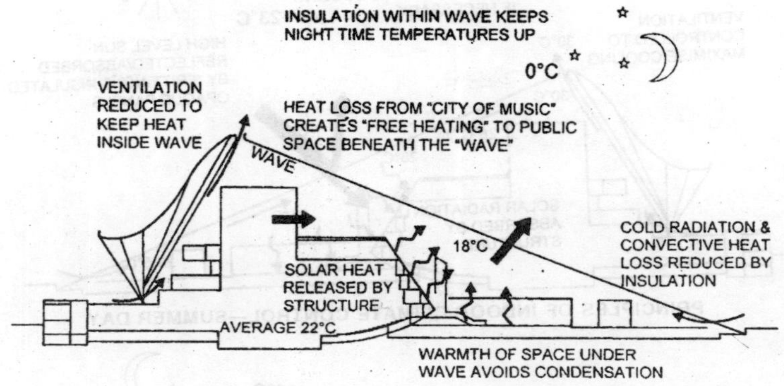

PRINCIPLES OF INDOOR CLIMATE CONTROL—WINTER NIGHT

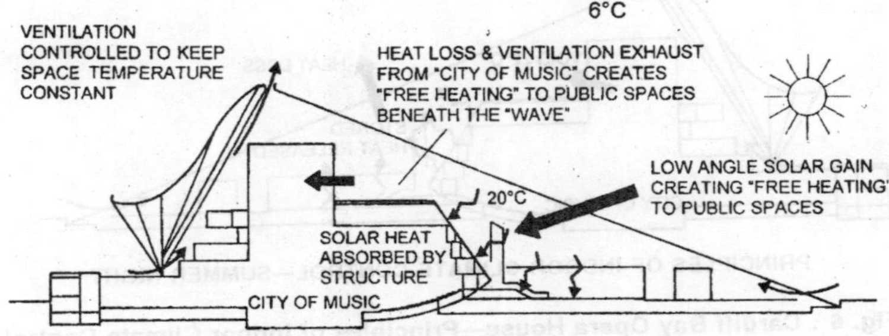

PRINCIPLES OF INDOOR CLIMATE CONTROL—WINTER DAY

Fig. 6 : Cardiff Bay Opera House—Principles of Indoor Climate Control

During heat waves, the micro-climate will not get uncomfortable: firstly because the space is so tall, warm air will rise to top levels. Secondly, part of the enclosures is opaque and the glass is partly obscured with fritting. At low level the foyer space will be at the same temperature as outside and substantially less than any normal non air-conditioned building.

During the winter in order to achieve a reasonable temperature under the Wave, (average U value 2.7W/sq. m.ºC) we exploited the "green house effect" in daytime and the heat loss of the "buildings" of the City of Music in the night. Consequently, the Atrium space will be maintained at about 18ºC in cold evenings when the outside temperature is –4ºC. The whole heating requirement will be the same as it would be if the Wave were omitted entirely (Fig. 6.).

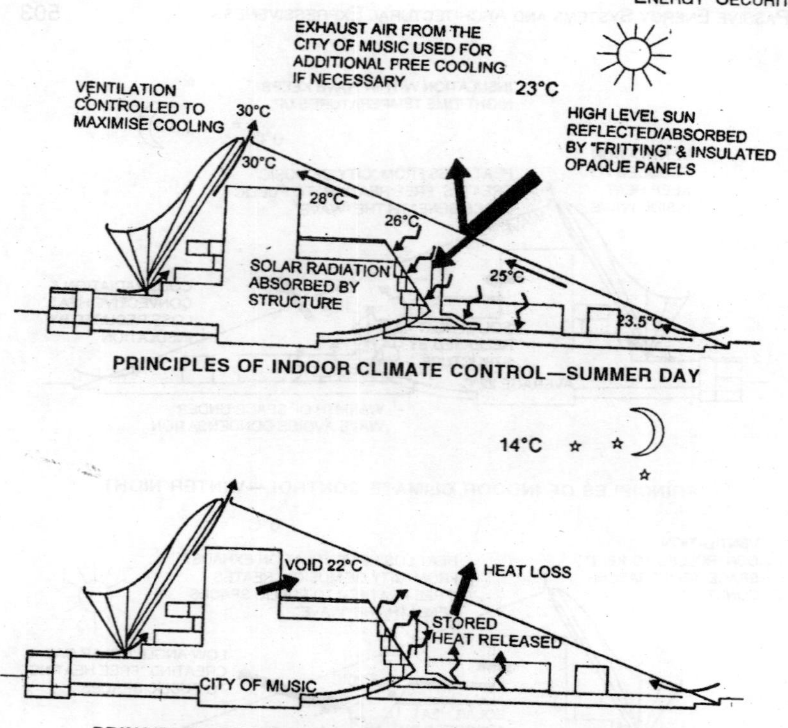

Fig. 6 : Cardiff Bay Opera House—Principles of Indoor Climate Control

CONTROLLING URBAN CLIMATE

The idea is to naturally "condition" a fragment of city in order to cool the individual building-elements that compose it. This energy theme was taken on in the designs submitted to the competitions for the construction of the **New Souks in Beirut**, for the new **Hall of Justice in Reggio Calabria** and in the **Extension of the Prado Museum** in Madrid. In the first prize internationally awarded design for Reggio a similar pergola of steel and aluminium panels covers an open space that determines, together with the existing buildings, a system of plazas to form the distributional node of the complex, whose buildings are in contact with a sensibly cooled mass of outside air.

In these cases an "artificial sky" was created, a sort of "pergola" capable of sheltering from the rain and shielding from

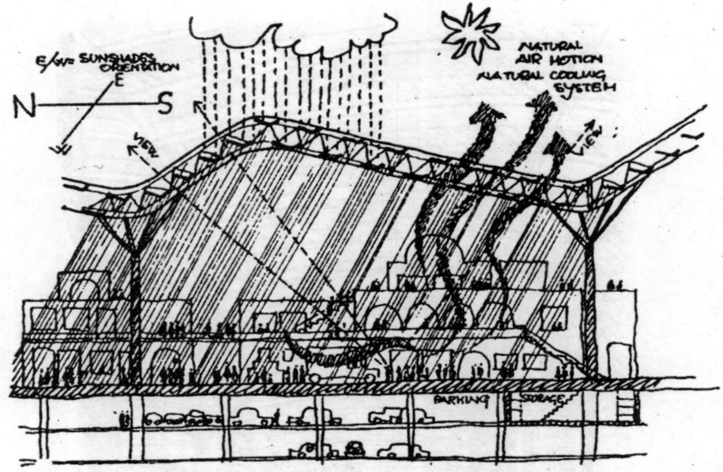

Fig. 7 : Reconstruction of the Souks of Beirut View of the Souks Protected by the Pergola

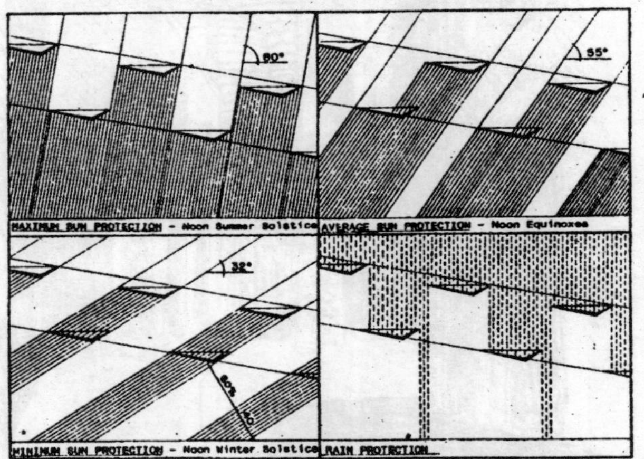

Fig. 7 : Reconstruction of the Souks of Beirut Functional Schemes of the Pergola

the sun's rays, while not obstructing the view towards the outside. Ascending movements of air are also activated by the considerable difference in temperature between the shaded areas on the ground and the structures of the sunshade roof and cold air is also drawn from the underground spaces.

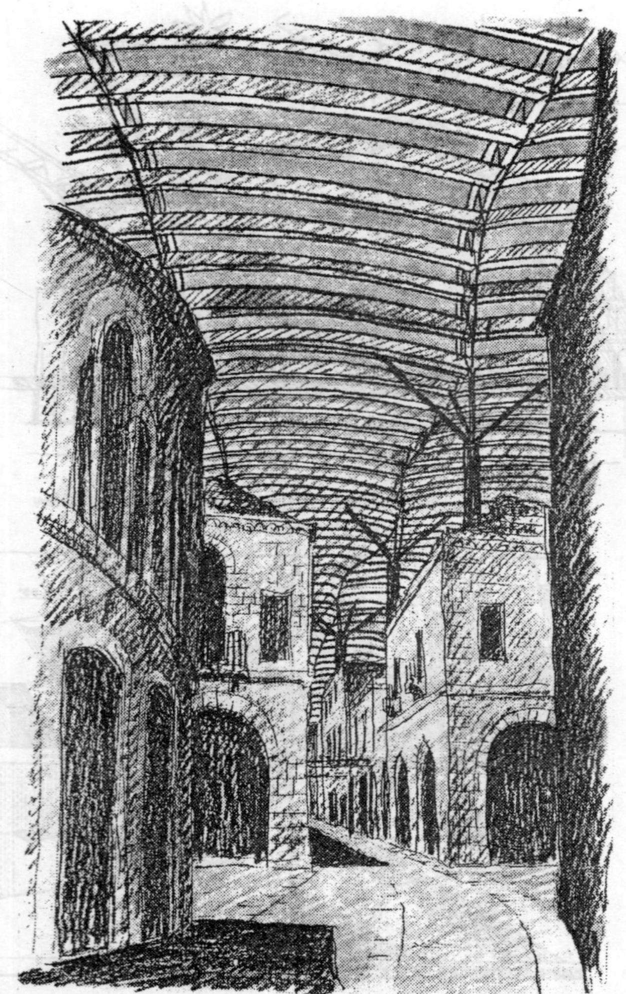

Fig. 7 : Reconstruction of the Souks of Beirut Urban Scheme with the Bioclimatic Pergola

In Beirut this strategy tends to restore spontaneity to the shops, which are independent of the "artificial sky" that protects them. Thus they are not constrained to follow pre-constituted building schemes but can rather pursue free and random forms as in the tradition of the ancient bazaars (Fig. 7). In the first prize awarded design for Reggio a similar pergola of steel and aluminium panels covers an open space that determines, together with the existing buildings, a system of plazas to form the distributional node of the complex, whose buildings are in contact with a sensibly cooled mass of outside air (Fig. 8).

Fig. 8 : Hall of Justice, Reggio Calabria—the Model

In Madrid the "Technological pergola" has as its purpose to unify the multiple articulations of the museum complex, which is broken up into several buildings, to shade the connecting urban spaces and to protect special glazed enclosures being adopted for some functions of the complex.

TALL BUILDINGS

The **Helicoidal Skyscraper** tackles the energy problem by minimising the quantities of material needed to built it. An aerodynamic form that opposes least resistance to the wind and a static system based on the use of steel under traction—and thus on the use of its strength most congenial to this material—form its rational basis. Inspired by the high structural efficiency of the limb of a mammal (a compressed central bone—external tensioned muscles) the 600m. high skyscraper morphology consists in three inhabited sails cambered and mutually wrapped, like those

Fig. 8 : Hall of Justice, Reggio Calabria—View of the . Internal Piaza Protected by the Bioclimatic Pergola

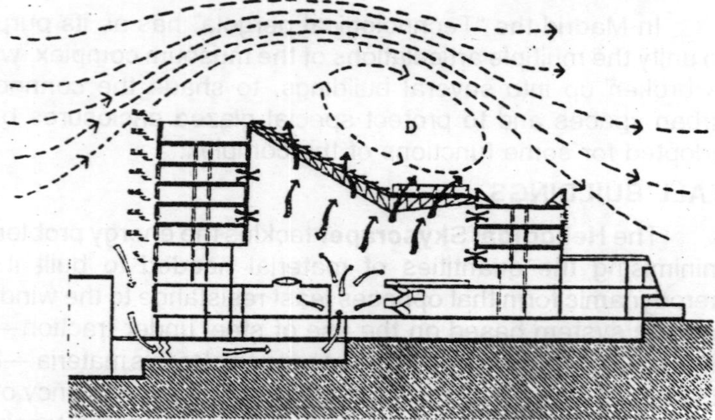

Fig. 8 : Hall of Justice, Reggio Calabria—Section (Effetto sul fabbricato del flusso dei venti provenienti da Nord; D=Zona di depressione)

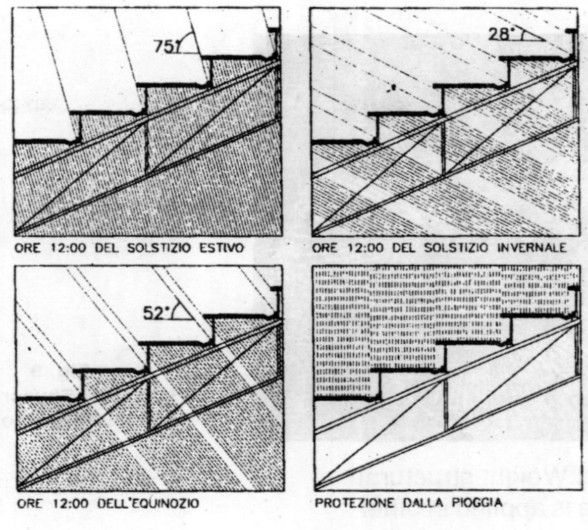

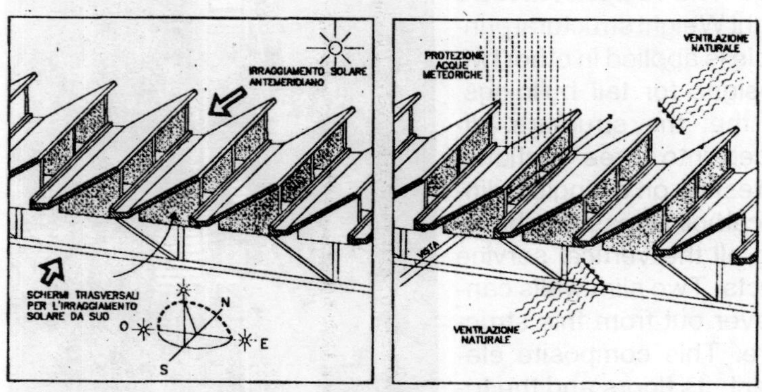

Fig. 8 : Hall of Justice, Reggio Calabria Functional Schemes of the Pergola

of a ship ready to "loose the wind", that is, an instant before the wind no longer has a hold in it. A bad form for sailing, but an excellent one for the static of a building. Each sail is anchored to one of the three hollow cylinders forming the "bone" of the system by a host of stay-cables compressing the floors to the core. The floors do not therefore represent a dead load but instead actively work with the statics of the organism as a whole.

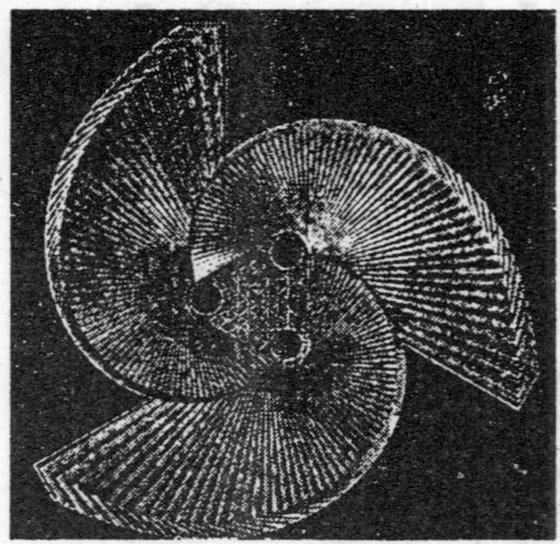

Fig. 9
Helicoidal Skyscraper—Plan and Elevation

The light Weight structural principle is applied in other two designs for tall buildings. The **Tower of Mestre** Light Weight structural principle is applied in other two designs for tall buildings. In the, the structure, divided into three sections, is resting on a single cylindrical hollow mastcore housing all the vertical service ducts. Two slim walls cantilever out from this structure. This composite element, its floors and the triangular modular steel grid of the facades, act together as a lightweight structural unit. The form itself provides both strength and rigidity (Fig. 10). The **Tower of Calabria**, the headquarters and symbol of the Calabria Region, stands in a vast, empty, rolling land-

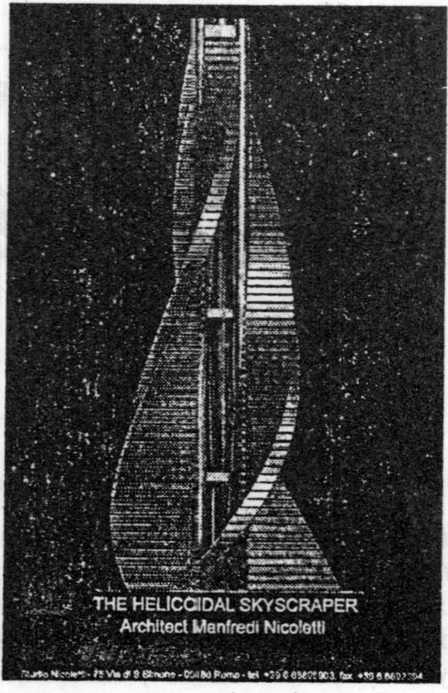

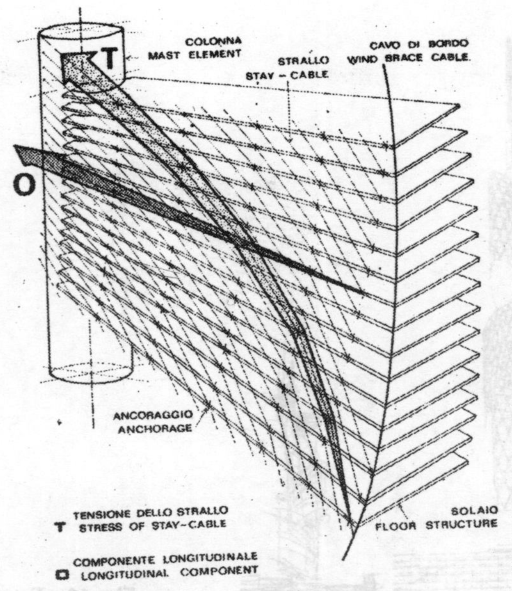

Fig. 9 : Helicoidal Skyscraper Structural Scheme

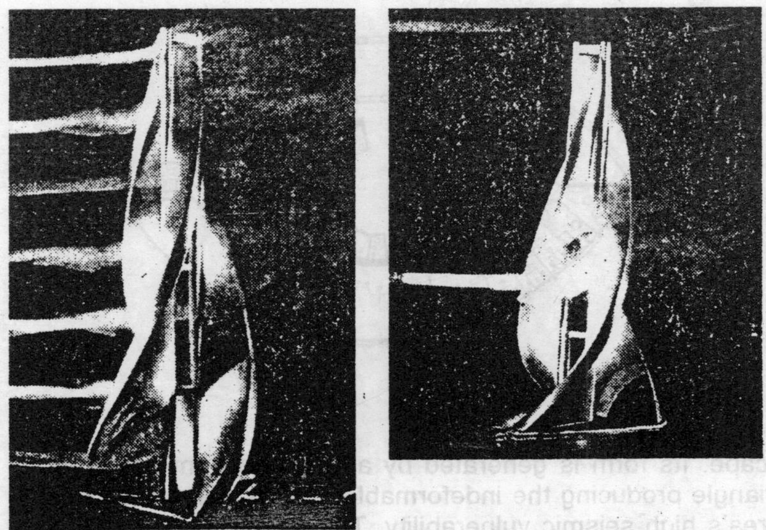

Fig. 9 : Helicoidal Skyscraper Aerodynamic Experiments

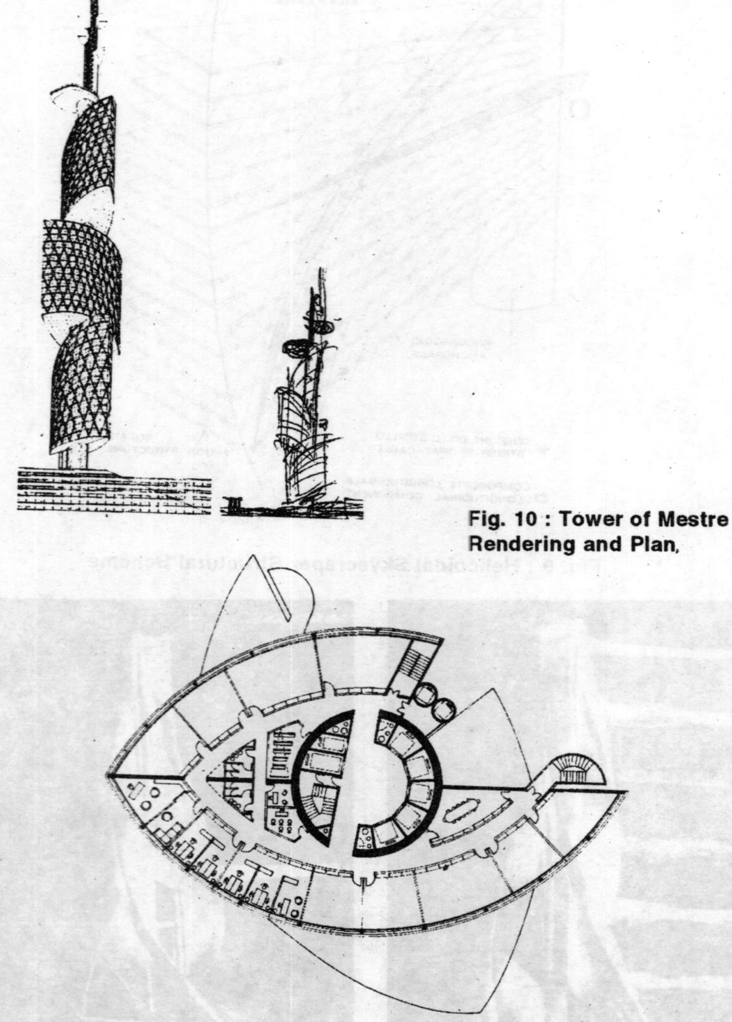

Fig. 10 : Tower of Mestre Rendering and Plan,

scape. Its form is generated by a space system based on a triangle producing the indeformable shapes demanded by the area's high seismic vulnerability. The triangular base and the triangular top of the building are rotated by 60° in respect of each

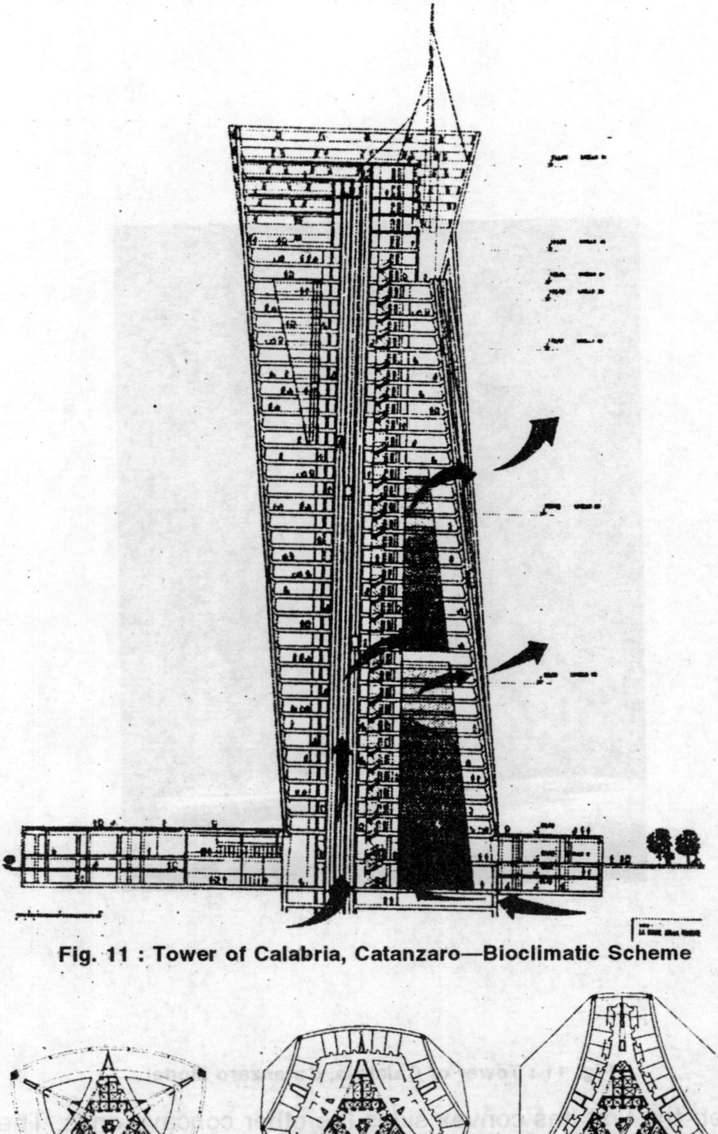

Fig. 11 : Tower of Calabria, Catanzaro—Bioclimatic Scheme

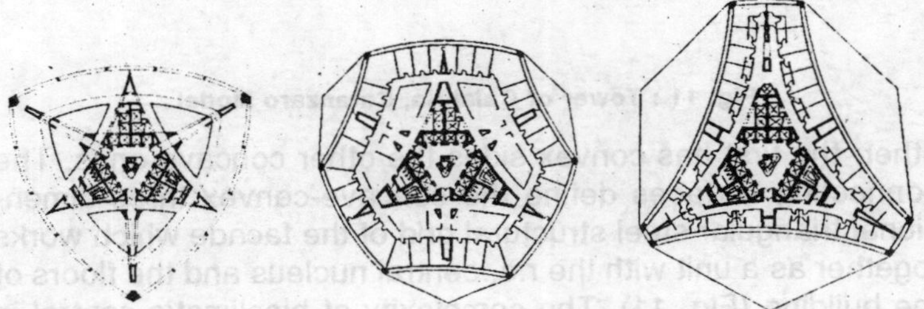

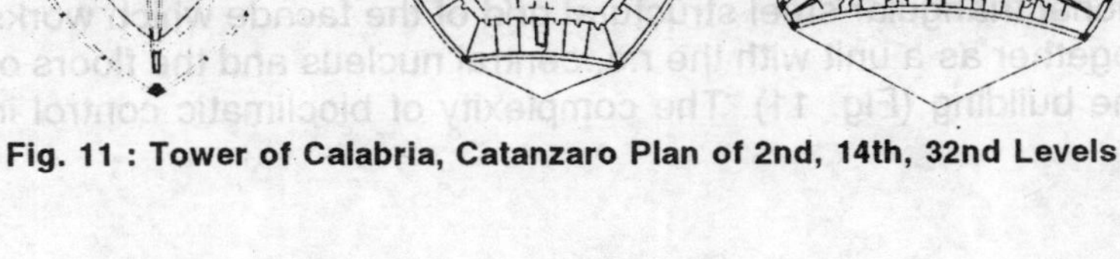

Fig. 11 : Tower of Calabria, Catanzaro Plan of 2nd, 14th, 32nd Levels

Fig. 11 : Tower of Calabria, Catanzaro Model

other: the first has convex sides the other concave ones. The connecting surfaces define the concave-convex three-dimensional triangular steel structural grid of the facade which works together as a unit with the r.c. central nucleus and the floors of the building (Fig. 11). The complexity of bioclimatic control in

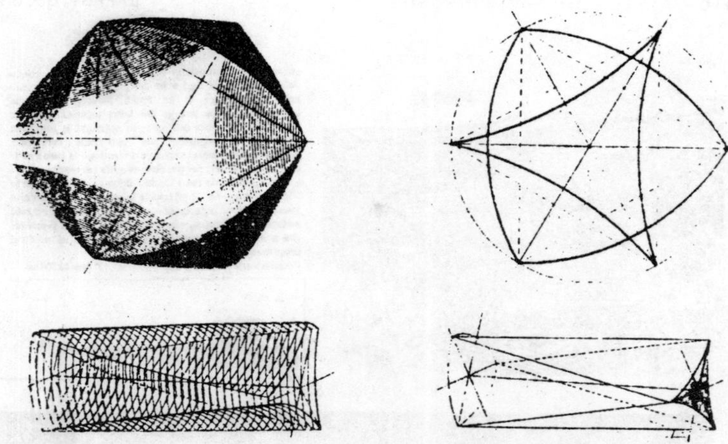

Fig. 11 : Tower of Calabria, Catanzaro Geometric Schemes

high rise structures mainly derives from the action of the wind and the difference in pressure and temperature between the top and the bottom of the building which generate convective flow of air. Direct exchange of air between the inside and the outside spaces is to be excluded due to the different pressure of air on the two opposite sides of the volume. In the Tower of Calabria, the first two lots of ten floors are characterised by inner courts with sealed glass walls. Cold air is conveyed from the basement floors and from the shafts of the central nucleus of the building and expelled through the radial empty elements of the "star" shaped system of the internal structure. Infra-red radiation is controlled by special heat reducing glass panels and by orientable inner "micro-louvers". Natural light is projected towards the office areas more distant from the windows by a system of "light shelves" in which artificial lighting is also incorporated.

THE ACROPOLIS MUSEUM

In this design one of the crucial issues was the study of the natural lighting system, and particularly its most important symbolic element: the "Eye" looking at the Acropolis. In this museum, built at the foot of the "holy rock", the direct penetration of sun rays was drastically excluded for their damaging effect on the marbles of the precious ancient sculptures. Through the eye the view of the Acropolis enter within the archaeologic exhibitions,

Ai piedi dell'Acropoli, il Museo racchiude ogni reperto dalla preistoria ad oggi e ne continua il profilo roccioso nella morfologia di un Podio percorso da "onde geologiche" da cui emerge una lastra inclinata perforata dall'Occhio. La vista dall'Acropoli organizza le collezioni archeologiche seguendo una stratigrafia temporale. Attraverso una rampa i visitatori discendono in basso nelle epoche più remote, per poi risalire a quelle più recenti.
Coperto dalla lastra con l'Occhio, il fulcro del progetto è lo "Spazio Partenone", il principale luogo espositivo in cui il vuoto riproduce le dimensioni del Tempio. Intorno ad esso metope, fregi, frontoni sono collocati nella esatta posizione che avevano sul Partenone, riacquistando in tal modo il proprio significato architettonico.
Consistenza dell' Opera: 1ª fase 30.000 m², 2ª fase 42.700 m².

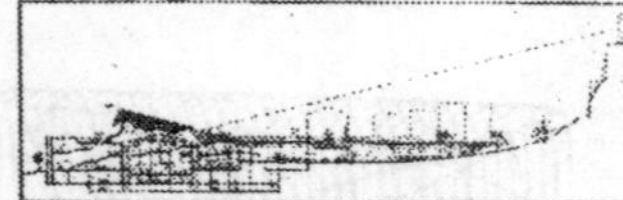

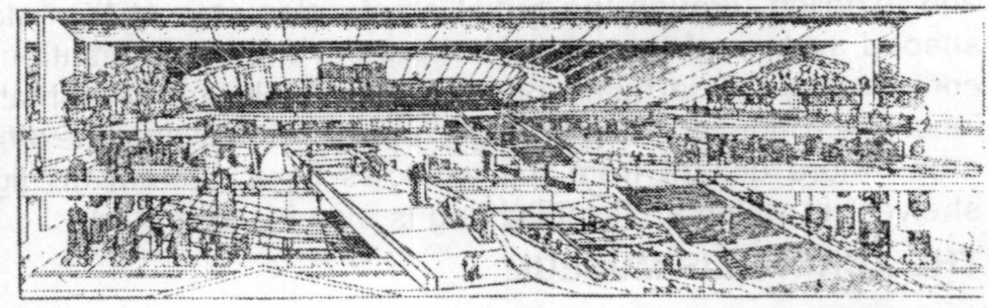

Fig. 12 : The New Acropolis Museum, Athens (M. Nicoletti-L. Passarelli)—View from the Acropolis, the Parthenon Space

giving to them a sense of concreteness and becoming an integral part of museographic system. Due to its complexity, several different passive method were applied in this museum.

The heart-core of the design is the "Parthenon Space" a void reproducing the dimensions of the temple and also its

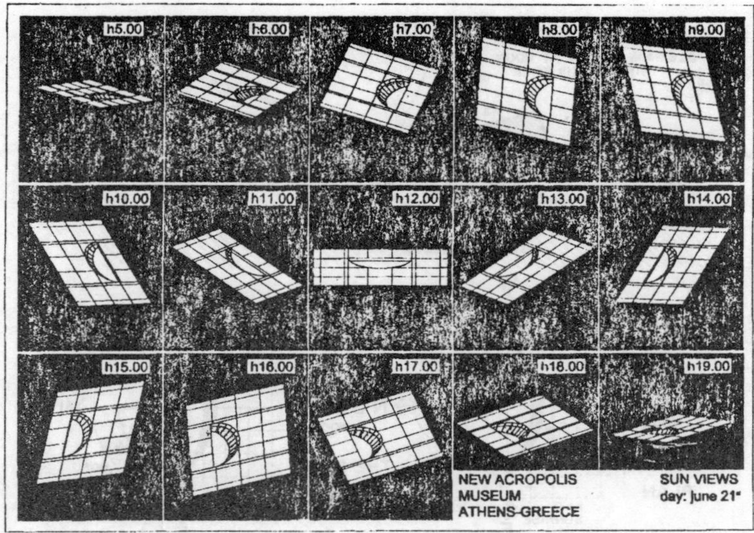

Fig. 12 : The New Acropolis Museum, Athens (M. Nicoletti–L. Passarelli)—Sunviews of the "eye"

orientation according to the cardinal points. In this space the metopes, friezes and pediments are exhibited in the exact position they had on the temple site, thus preserving their original architectural meaning. The building skin reacts to the environment to ensure the best internal conditions for the artworks and the expected two million visitors per year between May and September. Facing North, the glazed surface of the Eye avoids direct solar radiation but will be skimmed by them for brief moments every year. The external envelope of the 40 per cent hypogeous structure is naturally cooled by an epidermic ventilation system activated by double walls and the cladding. Natural light can penetrate only through a very limited typology of perforations besides the Eye: roof slits, lightwells and the lateral glazed areas of the Parthenon Space, protected by very large light-shelves. An extensive controlled use of daylighting will improve the quality of perception while saving energy. In collaboration with the Fraunhofer Institute fur Solare Energie system our team developed the simulation models to measure the different temperatures and lighting levels determined by the sun radiations in the various exhibition areas of the museum, in different

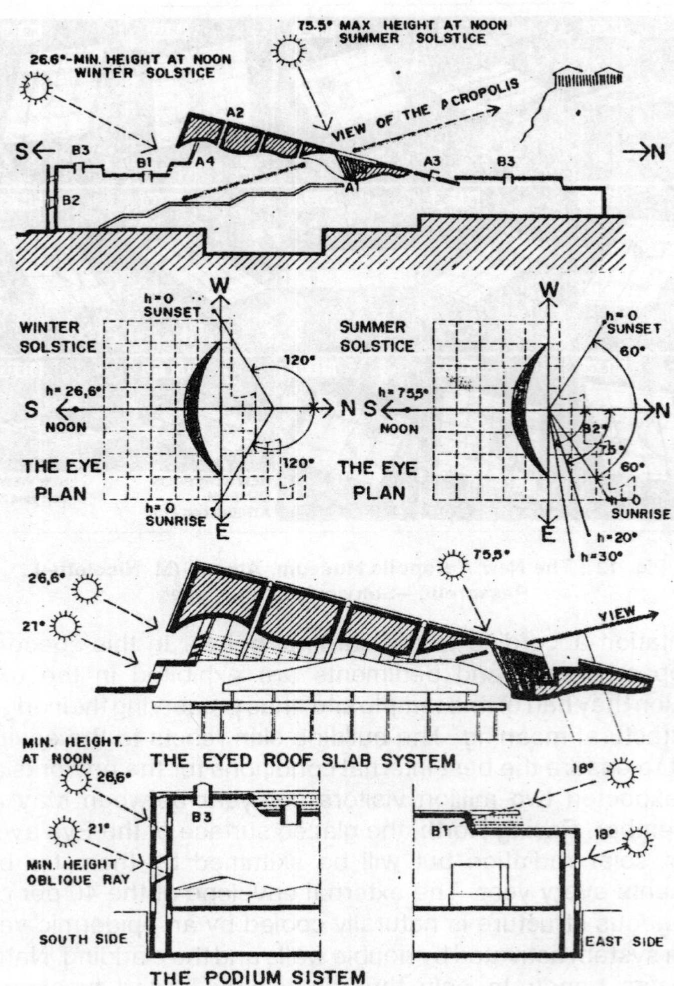

Fig. 12 : The New Acropolis Museum, Athens (M. Nicoletti–L. Passarelli)—Solar Scheme

hours, seasons and sky conditions. This last data were provided by the University of Athens. The study was financed by the European Community (Fig. 12).

32

A Bioclimatic Approach to Building in the Mediterranean Climate

L. TIRONE
Portugal

ABSTRACT

In the Mediterranean climate an improved building envelope via the application of passive solar technologies, can reduce the user's energy bill by over 80 per cent while increasing thermal, air quantity and visual comfort levels.

In the residential sector two types of buildings are presented incorporating those best available technologies: A wide variety of houses integrated in the ecologically planned developments Jade and Quinta Verde which are set in rural contexts and the Torre Verde multistorey residential building set in a urban city context, both in Portugal.

INTRODUCTION

There is no doubt that the quality of life our descendants will inherit from us, depends heavily on our effort in delivering more environmentally friendly solutions for the biggest energy spending sector world-wide: the buildings that provide shelter for us.

Since it is a proven fact that (especially in our mild Mediterranean climate) with the application of passive solar technologies, buildings can be more comfortable being significantly less energivorous than their conventional counterparts, the challenge lies in overcoming the inertia of this conventionality, and introduce the application of these technologies by generalising them into the construction industry, as a common way of building.

JADE AND QUINTA VERDE

The development JADE of 6 bioclimatic houses and the

development Quinta Verde which consists of 90 houses set in an ecologically planned development, both define the south limits of the village of Nafarros, Sintra. In the first, 4 typologies were developed focussing the design approach for a bio-climatic dwelling and in the second further 10 typologies emerged, based on the following design criteria.

Thermal comfort considerations as well as the integration of Passive Solar Technologies were weaved into the design of each dwelling from the very first design stages, by assimilating the thermal consultants' comments and their calculations as part of the design process in a regular and open dialogue. The challenge of incorporating more and more quality into our projects brings, hand in hand with it, the necessity for this dialogue with specialists at early phases in the design.

More than other consultants, the thermal consultant has a very important role in the primary design stage of a bioclimatic building since his qualitative input is fundamental for the definition of the following:

- Position and sizing windows correctly for solar gains and ventilation paths.
- Position and sizing of mass walls (non-ventilated Trombe walls) in proportion with the spaces they serve.
- Specify correct insulation system and quantity (thicknesses).

As the project begins to take shape and the building envelope appears, structural and services engineering being to have a determining input in the design while the thermal engineer then focuses on the tuning of the active solar system.

This intimate dialogue with the specialists makes it possible to include more and more quality and the best available technologies into the projects we create.

In the thermal consultant's opinion the project Quinta Verde is a unique project in Portugal in as much as it set out to construct 90 dwellings in which the architect devoted the necessary attention to climatic, technological and environmental factors during the design process. The role of the thermal consultant, in an effort to make the most of the inter-disciplinary context, has been to analyse the given options of the projects for the individual typologies with the available software means, evaluate the results and point out the available options. The following areas benefited from the dialogue with thermal consultant:

- Appraisal of the thermal behaviour for the building, including the compliance with the current building regulations for energy consumption buildings (Law 40/90). The figures representing the thermal performance of the different typologies on the table relate to the above regulations, which work as an appeal to the improvement of building solutions rather than a mere bureaucratic instrument.

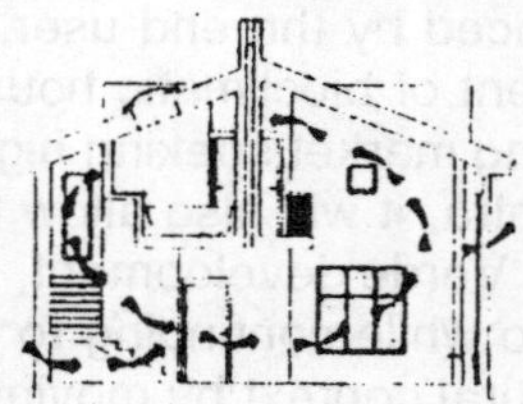

- The calculation of the contribution of the passive solar components, using software programmes which are adapted to Portugal. This kind of exercise gives an idea of the potential heating requirements which can be satisfied by glazed areas and by "mass walls" (non ventilated Tombe walls), although the effective results can be strongly influenced by the end user's behaviour.

Nonetheless the table shows the enormous potential of passive solar technologies with regard to indoor heating. To adapt the heat storage components to the architecture, it was possible to conceive some innovative elements, allowing not only the heat transferral function but also that of a comfortable window seat.

- It is also important to point out that in these local climatic conditions it is possible to attain indoor comfort in summer by guaranteeing adequate external overshadowing of windows and continuous external thermal insulation.
- The calculation of the central heating system which may eventually be installed in these houses.

Once again and according to the calculations, the energy saving for the purpose of attaining thermal comfort, is nearing the 90 per cent less in comparison to what it would be a conventional house.

- Further, the houses will have solar panels for domestic hot water supply. The system gives priority to the solar energy captured in the solar panels.

- This project was awarded a EU grant as part of the THERMIE' 93 programme and the developer is the Quinta Verde Cooperative which is constituted by Tirone Nunes Urbanismo, Lda and C.H.E.S.M.A.S., a local established Housing Cooperative. All houses were sold before construction started on site and thus, the construction of the houses was financed by the end-user.

A new development of bioclimatic houses was conceived in response to a growing market seeking high quality houses in the rural context of Sintra, it will also allow the first generation of clients of the Quinta Verde development, who aim to enlarge their typology, to do so while continuing to benefit from bioclimatic quality and the rural context by moving to a development nearby.

By reducing up to 90 per cent of the energy needed to make these houses comfortable in comparison with a conventional construction, Quinta Verde fulfills its exemplary role on the building scale by applying the best available technologies for a sustainable future.

TORRE VERDE

Torre Verde is the Bioclimatic Residential Building of the Expo' 98's Intervention Area:

Torre Verde, with the urban park to the North, the Tagus river estuary to the East and the new city to the South, contains all the desirable ingredients which guarantee the high quality of life we all seek while living in the city.

The challenge and objective of this project was to demonstrate that the application of passive solar technologies in residential buildings, even in a high density urban context such as this new quarter of Lisbon, offer very high levels of thermal comfort can be achieved all year round with a minimal dependency on non-removable energies.

Tirone Nunes Urbanismo, Lda. applied for and won a EU grant within a group of projects named "Expo-Cities", including projects from Expo' 98 Lisbon, Expo 2000 Hannover, Utrecht and Mallorca, and the European Commission has thus, once again, set a Tirone Nunes Urbanismo, Lda. project on the international scene for its environmental friendly qualities.

The building, with its 42 apartments, is under construction and is due for completion before Expo' 98.

The technical aspects which contribute to the comfort conditions of the apartments in the Torre Verde project are as follows:

In Torre Verde the environmental pre-occupations were an integral part of the design process, thus involving the thermal, the acoustic and the structural engineers, from the very outset.

Here again, we applied the best available technologies in order to attain thermal comfort, acoustic comfort and the best indoor air quality conditions.

Torre Verde is a multistorey residential tower, set in a square plot where the main facade faces exactly South over a raised ground floor condominium garden.

The design team relied on the dialogue with the thermal engineer in order to design the building envelope and all its components. In the field of passive solar technologies, the following design measures were taken:

- Positioning and sizing of windows correctly to guarantee the solar gains and the ventilation paths necessary to attain the desidered comfort levels in each apartment.
- Positioning and sizing of mass walls (non-ventilated Trombe walls) correctly in proportion with the spaces they serve, in order to optimise indirect gains during cold evenings and to reduce the consumption of conventional energy for heating purposes.
- Dimensioning correctly the continuous, externally applied insulation system of the building, in order to minimise thermal loss during the cold season and minimise thermal gains during the hot season.
- External shading with movable blinds and fixed overhangs to contribute to the exclusion of excessive solar gains during the hot season.

Since it is a twelve-storey building with 42 apartments, more than in Qunta Verde it was necessary in Torre Verde to carefully contemplate a series of innovative factors in order to offer the best available services to the end users. In the field of active solar technologies, the following measures were taken:

- Solar panels for domestic hot water will be installed on the roof of the building, with a centralised gas boiler back up system. This system offers each client the hot water service from predominantly renewable sources.

- All those spaces that cannot be naturally ventilated like bathrooms and kitchens are serviced by mechanical ventilation system, which has had to be carefully turned in order not to interfere with the passive technologies applied.
- Due to the pressurisation of the common circulation areas in this twelve-storey storey building, the natural ventilation contributing to the passive cooling of the building had to be carefully considered.

In comparison with other European countries, Portugal has a very demanding "noise in buildings" regulation and Torre Verde has complied with it.

With a sufficiently air tight building envelope, acoustically well insulated (including double glazing with thickness of 6 m and 8 m), glasses with continuous resilient membranes underneath floor finishes in the apartments and double walls between apartments it is possible to guarantee a high level of acoustic comfort and privacy in each apartment.

Torre Verde distinguishes itself by the diversity of its volumes and its aesthetic form, being part of the most sought after residential area in Lisbon, in anticipation of '98 World Fair.

But, above all, what makes this building special is the invisible fact that, for all the reasons mentioned above, it is the building which most positively contributes to the quality of life of future generations.

PROJECTS FOR THE NEW MILLENNIUM

Tirone Nunes Urbanismo, Lda. continues to invest in the EXPO' 98 invention area which, from the end of this year, will be a series of new urban quarters of Lisbon and Loures.

Our aim with this new group of projects is to close the northern urban edge of EXPO with a strong and clear statement on sustainable building practice, since in the Mediterranean climate passive solar technologies can guarantee the highest comfort levels and be ideal low emission neighbours to the urban Park.

Five twelve-storey residential towers and one five storey hotel will mark the end of the urban tissue and the beginning of the largest Iberian urban park—Parque do Tejo e Trancão, on the right shore of the Tagus river estuary.

The towers will have between 42 and 46 apartments each, will all face the South, respecting passive solar principles, and

East reaping the most benefit of the spectacular views over the river estuary.

The hotel is to have presently 120 rooms spread over 5 storeys and all with an ideal solar orientation and panoramic views over the river estuary.

Its classification will be 3 or 4 star and the clients will be predominantly professional.

These projects will be a positive contribution to the new millennium and we hope that the global impact of EXPO' 98 will support the dissemination of these sustainable technologies which respect both the planet and future generations.

Both Quinta Verde and Torre Verde have already hosted a great number of visitors, including; among many other, groups of university students and local authority technicians.

Those innovative projects have also been presented by TNUL in conferences, meetings, and seminaries in Brussels, Paris, Milan, Turin, Copenhagen, Winterswijk, Milton Keynes, Swansea, Jerusalem, Madrid, Praia, Porto Santo, Porto, Lisbon, Estoril and Sintra.

TNUL have launched a group of new projects which so beyond the residential sector into the hotel sector.

Each new project is, for us, a new challenge and while the Passive Solar tools becomes less and less visible the positive results become more and more recognised.

PART — 7

FORUM OF THE UNESCO INTERNATIONAL SCHOOL OF SCIENCE FOR PEACE

"ENERGY SECURITY IN THE THIRD MILLENNIUM: SCIENTIFIC AND TECHNOLOGICAL ISSUES"

33

The International Forum on "Energy Security in the Third Millennium: Scientific and Technological Issues"

M. CANEPA,
U. FARINELLI,
V. KOUZMINOV,
M. MARTELLINI
and
L. PAGLIANO

An important forum of the UNESCO International School of Science for Peace organised by the Landau Network–Centro Volta (LNCV), Russian Centre for Energy Policy (CEP), UNESCO Venice Office (UVO), ENEA and Italian Ministry of Foreign Affairs, was held in Como, Italy (14-16 May 1998) at the Centre for Scientific Culture "A. Volta". The Forum aimed at analysing the most relevant issues facing the global sector over the next two decades as renewables, general energy policies, market accesses, sustainable energy for the developing countries and highly efficient-environmentally safety new technologies. The participants identified the following important messages to be addressed to the governments and world's leading energy companies and institutions.

1. EVOLUTION OF THE ENERGY SITUATION

The world's outlook on energy has deeply changed in the last decade. The "Cold War" is over, even if local conflicts are not. Resources of fossil fuels have been increasing with time, rather than decreasing, due to new discoveries and to re-evaluation of known deposits. International energy prices, although still fluctuating, seem less volatile than in the past. Preoccupations on the availability of energy and on its price have therefore decreased both in the attention of the governments and in the focus of media. On the contrary, preoccupations have shifted toward the environmental effects of the energy cycle and the threat it poses on the stability of global climate. The concept of sustainable development, from a specialised and cryptic idea, has become a frequent presence on the first pages of newspapers. The governments of industrialised countries have started committing themselves towards changing their energy policies in the direction of sustainability. Also developing countries, without entering precise engagements, have shown concerns in this respect, while already suffering severe local and regional environmental problems deriving from energy (or transport).

But also the way in which energy policy is made is rapidly changing. In most (not all) countries, until some time ago, energy policies were implemented through a command-and-control chain, according to top-down plans, and making use of national (state-owned and operated) utilities and energy companies. At present, it is recognised that the use of market mechanisms are required for an effective implementation, since only market forces can ensure an optimised allocation of resources. At the same time, market by itself will respond only to short-term signals, and it will not by itself aim at sustainability, by definition a long-term goal. The role of governments is therefore to introduce the corrections that are needed to account for these goals.

Other important changes have occurred. Some concern the level at which decisions relevant for energy policies are taken. On the one hand, international or supranational conditioning acquire increasing importance: the Kyoto protocol is one example; and for the European countries, the directives on the liberalisation of the energy markets by the European Union are changing very deeply the structures and the operating rules of the energy systems of member states. In the opposite direction, the process of decentralisation is progressing in many countries, and many duties and competencies in the energy and environment field are gradually passing from national governments to regional and local authorities.

The changes in structure of the energy system are having as a consequence a great increase of number and diversification of the actors on the energy scene. The liberalisation of the energy market is producing a variety of independent power producers, of self-producers, of distributors, of energy traders and brokers, of energy service companies, of new financial subjects for project financing, leasing, third party financing etc. Moreover, the attention on the energy cycle, which was previously concentrated nearly exclusively on the supply side, is now shifting more and more on the efficiency of the final uses of energy. This means that every entity that utilises energy, from an industry to the individual consumer, becomes an actor of the energy cycle.

The objectives of the energy policy of single countries or groups of countries remain broadly the same, with changing emphasis and priorities: making enough energy available at affordable prices to sustain economic growth and social promotion, while respecting the environment and the stability of global climate.

The technological lines along which the new energy paradigm should be constructed are also broadly shared: they are the increase of efficiency in all the sectors of the final uses of energy; the introduction and diffusion of renewable sources of energy; and a more sustainable conversion and utilisation of the traditional forms of energy, in particular fossil fuels. Progress along all these lines has been continuous, mostly by incremental improvements, although the investments in research and development have regrettably been steadily decreasing in the recent years, mainly in the public sector.

Finding the right way for attracting private investments shall be essential in the future rather than further research; indeed large scale private investments will reduce costs and improve technologies. One example is the Japanese leadership on low consumption motor cars. Here is private investment driving new technologies. The key point is to find a good compromise between public and private investments in the energy sector in order to cover the large energy demand expected in the next decades.

The success of an energy policy rests on two legs: on one side the availability of the right technologies; on the other hand, the policy conditions that allow these technologies to be implemented and diffused. Since both technologies and policy instruments, as we have seen, are evolving, and sometime rather rapidly, it seemed appropriate at this time to review the situation by choosing at least a few of what were considered key issues: and this is what the Forum has attempted to do. In line with the scientific character of the Landau Network, these issues are mostly of a technological nature: however, technology and policy often have strict interconnections, so that in some instances both sides have to be considered in parallel.

2. ENERGY SECURITY IN THE GLOBAL MARKET

From 1980 to 1995, energy consumption at the world level[1] has increased from 7021 Mtoe to 8852 Mtoe (+26%). However, this growth is compounded very unevenly: an increase of 18.3% in OECD countries, a decrease of 21.7 per cent in the countries with economies in transition (due to the serious economic crisis), an increase of 108 per cent in non-OECD Asia (including Middle East countries) and increases of 43 per cent and 61 per cent in Latin America and Africa respectively. It is perhaps even more instructive to look at the per capita energy consumption: this has actually remained constant in Africa at a level of 0.50 toe/person.year (against a world average of 1.56) and even decreased in sub-Saharan Africa (from 0.465 to 0.456) due to the massive population growth, while in China energy consumption per capita has grown from 0.54 to 0.83 and in India from 0.21 to 0.33. Differences in rates of development are thus reflected in per capita energy availability.

At the world level, the economy has grown more rapidly than energy consumption: correspondingly, energy intensity—defined as the ratio between (commercial) energy consumption and the gross national product in a given country—has decreased from 539 toe/MECU (1985) in 1980 to 477 in 1995 (−11.5%). However, data are very different from one region to another. The energy intensity in the same has decreased by 17.1 per cent from 291 to 241 in the European Union, from 550 to 446 in North America (−18.9%) but it has increased elsewhere, from 1923 to 2349 in the former Soviet Union, where the breakdown of the economy has been greater than that of energy consumption (and where the extremely high values point to a chronic inefficiency); in developing countries, it has dropped from 1469 to 1066 in Asia, but it has increased from 770 to 997 in Africa, and from 463 to 511 in Latin America.

1. Data in this section are taken from: European Commission, Directorate General for Energy DG-XVII, "Energy in Europe, 1997—Annual Energy Review", Special Issue September 1997.

Even if the total energy consumption in the world has greatly increased, population has increased just as much, so that the per-capita energy consumption has remained essentially constant: 1.54 toe (tonnes oil equivalent) per person per year in 1995 against 1.56 in 1980. However, this derives from a substantial increase in Asia and a substantial decrease in the economies in transition, and a nearly constant value elsewhere.

As we have mentioned, there is no short-term preoccupation of physical exhaustion of energy resources. Industrialised countries have shown a sufficient capability of finding alternatives to the oil supply of single countries or even regions, so that unilateral initiatives of suppliers, as in 1973 or 1979, are discouraged. However, to assume that further energy crises will not occur in the future would be unjustifiably optimistic. Scenarios of political instability in some of the main oil or gas production countries are conceivable, such as to cut for not too short a time the supply of oil or gas to the outside. Even if the difficulty could be overcome, the impact on economy could be severe, especially for countries which depend nearly entirely on energy imports.

Even more serious are preoccupations of a structural nature deriving from a production capacity unable to meet demand. It is actually possible to imagine scenarios of demand growth which are greater than currently assumed: for instance, that countries in rapid growth, like China and India, introduce accelerated individual motorisation, like the Europe of the 1950's and 1960's, thus greatly increasing the demand for oil products. A delay in the investments needed to prepare the supply in time for the increased demand would result in tensions and possibly price shocks. A similar preoccupation could concern the investments for gas pipelines.

As evidenced in the past, energy is a major cause of potential conflicts. Preventing energy crises is an important contribution to world peace.

3. SUSTAINABLE ENERGY FOR DEVELOPMENT

The importance of energy for development is certainly not new, but it may have failed to be recognised explicitly in the past. Consequently, an important part (between 25% and 30%) of all the money devoted to aid to development programmes, both as grants and as loans, by bilateral, multilateral or international institutions as well as non-governmental organisations (NGO) go to energy-related projects.

However, what has been missing until some time ago has been the link between energy, development and sustainability. Most of the energy projects that have been supported have been of the traditional type: first of all, they were supply oriented (i.e. directed to providing energy rather than to using it effectively); second, they were mostly directed to large-scale plants based on traditional technologies (fossil fuels or large hydro-electric plants); finally, environmental assessments were carried out for most projects, at least in the recent times, but a general evaluation and prioritisation of projects in terms of sustainable development has generally been missing. It has been calculated[2] that between 1980 and 1991 only about 2.5 per cent of official development assistance for energy from all donors has gone for renewable energy development.

The challenge is to turn now to sustainable energy schemes. Sustainability has three aspects: economic, environmental and social. If industrialised countries are today particularly interested in the environmental (and climatic) aspects, developing countries are understandably particularly interested in the aspect of economic growth. Social aspects, and quality of life, are equally important. Energy availability has been shown to influence demography and population growth; gender roles and the promotion of women; agriculture and food; water and health; education, job creation, land degradation etc. Such connections have been recently identified[3] and studied[4]. Some concrete links have also been presented in this Forum.

New possibilities of obtaining energy services in a sustainable way have now become available as the result of the development of new technologies, in particular based on renewable sources of energy. New, innovative approaches are needed, and developing countries must lead the way. They should be able to make the best use of the fact that they are not committed

2. K. Kozloff and O. Shobowale, "Rethinking Development Assistance for Renewable Electricity", World Resources Institute, Washington D.C., 1994.

3. United Nations Development Programme (UNDP), "Energy as an Instrument of Socio-Economic Development" (1995).

4. United Nations Development Programme (UNDP), "Energy after Rio", (1997)

to the solution adopted by industrialised countries in the past, because they do not have all the infrastructures in place and do not have to exploit the large past investments: they can jump–or "leapfrog" —to new, advanced solutions more adapted to their situations, needs and culture.

4. INTRODUCING SUSTAINABLE ENERGY IN DEVELOPING COUNTRIES

The diffusion of sustainable energy forms meets with many barriers, which in a certain sense can be referred to an insufficient or improper functioning of the market. One of the most important barriers is the subsidy given by many governments to traditional forms of energy that inhibit sustainable development. Once recognised the importance of energy for development, many governments subsidise electricity or various fuels, so that their price to the final consumer is lower than the cost of production and delivery. In many developing countries, energy prices and tariffs are much lower than in industrialised countries, although the cost of producing and delivering energy is by no means lower. This has the double effect of discouraging energy conservation, by making interventions to increase efficiency artificially more expensive than the energy which is saved, and of creating a barrier to the introduction of new forms of energy, renewables in particular, which are not equally subsidised.

Another barrier—which to some extent is common to industrialised countries—is the dispersion of interventions in energy efficiency and distributed energy generation, in particular by renewable sources. The financing of very large scale supply-side projects (a gigawatt-size hydroelectric or coal-powered or nuclear plant) can find capital for investment at much lower interest and longer return times than the corresponding hundred thousand small projects of micro-hydro or wind installation or efficiency improvement.

Finally, as in most industrialised countries, externalities are not reflected in the price of energy. Although there is no clear consensus on the way in which this "internalisation" should occur, and that it would not be fair to expect developing countries to precede advanced countries on this route, it is clear that a gradual introduction of externalities would greatly improve the prospects of increased energy efficiency and of renewable energy sources.

Last but not least, the new market instruments "emission trading" could modify the future of energy structures in favour of energy efficiency and renewable energy, in particular in transition economies and developing countries; this could help to achieve global targets of CO_2 emission reduction at a lower cost for everyone.

Finally the Climate Change Convention which was elaborated in Kyoto at the end of 1997 has opened new possibilities for the developing countries through the implementation of feasibility instruments: *emission trading, joint implementation* and new mechanisms development. These instruments which would take place simultaneously before 2008 would change considerably the strategies in terms of transfer of clean and efficient technologies. This has to be already taken into account in the energy planning of the developing countries.

5. ENERGY POLICY IN THE COMMONWEALTH OF INDEPENDENT STATES

The countries of the Commonwealth of Independent States (CIS), in comparison with the western industrialised countries, are characterised by a very high energy intensity, i.e. the ratio of energy consumption (primary or final) to GDP. The situation varies from country to country but, as a general rule, the energy intensities of the CIS countries are two to three times higher than those of the EU countries. Furthermore the CIS countries are the world's largest energy related CO_2/GDP emitters. Unfortunately there is no legislative and economic background for implementation of the energy saving potential in Russia. The economic instability and lack of investments do not enable to foresee the perspective of energy efficiency. Thinking about financing energy efficiency in CIS countries, we can point out two mechanisms:

a) **energy efficiency joint implementation** (EEJI) projects. The term joint implementation is used to describe a wide range of possible arrangements between interest in two or more countries (in the present case countries belonging to EU and CIS), leading to the implementation of co-operative development projects that seek to reduce or sequester greenhouse gas emissions. Attractiveness of investments in EEJI projects, mainly at the end use, are quite perspective in terms of a wide energy saving equipment and service market, small costs on projects and hence low financial risk, potentially quick playback and State schemes (in easing credits and tax reduction) of reimbursement of energy efficiency investments due to saved energy. We acknowledge in this regard the creation of energy service companies in Russia and Ukraine undertaken by the EBRD.

b) **carbon emission credit (CEC).** Due to the strong economic decline taking place in CIS countries since the disintegration of the Soviet Union in 1990—i.e. in 1996 the Russian GDP was only 59 per cent of 1990 level—a permanent carbon emission mitigation is having place. Forecasts for the period 1990–2010 cumulative reduction of carbon emission compared to 1990 levels would be about 2 billion tons only for Russia. Meanwhile, according to Kyoto Protocol, Russia is committed to return in 2008–2012 to 1990 level. Therefore Russia, as well as other CIS countries, have CEC which they could use for trading mechanisms as formulated in Article 17 of the Kyoto Protocol. We recommend that such "carbon income" for the CIS countries should be primarily allocated for EEJI's on the project base (see point a).

6. ENERGY AND ECO-EFFICIENCY

While there has been progress in material and energy efficiency, particularly with reference to non-renewable resources, overall trends remain unsustainable. As a result, increasing levels of pollution threaten to exceed the capacity of the global environment to absorb them, increasing the potential obstacles to economic and social development.

Unsustainable patterns of production and consumption are the major cause of continued deterioration of the global environment. The development and further elaboration of national policies and strategies are needed to encourage changes in unsustainable consumption and production patterns, while strengthening, as appropriate, international approaches and policies that promote sustainable consumption patterns on the basis of the principle of common but differentiated responsibilities, applying the polluter pays principle, and encouraging producer responsibility and greater consumer awareness. Eco-efficiency, cost internalisation and product policies are also important tools for making consumption and production patterns more sustainable.

New and renewable sources of energy (NRSE) represent an attractive area for research and development in meeting constantly growing energy demands which should fully satisfy newly established requirements for sustainable development with a particular emphasis on the improvement of the quality of environment.

In this case all initiatives to foster international co-operation in energy R&D related to NRSE should be promoted and welcome. It was noticed with a great satisfaction that such prestigious international organisations as UNESCO are giving a high priority in their working agendas to this particular matter and the World Solar Programme initiated by UNESCO is focused on achieving these objectives.

In order to achieve a sustainable energy strategy, priority must be given to energy conservation, the improvement of energy efficiency and the use of renewable sources of energy. Action in this area should focus on:

a) Promoting international and national programmes for energy and material efficiency with time-tables for their implementation, as appropriate. In this regard, attention should be given to studies that propose to improve the efficiency of resource use, including consideration of a ten-fold improvement in resource productivity in industrialised countries in the long term and a possible factor-four increase in industrialised countries in the next two or three decades. Clean and efficient technologies will play a crucial role in the sustainable development. Nevertheless, they will not solve alone energy and environmental issues. General policies have to be implemented in parallel, based on economic instruments which should both orient the behaviour and support the deployment of new technologies in a consistent, cost-effective and acceptable way. Ways to internalise externalities have to be found and implemented.
b) Further research is required to study the feasibility of these goals and the practical measures needed for their implementation. Industrialised countries will have a special responsibility and must take the lead in this respect. In particular, these countries must promote efforts in re-

search, development and use of renewable energy technologies at the international and national levels.

c) The UN-FCCC (Framework Convention Climate Changes) evolution will support the implementation of new feasibility instruments which should favour energy efficiency and renewable energies, in particular in transition economies and developing countries. This major opportunity which will take place in the coming decade has to be organised as soon as possible; objectives of greenhouse gas emission reduction compatible with an economic sustainable development have to be defined consistently.

d) The UN-Commission on Sustainable Development should consider in the coming years policies and measures necessary to implement eco-efficiency and, for this purpose, encourage the relevant bodies to adopt measures aimed at assisting developing countries in improving energy and material efficiency through the promotion of their endogenous capacity-building and economic development with enhanced and effective international support.

e) Promoting measures favouring eco-efficiency; however, developed countries should pay special attention to the needs of developing countries, in particular by encouraging positive impacts, and to the need to avoid negative impacts on export opportunities and on market access for developing countries and, as appropriate, for countries with economies in transition. These considerations become particularly important in the context of the joint implementation and the clean developments mechanisms which take place within the climate change strategy.

f) The Commission of the European Union should contribute to the above international efforts and should place those efforts within the centre of their future policies, especially to support research in this areas as proposed as one of the three major initiatives of the European Union at the Special Session of the General Assembly.

g) Special efforts should be undertaken in order to assure technological developments in the field of NRSE in order to provide international and national markets with competitive installations for the production, accumulation and distribution of different types of energy extracted from and generated with NRSE which should be acceptable for all countries regardless of their levels of economic development.

The increased use of NRSE should also contribute to the reduction of green house gas emissions and therefore to the meeting of objectives established by recent international agreements mentioned above.

7. ENERGY GENERATION AND ENERGY-SAVING SERVICE MARKETS: THE STAKE OF HARMONISATION

Energy market restructuring is expected to capture economic efficiencies resulting from price setting by market forces. Market-based pricing, if it is not unduly affected by abuses of market power or other influences, could lead to prices closer to the marginal cost of production. This would begin to address an early rationale for government intervention (e.g. through energy efficiency and renewable energy programmes), which was that regulated prices might not accurately reflect the true marginal cost of production, leading to inefficient production and consumption decisions. However, a question arises whether the market would be able to capture long term marginal costs, and even in this case many other market barriers not related to incorrect price signals restrict the realisation of a cost effective energy efficiency potential. It seems unlikely that energy sector restructuring which aims to increase competition on the generation side will, by itself, address these market failures, unless a systematic and co-ordinated approach targeted at removing/lowering these failures/barriers is put in place. Many studies and efficiency programmes in the field confirm that a large cost-effective potential to increase end-use efficiency exists both in developed and developing countries. The costs of these energy efficiency resources are between 40 and 80 per cent lower than the cost of production by new plants. It is estimated that with well-designed policy programmes, the bill for electricity services in Western Europe in the year 2020 (including all investments in energy efficiency measures) could be reduced by 15–35 per cent percent below the levels expected under business-as-usual policies. If market liberalisation were combined with these demand side opportunities, economic benefits would be three to five times larger than those from generation side restructuring alone.

The States should therefore establish market-based legal and regulatory frameworks, which explicitly harmonise objectives of economic efficiency through generation liberalisation, with those of security of supply, end-use efficiency, renewables exploitation, environmental protection, and R&D in these key areas. A correct formulation of the ground rules will ensure that the various actors are encouraged to participate actively in the creation and development of the new energy markets both on the generation and the demand side, and are fully enabled to participate to the added value that will thus be created. Such a formulation will reduce the need for government intervention and will allow all actors free to seek their own best interest within a socially beneficial framework. Long-term vision, legal and regulatory stability should provide a favourable climate for action for all the actors on the scene: customers, energy service companies, utilities, manufacturers, financing bodies, designers, research groups.

To ensure the proper functioning of energy markets, customers should have access to clear and certified information concerning price structure, contract arrangement, environmental quality of energy supply, and energy efficiency levels of end-use devices. It is therefore urgent to implement a co-ordinated approach to accelerate the introduction of energy labelling and minimum efficiency standards for various end-use devices (electric motors, light sources, domestic and office appliances, etc.).

Most commonly, the regulation of those parts of business which continue to be technical monopolies or are not yet open to competition, fixes tariffs in such a way that utilities receive a huge incentive at expanding sales even when energy efficiency resources are less expensive than new production. It is, therefore, of the utmost importance that the EU Ministries rapidly approve the proposal for a "Council Directive to introduce rational planning techniques in the electricity and gas distribution sector", which was passed with a large majority in November 1996 by the European Parliament. It is equally important that other countries adopt similar resolutions in order to remove barriers to the creation of open, functioning markets for energy related services. States should:

- set conditions (e.g. involving or guarantee funds, access to customers information, etc.) to remove entry barriers to the activities of energy service companies, either large or small, independent or affiliated to utilities.

- establish procedures whereby electricity and gas distribution/supply companies periodically present integrated resource plans which shall evaluate all resource alternatives (including Demand Side Management—DSM) on an equal economic basis.
- remove present disincentives to energy efficiency activities for distribution/supply companies through de-coupling profits from sales volumes.
- make compulsory the integration of options regarding DSM into capacity tendering/contracting procedures.

Activities of manufacturers and investments of financing bodies will not develop without a clear picture of the environment in which they will be asked to act. Market transformation actions through technology procurement can play a relevant role in this direction. Through the facilitation of energy experts, groups of large or associated customers develop functional specifications and an aggregated demand for more efficient products. The existence of an informed buyer group and the prospect of large deliveries reduce the risk for manufacturers to develop and produce goods based on new, more efficient technologies/concepts. On the basis of the successful experience in Northern Europe and of the studies conducted by EU and IEA, market transformation actions should be performed for a series of new/improved energy efficiency technologies.

Frequently building designers receive retribution in proportion to the size and capacity of the climate controlling equipment, a practice that provides a direct incentive for over-sizing and inaccurate design. Careful, good quality design should be suitably paid for through correction mechanisms that account for the energy and environmental performances and comfort levels of the complete building and installed plant. Energy certification of buildings should be introduced in order to allow an informed customer choice and remove present distortions in the real estate market.

Long term targeted research programmes (such as the European SAVE, Thermie, etc.), an adequate level of funding, and independent quality control procedures, will enable research groups to fully bring their contribution to increase national productivity levels in line with environmental protection. Public interest activities, like R&D, lower return energy efficiency programmes, low income customers programmes, renewables, should be ensured in the new market situation, through the creation of funds fuelled by a non by-passable levy at the distribution level.

8. BIOCLIMATIC ARCHITECTURE

Among energy saving strategies, a significant role can be played by bioclimatic architecture. If one thinks that both in Europe and in the United States residential and commercial buildings make up for about one third of total energy consumption (and in Italy, in particular, they account for 30–40 per cent of CO_2 and CFC emissions) one can easily understand why more extensive studies should be conducted in this field.

Unfortunately, Italy ranks first in Europe in terms of energy consumption per building, with 2,02 toe (tons oil equivalent) burnt each year, as against 0,72 in Denmark.

Bioclimatic architecture, which may also be defined "sustainable architecture", has been used many times in the past by ancient civilisations, such as the Indians who built the Mesa Verde settlement in Colorado in the 13th century. The village is located in a horizontal cut in the rock and is made of dwellings representing a perfect compromise between a cave and a building. It benefits from a constant microclimate throughout the year, thus avoiding a considerable temperature difference between very cold winters and torrid summers.

Other examples of bioclimatic architecture include the so-called "Wind Towers" in Iran, and in Italy: Matera's rocks, Pantelleria's "Dammusi", houses in the Chiavenna valley, Alberobello's "Trulli" dwellings, the more sophisticated renaissance villas in Costozza, also mentioned by the Italian 16th century architect Palladio and the passive solar energy solution adopted by Rafael in Villa Madama, Rome.

In fact, bioclimatic architecture is not a new invention, but rather a rational and scientific application of ancient knowledge and common sense, with the aim of obtaining maximum physical and psychological comfort and the best lighting, heating and cooling conditions, while adopting the most energy efficient and environmentally friendly techniques and technological solutions.

The study of local micro-climates, following the example of Le Corbusier's *climatic grid*, the correct exposure of buildings, the use of suitable materials and other technical solutions are all carefully combined in a design granting a reduction in energy consumption up to 40 per cent.

Modern examples of bioclimatic buildings were presented in a special session of the Forum and were shown at the same time at the ENEA "Bioarchitecture" Exhibition (at Villa Erba, Cernobbio-Como). The Exhibition was displayed in 30 European countries.

Experts have stressed the need for a larger public discussion on this subject and for large scale practical applications, in order to tackle growing population numbers and escalating energy needs (as many as 2 billion people have no access to energy sources today).

Since such problems cannot be the responsibility of single citizens alone, experts have emphasised the need for a direct commitment on the part of public authorities in encouraging bioclimatic architecture in city planning schemes and improving energy efficiency and environmental quality. Therefore city planning will be a fundamental tool to enhance the application of bioarchitecture technologies for the improvement of environmental quality.

Finally, R&D in the field of architecture should receive more incentives (also from the European Union) so as to foster and disseminate active and passive energy saving methods, while at the same time improving the quality of life in urban as in rural settlements.

9. NEW ENERGY SAVING TECHNOLOGIES: FUEL CELLS AND NON-CONVENTIONAL TRANSPORTATION FUELS

The development of new highly efficient and climate-friendly energy technologies and processes of the 21st century is an essential part of a long-term strategy to secure an integrated source approach to the growing future energy demand and to reduce greenhouse gas emissions. Among the distinguished technological options priorities must be given to those which can substantially change the existing partners fo energy supply facilties and systems such as: i) fuel cells and ii) oil-substitute alternative transportation fuels.

i) Fuel cells will have an always increasing space in a scenario for the next decades, where will coexist:

- liberalisation of the energy market and distributed generation (DG), with independent power producers, energy traders, brokers, self-producers and distributors
- abundant availability of fossil fuels
- accounting of cost internalization for emissions
- CO_2 market development, where systems allowing an easier CO_2 fixation will have a greater value
- availability of abundant hydrogen from renewable sources

In fact, fuel cells, as direct means for chemical energy conversion into electricity, are—independently from the technology considered—a valid way for reducing the amount of primary energy necessary to produce electrical energy. As electrochemical generators, they directly transform into electrical energy the free-energy variation, associated to an overall chemical reaction running through two electro-chemical processes taking place on separated anodic and cathodic regions.

In addition to high efficiency, fuel cells offer additional advantages in terms of good performance at base and partial load, very low emissions, modularity (easy adjustment of plant capacity to power-demand increase), and reduced plant erection time.

FC power plants, if compared to conventional power plants (traditional steam-water and combined cycles), are delivering to atmosphere almost no SO_x, NO_x and 1/3 of the standard CO_2. A direct consequence of this is, in a decentralised electrical energy generation system, the possibility of setting this type of plants almost everywhere, in a very short time (preassembled or skid mounted) and radically decreasing the transportation and distribution losses.

The full value given to the active and reactive power control in the grid knots, the possibility of taking local advantage from the available heat, as well as the internalisation of energy production induced costs, will give a boosting effect on this technology in the next decades.

By the way, a great potential for distributed generation is offered by the emerging countries having not yet strongly invested in the transmission and distribution systems associated with centralised generation.

Furthermore, the current transport mechanism inside the cell matrix implies the supplying proper flow rates of carbon dioxide (to the cathode side) or the recycling CO_2 (to the anode). These two peculiarities allow choosing a suitable plant configuration and giving the opportunity for an easy recovery of CO_2 from the anode exhaust for subsequent fixation processes if a gas turbine retrofits the fuel cell system.

Distributed generation share is expected to increase significantly in the next decades up to 15–20 per cent reaching 100 GW per year and, probably, over. A significant share of this market sector, up to about 10 per cent, seems to be approachable for fuel cells even if their utilisation in larger size power plants will be feasible later, when fuel cell technology will reach a higher level of maturity. In the medium term, as an example, Molten Carbonate Fuel Cells (MCFC) will be suitable for plant size in the range 500 kW–5 MW, the upper one being easily obtainable exploiting the fuel cell modularity.

To summarize, it is expected that FC can be used variously, for example, as stationary power plants, dispersed power supplies used by households, hotels, hospitals, etc., and hybrid power supplies of electric motor vehicles. So, their earlier practical application is strongly desired FC in many countries, and particularly in the developing ones, can be identified as one of the main national priorities for energy strategy to be urgently promoted.

ii) Energy consumption to fulfil mobility needs has closely followed the world economic output. Despite the achieved efficiency in transportation sector and evolution of oil markets, oil security remains a serious concern, particularly given the prospects of increasing oil dependence from the depleting oil reserves in the Middle East. The advanced technologies for synthesis of liquid motor fuels from natural gas and biomass are good options to reduce dependence on depleting oil resources and to diminish local and global pollution. For instance, it was recognised that biodiesel is:

- a renewable form of energy that may become the flagship of all non-food projects, due to an expensive agricultural over-production in Europe and a surplus in soybean oil in the US
- climate-friendly since it contributes to the reduction of

greenhouse gases by at least 3.2 kg CO_2 –equivalent per 1 kg biodiesel

- a compound with a very low toxicity being its biodegradability of more than 90 per cent within 3 weeks. A further benefit concerns the creation of additional jobs through local production of bio-energy as liquid fuel. In this regard, the Brazilian Ethanol Programme was seen as successful even if it is not yet cost effective.

Finally, from a market point of view, the exercise of internalising all external costs gives i) job-creation-bonus ii) less-global-warning-bonus and and iii) energy-supply-bonus for the alternative motor fuels such as compressed natural gas, liquid petroleum gas and biofuels.

10. CONCLUSIONS AND RECOMMENDATIONS

Energy is an essential factor of economic development. Although reserves of fossil fuels are not yet approaching physical exhaustion, and energy prices remain low on the international market, it is unduly optimistic to assume that energy crises will not occur in the future. The rapid growth of energy demand, especially in the developing countries, is likely to bring to new tensions in the market, unless appropriate preventive measures are taken. As proven by past history, energy is a major cause of potential conflicts. Preventing crises by ensuring energy security is an important contribution to world peace.

At the same time, preoccupations on the environmental and climatic effects deriving from the increased use of fossil fuels is mounting. Although much remains to be explored and proven as concerns climate change, evidence is enough to justify a preventive action, especially as such action also meets other objectives, such as the reduction of local and regional environmental effects and the conservation of finite, non renewable resources.

Sustainable energy schemes meet the objectives of both energy security and environmental protection. They are based on increased efficiency in all the chains of energy use, especially at the final consumption, on the utilisation of renewables. Sustainable energy is essential to support development and to supply modern energy services to the two billion people in rural area who until now have no access to commercial energy. In this connection, the availability of energy is also a powerful motor of social promotion. The shift toward more sustainable energy cycles in all countries, for different applications, in urban as well as in rural contexts, has been advocated by governments of most countries and by public opinion.

Improved and new technologies are needed to meet growing demand and environmental challenges. This requires an increased effort of research and development (R&D). Unfortunately, investment in energy research has been decreasing both in the public and in the private sector; medium- and long-term programmes have particularly been affected.

Together with technologies, policy instruments are needed in order to implement and diffuse sustainable energy schemes. Such policies must make use of market forces to optimise the allocation of resources, but the market is not sufficient to bring about sustainable development, unless long-term signals are introduced to correct its limited vision. This is a task for governments. Innovation in energy policy instruments to correct market forces is just as important as technological innovation. Market "pull" and technology "push" are the prerequisites of the proper time deployment new technological options.

International cooperation is essential in bringing about this transition. The instruments being set up in the frame of the Convention on Climate Change—emission trading, joint implementation, clean development mechanism—may be very useful in bringing about the desired change in energy paradigm, provided they are implemented in a way apt to stimulate the transfer and the large scale deployment of the most efficient and effective technologies with high environmental performance.

The Forum of the UNESCO International School of Science for Peace and Energy Security for the Third Millennium at Como (14–16 May 1998), organised by Landau Network–Centro Volta, Russian Centre for Energy Policy, UNESCO Venice Office, Italian Ministry of Foreign Affairs and ENEA has examined various technical aspects connected with this problematique. It has underlined that satisfactory progress is being achieved with emerging technologies of the highest interest, such as fuel cells for both stationary and transport applications; renewable energy technology and biofuels; high efficiency end-use equipment and appliances, bioarchitecture technologies in building, etc. The satisfactory progress of the World Solar Programme, process initiated by UNESCO, has been recognised. Perspectives for safer, and also smaller, nuclear systems have also been presented, particularly for needs of developing countries.

The Forum has also examined the particular problems of two regions of critical importance for the future energy scenarios: developing countries and economies in transition. For developing countries it has been emphasised that energy projects by donor and lending agencies, until now concentrated on large-scale, supply-side traditional plants or transmission lines, should give priority to demand-side interventions and to distributed energy generation, in particular by renewable sources. It has also been stressed that the direct involvement of the local authorities and of private business is a pre-requisite for success, and that the creation of the right conditions is essential, including removal of incentives to traditional energy, capacity building for installation, maintenance and marketing, updating of norms, setting up of local institutions etc.

For countries in transition, the great potential of improvement in energy efficiency should be considered as mean to enhance the energy security at the regional and global levels, to raise national economy and protect environment. The cost effective part of energy efficiency should be kept in mind as allocating priorities in the no regret investments for greenhouse gas emissions reductions.

In conclusion, the Forum has expressed the following recommendations:

1. Governments and institutions should invest more in sustainable energy R&D, and encourage the private sector to do the same. Support for long-term energy R&D does not necessarily imply equal support for all technologies. Periodic assessments of resources allocation and the effectiveness of R&D strategies are required.
2. R&D effort could be made more effective by enhanced international cooperation and by a more pointed identification of critical technology to be sought by a joint assessment effort.
3. Studies and experimentation of institutional innovations, especially in developing countries, should be encouraged.
4. End-use energy efficiency should be encouraged, in particular through integrated resource planning and demand-